Ecology: Concepts and Applications

Ecology: Concepts and Applications

Editor: Madison Morgan

R CALLISTO
REFERENCE

www.callistoreference.com

Callisto Reference,
118-35 Queens Blvd., Suite 400,
Forest Hills, NY 11375, USA

Visit us on the World Wide Web at:
www.callistoreference.com

ISBN: 978-1-64116-062-9 (Hardback)

Cataloging-in-Publication Data

Ecology : concepts and applications / edited by Madison Morgan.
    p. cm.
Includes bibliographical references and index.
ISBN 978-1-64116-062-9
1. Ecology. 2. Population biology. 3. Environmental sciences. I. Morgan, Madison.
QH541 .E26 2019
577--dc23

Table of Contents

Preface

The branch of biology, which involves the study of organisms and their interactions with their environment, is called ecology. It focuses on the interactions between organisms and with the abiotic factors of the environment. Some of the significant topics studied under this field include competition and cooperation between species, biodiversity, population of organisms, biomass and distribution of organisms. Ecology has applications in diverse fields like forestry, conservation biology, community health, wetland management, human social interaction, etc. The understanding of the principles of ecology is fundamental to the development of behavioral ecology, social ecology, cognitive ecology, biogeography, molecular ecology and social ecology. From theories to research to practical applications, case studies related to all contemporary topics of relevance to this field have been included in this book. Some of the diverse topics covered herein address the varied branches that fall under this category. This book includes contributions of experts and scientists, which will provide innovative insights into the field of ecology.

All of the data presented henceforth, was collaborated in the wake of recent advancements in the field. The aim of this book is to present the diversified developments from across the globe in a comprehensible manner. The opinions expressed in each chapter belong solely to the contributing authors. Their interpretations of the topics are the integral part of this book, which I have carefully compiled for a better understanding of the readers.

At the end, I would like to thank all those who dedicated their time and efforts for the successful completion of this book. I also wish to convey my gratitude towards my friends and family who supported me at every step.

Editor

Modifications of nutrient regime, chlorophyll-a, and trophic state relations in Daechung Reservoir after the construction of an upper dam

Neha P. Ingole and Kwang-Guk An[*]

Abstract

Background: Previous numerous studies on watershed scale demonstrated that the constructions of upper dams may influence the below dams due to modifications of flow regime and nutrient inputs. Little is known about how the dam constructions influence the downstream lakes or reservoirs in the regional scale. This study demonstrates how the construction of upper dam (i.e., Yongdam Dam) influences nutrient regime, trophic relations, and empirical models in Daechung Reservoir (DR). Yongdam Dam was constructed at the upstream region of DR in year 2000.

Results: The analysis of hydrological variables showed that inflow and discharge in the DR were largely reduced after the year 2000. The construction of upper dam construction also resulted in increases of water temperature, pH and conductivity (as an indicator of ionic content) in the DR. Empirical models of TP-CHL and N:P ratio-CHL suggested that stronger responses of CHL to the phosphorus were evident after the upper dam construction, indicating that algal production at a unit phosphorus increased after the upper dam construction. Mann-Kendall tests on the relations of N:P ratios to TN showed weak or no relations ($t_{au} = -0.143$, $z = -0.371$, $p = 0.7105$) before the dam construction, while the relation of N:P ratios to TP showed strong in the periods of before- ($t_{au} = -0714$, $z = -2.351$, $p = 0.0187$) and after the construction ($t_{au} = -0.868$, $z = -4.270$, $p = 0.0000$). This outcome indicates that TP is key determinant on N:P ratios in the reservoir. Scatter Plots on Trophic State Index Deviations (TSIDs) of "TSI(SD) - TSI(CHL)" against "TSI(TP) - TSI(CHL)" showed that the dominance of clay turbidity or light limitation was evident before the upper dam construction [TSI(TP) - TSI(CHL) > 0 and TSI(SD) - TSI(CHL) > 0] and phosphorus limitation became stronger after the dam construction [(TSI(TP) - TSI(CHL) < 0 and TSI(SD) - TSI(CHL) > 0].

Conclusions: Overall, our analysis suggests that the upper dam construction modified the response of trophic components (phytoplankton) to the nutrients or nutrient ratios through the alteration of flow regime, resulting in modifications of ecological functions and trophic relations in the low trophic levels.

Keywords: Upper dam construction, Trophic state deviation, Nutrient regime, Seasonality

Background

The Daechung Reservoir is located on the upper part of the Geum River in the central region of South Korea and has a surface area of 6.8×107 m^2, a volume of 14.3×108 m^3, a mean depth of 21.2 m, and a maximum depth of 69 m at an elevation of 80 m above mean sea level (MSL). This reservoir was formed by the construction of a multipurpose dam in 1980 to conserve water resources for drinking, agricultural, and industrial use and for electric power supply. This reservoir supplies water to several central regions including Daejeon,

Chongju, and Chonahn cities. This reservoir is a large branch-type lake with a 72-m-high dam and a gross storage capacity of 1490 Mm^3 (Oh et al. 2001). Previous studies of reservoirs (artificial lakes; Kimmel and Groeger 1984; Kennedy et al. 1985; Cole and Hannan 1990) pointed out that spatial heterogeneities in physical structure, chemical water quality, and biological components are large and temporal variations are large due to large fluctuations of rainfall and/ or runoff from the watershed. Such heterogeneity is mainly attributed to greater flushing rate than natural lakes (Canfield and Bachmann 1981; Straskraba 1996).

Reservoir ecosystems typically have prominent longitudinal heterogeneities in water quality from the

* Correspondence: kgan@cnu.ac.kr
Department of Biological Sciences, College of Biological Sciences and Biotechnology, Chungnam National University, Daejeon 34134, South Korea

headwaters to the dam (Kennedy et al. 1982, 1985; Kimmel et al. 1990). Spatial and temporal variabilities of nutrients, trophic state, and algal productions are large in reservoir ecosystems, and the variabilities are especially greater in monsoon regions (Asia) than non-monsoon regions (Park et al. 2010). Typical longitudinal gradients in the Daechung Reservoir were evident in nutrients (N, P), water transparency, suspended solids, and algal biomass (An and Park 2002; An and Jones 2000); thus, the functional zones were divided into three reaches along the main axis of the reservoir from the headwaters to the dam (An and Park 2002; Sedell et al. 1990). The riverine zone (running water) which is mainly influenced by external inputs and is frequently light limited showed high total phosphorus (TP), high inorganic turbidity, and low Secchi depth (Oberholster et al. 2013; Irigoien and Castel 1997; O'Boyle et al. 2015), and the length was maximized in the flooding monsoon (Puckridge et al. 2000).

Although phosphorus is a primary limiting nutrient for algal growth in temperate regions (Correll 1999), light limitation is frequently found in the riverine zone during the flooding season (Lehman et al. 2007). Under these circumstances, nitrogen or phosphorus may not be a key factor regulating the primary production in the reservoir (Sterner 2008) In contrast, the lacustrine zone (stagnant water) which is lake-like and influenced by internal processes showed low nutrients (N, P), high availability of underwater light (Havens et al. 2003), and low primary productivity as chlorophyll-a. The transition zone

is between the riverine and lacustrine zones, and the characteristics were intermediate, compared to the riverine and lacustrine zones. Also, the Daechung Reservoir has high temporal variations of limnological parameters seasonally due to intense Asian monsoon rain during the short period of July–August. Thus, longitudinal characteristics of the three zones were largely modified by the monsoon flow. Ionic contents in the lake water were diluted by the rainwater and most pronounced in the riverine zone by high flow (Moss 1998). An and Park (2002) found that the relations in empirical models of chlorophyll-a (CHL)-TP were largely modified depending on the location of the functional zones, and the light limitation is most pronounced in the riverine zone during the monsoon.

Under the high spatial and temporal variabilities of limnological conditions in the Daechung Reservoir (DR), the Yongdam Dam (YD) was constructed at the upstream region of the DR in 2000. Thus, flow regime and hydrology in the DR were probably modified due to the constructions of the YD. Serial discontinuity concepts (SDCs; Ward and Stanford 1983) hypothesized that river nutrient cycling is strongly altered by constructions of the upper dam, especially in low to mid-region waterbodies, even though empirical evidence of such a pattern is not strong. This hypothesis implies that the construction of the YD may alter the nutrient regimes and trophic state in the DR, resulting in modifications of algal response to nutrients (N, P).

Currently, little is known about research outcomes on how the constructions of the upper dam influence the downstream waterbody of the DR. Such construction

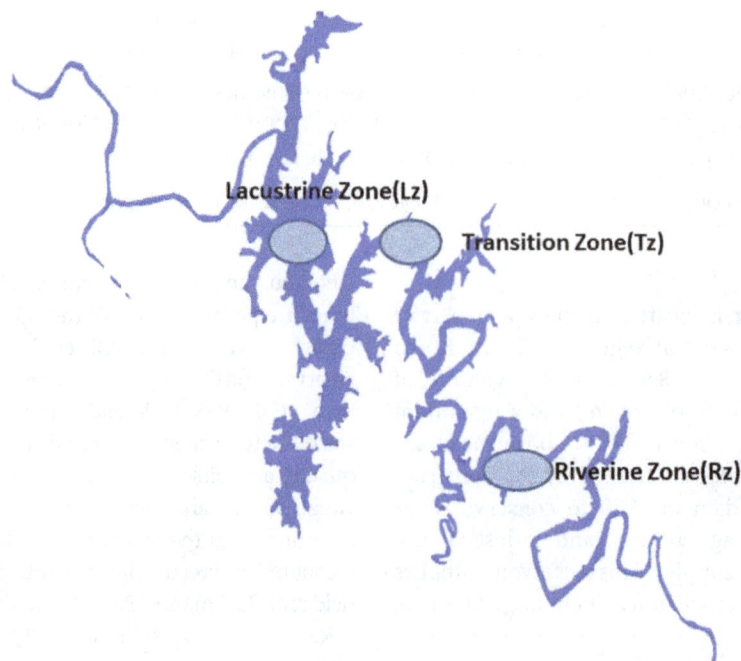

Fig. 1 Map of the Daechung Reservoir showing the three sampling sites of riverine (*Rz*), transition (*Tz*), and lacustrine zones (*Lz*)

may reduce inflows and outflows to DR, and these hydrological factors, in turn, may directly influence the N:P ratios and yields of chlorophyll per unit of nutrients (N or P), resulting in modifications of the eutrophication processes in DR. The objective of the present study was to evaluate the influence of the YD on nutrient regime, algal response, light availability in the DR with an emphasis on how flow regime influenced spatially (riverine, transition, and lacustrine zone) the water chemistry along the length of this morphologically complex reservoir and how the conditions vary seasonally (premonsoon, monsoon, and postmonsoon) in relation to the Asian monsoon. For the analysis, empirical relations of CHL-nutrients and Trophic State Index Deviations (TSIDs) were demonstrated in this study.

Methods
Site description and data collection
The DR is located in the middle of Geum River, Chungbuk and Chungnam Provinces, Korea (36° 50 N, 127° 50 E) and was formed in December 1980 by construction of a dam (Fig. 1). The reservoir is identified as a dendritic-type waterbody with longitudinal gradients from the headwaters to

the dam. The reservoir has two intake towers of Muneu and Dongmeon areas for drinking water supplies and has a surface area of 6.8×10^7 m^2, a volume of 14.3×10^8 m^3, a mean depth of 21.2 m, and a maximum depth of 69 m.

In this reservoir, three sampling sites were designated in order to cover the longitudinal gradients of the reservoir depth. The longitudinal zones in the reservoir were divided into three categories of riverine (Rz), transition (Tz), and lacustrine zones (Lz; Fig. 1). The Rz reflects the headwater zone, and the Lz reflects the down lake near the dam region. The Tz reflects the intermediate characteristics between the Rz and Lz. Seasonal variations were also considered in the analysis; the terms of premonsoon (January–June), monsoon (July–August), and postmonsoon (September–December) were used in describing the temporal conditions.

Chemical variables and trophic state analysis
The seasonal and spatial variations were total nitrogen (TN) and TP in DR along with total suspended solids (TSS) and CHL. Long-term limnological data from 1992 to 2013 were analyzed, and this was obtained from the Water Information System, Ministry of Environment,

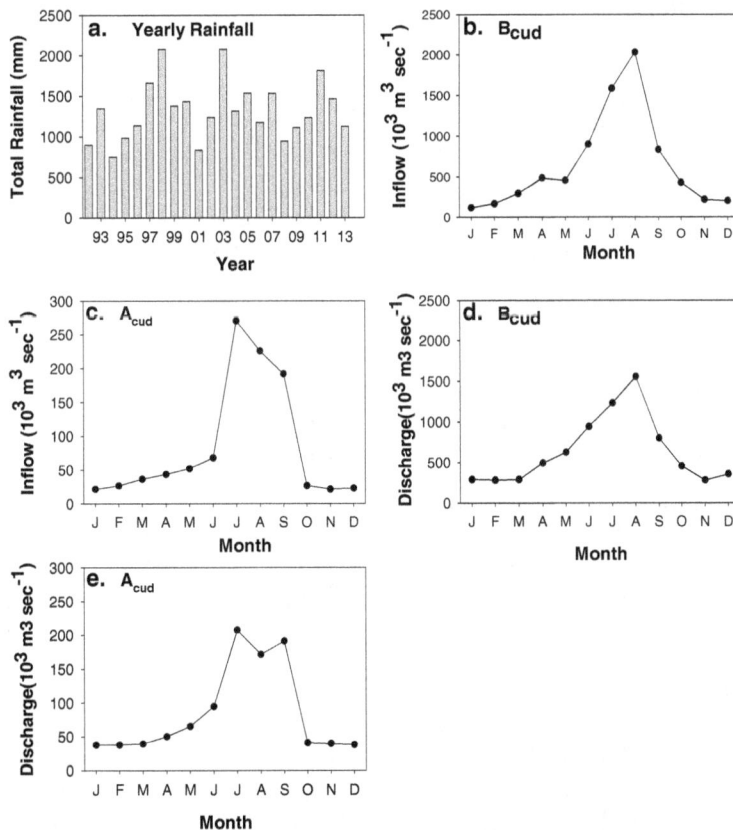

Fig. 2 Seasonal rainfall and flow regime in the Daechung Reservoir during 1992–2013. In the figure, total rainfall (a), inflow volume before the construction of the upper dam (B-CUD, b), inflow volume after the construction of the upper dam (A-CUD, c), the discharge B-CUD (d), and the discharge volume A-CUD (e) were depicted, based on a monthly mean of 12 years

South Korea. TP was determined using the ascorbic acid method after persulfate oxidation (Prepas and Rigler 1982), and TSS were filtered by GF/C filters. CHL concentration was measured by using a spectrophotometer after extraction in hot ethanol (Sartory and Grobbelaar 1984). Secchi transparency (i.e., Secchi depth) was estimated from the empirical equation of TSS. Annual, monthly, and seasonal mean data were log-transformed in order to follow the regression analysis requirements. Statistical analyses were performed using a Sigma Plot (Systat Software Inc.). The calculations of Trophic State Index (TSI) were followed by the approach of Carlson (1977), and the values of trophic parameters were calculated in the three zones and seasons using the three equations as follows:

$$TSI(TP) = 14.42\,Ln(TP) + 4.15$$
$$TSI(TN) = 14.43\,Ln(TN) + 54.45$$
$$TSI(SD) = -14.41\,Ln(SD) + 60$$
$$TSI(Chl) = 9.81\,Ln(Chl) + 30.6$$

The criteria of TSI were followed by the approach of Carlson (1977); values of TSI less than 40 were grouped into oligotrophic state, and the values of 40–50 and 50–70 were categorized as mesotrophic and eutrophic states, respectively. Values higher than 70 are considered as hypertrophic state (Carlson 1977). Non-algal light attenuation $coefficient$ (K_{na}) was estimated in order to evaluate the mechanisms controlling light attenuation in the water column. Non-algal light attenuation was calculated using the following formula:

$$K_{na} = 1/SD - 0.025\,Chl$$

Statistical analysis

Annual and seasonal mean data of trophic variables such as TN, TP, Secchi depth (SD), and CHL were log-transformed to normalize for the requirements of parametric regression analysis. In addition, we identified the annual trend and the trend analysis over the long-term period was performed by an approach of Mann-Kendall statistical test.

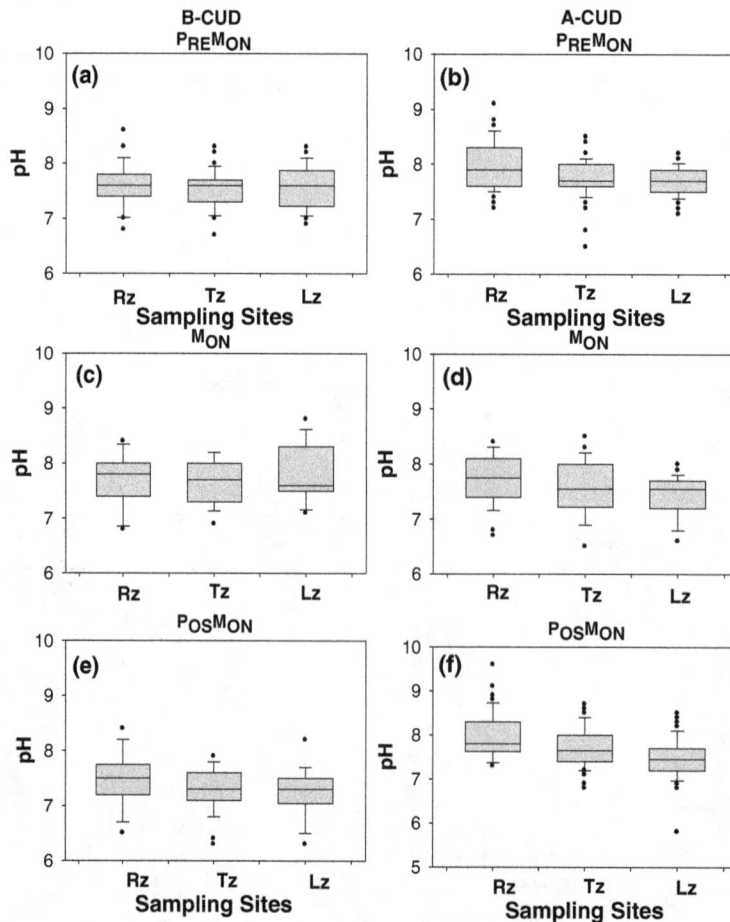

Fig. 3 Seasonal variation of pH among the three zones of the Daechung Reservoir before the construction (B-CUD) and after the construction of the upper dam (A-CUD) during the premonsoon (P_{RE}; **a, b**), monsoon (M_{ON}; **c, d**), and postmonsoon (P_{OS}; **e, f**)

Results and discussion

Modifications of flow regime after the upper dam construction

Annual and seasonal flow regime in DR is directly determined by rainfall patterns. During the study period of 1992–2013, high rainfalls occurred in the years of 1998, 2003, and 2011 and the maximum of 1738 mm was found in 1998 (Fig. 2). These periods were evidently flood years. In contrast, low rainfalls occurred in the years of 1994, 2001, and 2008 (Fig. 2a) and the minimum of 638 mm was found in 1994, indicating that these periods were drought years. In the meantime, rainfalls in the remaining years had no big differences (Fig. 2a).

Monthly inflow and discharge volume showed large differences between the two periods before (B-CUD) and after the construction of the upper dam (A-CUD). The largest differences of inflow and discharge volume between the B-CUD and A-CUD occurred in the monsoon season of July–August; monthly inflow was 2032.38 m^3 B-CUD in the August vs. 225.51 m^3 A-CUD in the August, while monthly discharge was 1575.42 m^3 B-

CUD in the August vs. 171.39 m^3 A-CUD in the August. In other words, inflow volume in the B-CUD directly influenced by rainfall, but the response was really weak in the A-CUD. Our observation of inflow and discharge volume suggests that large inflow and outflow volumes were reduced after the upper dam construction, especially during the monsoon of July–August. These hydrological changes after the upper dam construction may modify the nutrient regime, solid dynamics, ionic contents, and algal growth in DR.

Spatio-temporal variation of pH, water temperature, and conductivity

The construction of the YD resulted in increases of pH in DR. Before the dam construction (B-CUD), the mean values of pH during the premonsoon were <7.5 in the all three zones of Rz, Tz, and Lz (Fig. 3a). But, after the dam construction (A-CUD), the pH values during the premonsoon were >7.5 in all the three zones (Fig. 3b). These results suggest that during the premonsoon, the mean pH was significantly greater after the upper dam

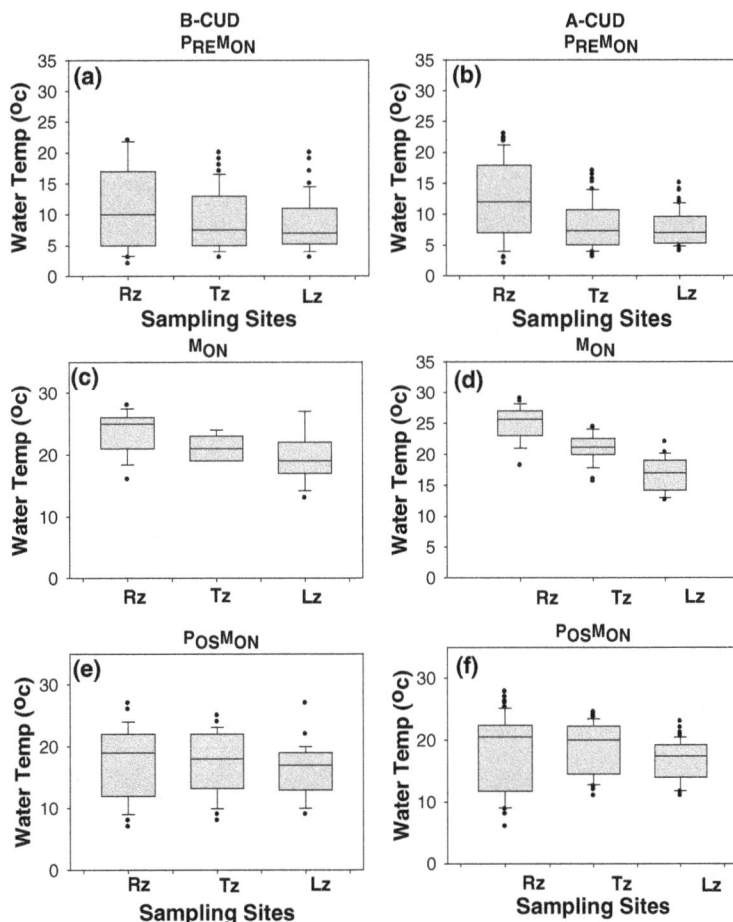

Fig. 4 Seasonal variation of water temperature among the three zones of the Daechung Reservoir before and after the construction of the upper dam during premonsoon (**a, b**), monsoon (**c, d**), and postmonsoon (**e, f**)

construction. During the monsoon, pH values, however, did not show a significant difference between the B-CUD and A-CUD (Fig. 3c, d). The differences in the mean pH between the B-CUD and A-CUD were most pronounced during the postmonsoon (Fig. 3e, f). We believe that pH increased after the upper dam construction.

The construction of the upper dam also resulted in increases of water temperature (W_t) in the DR, and this was similar to the pH. During the premonsoon, the mean W_t in the Rz was significantly greater in the A-CUD (9.8 °C) than the B-CUD (11.3 °C) while the mean W_t in the Tz and Lz was similar between the A-CUD and B-CUD (Fig. 4a, b). During the monsoon, the mean W_t, however, did not show a significant difference between the B-CUD and A-CUD (Fig. 4c, d). In the postmonsoon, the mean W_t in the two zones of Tz and Lz were greater in the A-CUD (>20 °C) than the B-CUD (<20 °C). This outcome indicates more increases of mean W_t in the Rz than the Lz and agrees with previous researches (Wetzel 2001) that water temperature increased after the dam construction. Similarly, specific

conductivity (at 25 °C) in the DR increased after the upper dam (YD) construction. This trend was evident in each premonsoon, monsoon, and postmonsoon season and also same as each zone of riverine, transition, and lacustrine (Fig. 5). The largest differences of conductivity between the B-CUD and A-CUD occurred in the Rz, shown in pH and water temperature.

Influence of the Yongdam Dam on empirical relations of trophic variables in the Daechung Reservoir

The empirical models of TN-CHL, TP-CHL, and N:P-CHL showed distinct differences between the B-CUD and A-CUD when we used all the sites (three sites at three zones plus other two sites = total five sites) in the DR (Fig. 6). There was no big difference in the relation of TN-CHL of the DR between the B-CUD and A-CUD. However, the relation of log-transformed TP-CHL in the A-CUD was modified in the DR; the relation was not significant ($p > 0.05$) before the dam construction but became very positively strong after the dam construction (Fig. 6c, d). Our analysis indicates that algal production

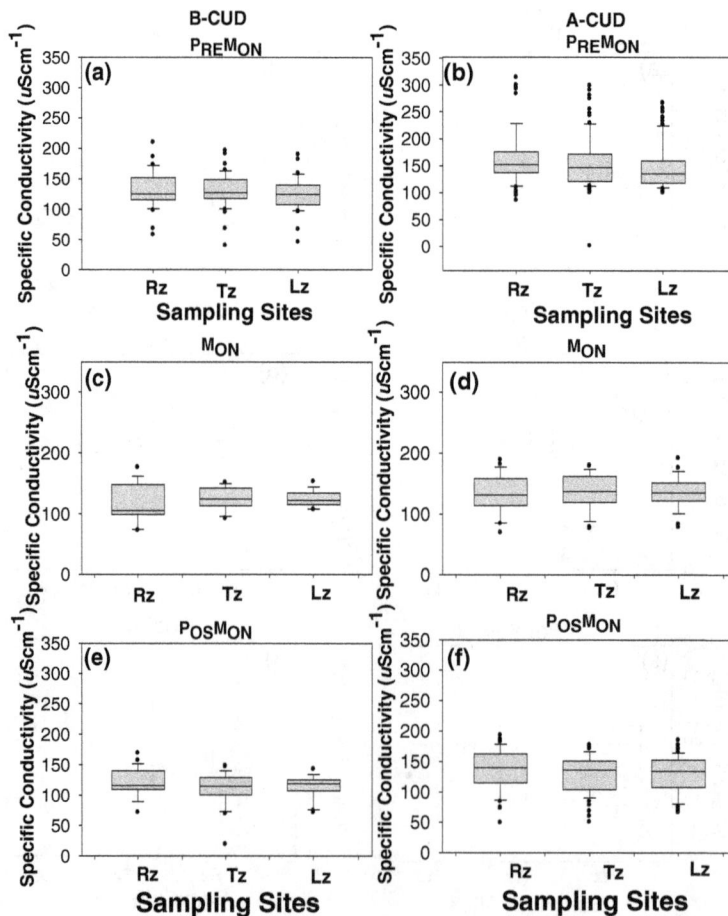

Fig. 5 Seasonal variation of specific conductivity among the three zones of the Daechung Reservoir before and after the construction of the upper dam during premonsoon (**a, b**), monsoon (**c, d**), and postmonsoon (**e, f**)

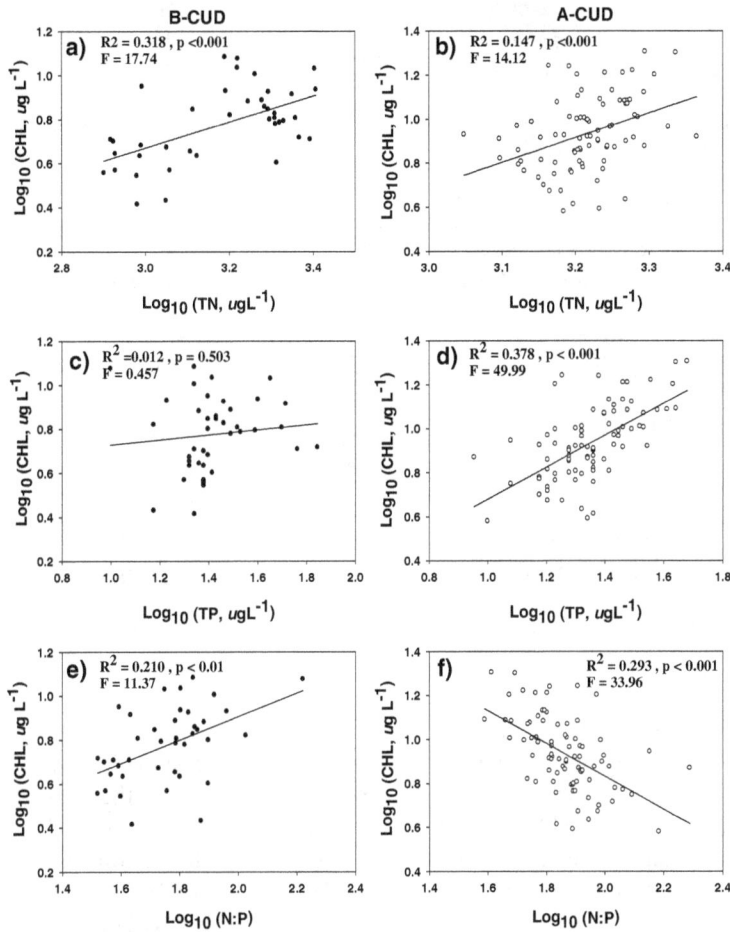

Fig. 6 The empirical models of TN-CHL, TP-CHL, and N:P-CHL in the Daechung Reservoir before the construction of the upper dam (**a**, **c**, **e**) and after the construction of the upper dam (**b**, **d**, **f**)

at a unit phosphorus increased after the upper dam construction and the input of phosphorus from the watershed may increase the eutrophication more rapidly, compared to the period B-CUD. This phenomenon was also supported by the relation of log-transformed CHL-N:P (Fig. 6e, f). The relation of N:P-CHL was positive during the period A-CUD but became negative after the upper dam construction. In other words, increases of N:P mass ratios resulted in less chlorophyll, so the CHL was directly associated with the magnitude of N:P ratios. These data suggests that the upper dam construction modified the ecological functions by the changes of trophic relations in the reservoir.

Relations of N:P ratios to nutrient regime and Mann-Kendall tests

Regression analysis of log-transformed N:P ratios on nutrient regime showed that functional relations of N:P were changed after the upper dam construction of the Yongdam Reservoir. We found that N:P ratios were

more strongly determined by phosphorus rather than nitrogen. Especially, the N:P ratios after the upper dam construction were not determined by TN. Thus, the regression model of log-transformed TN:TP against TN indicates stronger relation in the period after the dam construction than before the construction (Fig. 7a, b). The regression slope in the relation of N:P vs TN was 5.27 and 1.12 in the periods B-CUD and A-CUD, respectively (Fig. 7), and the relation became weak ($R^2 = 0.97$, $p < 0.001$) after the upper dam construction. In the mean time, N:P ratios were strongly influenced by phosphorus (Fig. 7c, d). This outcome suggests that the N:P ratios, frequently used as an index of nutrient limitation, were determined by the ambient concentrations of phosphorus.

In addition, Mann-Kendall tests on the relations of N:P ratios to TN showed weak or no relations in the periods B-CUD ($t_{au} = -0.143$, $z = -0.371$, $p = 0.7105$) and A-CUD ($t_{au} = -0.187$, $z = -0.876$, $p = 0.3811$; Table 1), indicating that the role of nitrogen to N:P was minor and

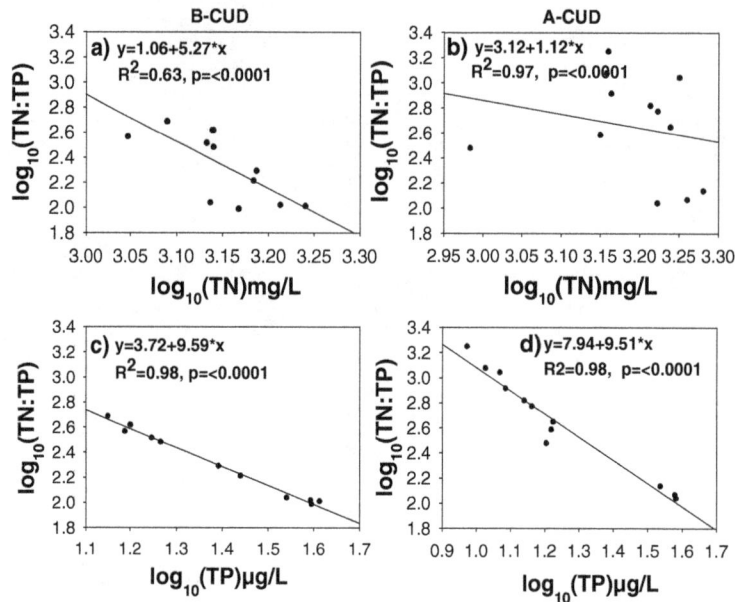

Fig. 7 The graph of regression analyses of log-transformed TN:TP against TN in the Daechung Reservoir B-CUD (**a**) and A-CUD (**b**) and regression analyses of log-transformed TN:TP against TP B-CUD (**c**) and A-CUD (**d**)

did not largely change. In the meantime, the relations of N:P ratios to TP were strong in the periods B-CUD (t_{au} = –0.714, z = –2.351, p = 0.0187) and A-CUD (t_{au} = –0.868, z = –4.270, p = 0.0000; Table 1). This result indicates that TP is an important determinant on N:P ratios, and the role did not change after the upper dam construction.

Relations of Trophic Sate Index (TSI) among the variables and the analysis of TSI deviation

A scatter plot of "TSI(SD) – TSI(CHL)" against "TSI(TP) – TSI(CHL)" before the upper dam construction indicated that about 93 % of the total observation fall into the section of TSI(TP) – TSI(CHL) > 0 and TSI(SD) – TSI(CHL) > 0 (Fig. 8a). These results suggest that the dominance of clay turbidity was evident before the upper dam construction. Thus, light limitation on algal growth dominated the

environment in the period before the upper dam construction, probably due to high current velocity or high clay particles. However, after the upper dam construction, turbidity of inorganic solids decreased largely, so the frequency of phosphorus limitation increased in the reservoir [TSI(TP) – TSI(CHL) < 0 and TSI(SD) – TSI(CHL) > 0]; (Fig. 8b). Thus, phosphorus limitation on algal growth dominated the environment after the upper dam construction. The changes of the environment is supported by the linear empirical models of TP-CHL and N:P ratio-CHL relations.

Conclusions

The present study demonstrates the evident influences of the upper dam construction (YD) on the DR. Empirical relations of log-transformed TP-CHL and N:P ratio-CHL after the upper dam construction were modified. In other words, phosphorus limitation

Table 1 Mann-Kendall statistical tests on the Log_{10} TN vs. Log_{10} TN:TP and Log_{10} TP vs. Log_{10} TN:TP between the two periods before (B-CUD) and after the construction of the upper dam (A-CUD)

(a) B-CUD					
B-CUD year (1992–1999)	t_{au}	s	z	p value	Linear model
Log_{10} TN vs. Log_{10} TN:TP	–0.143	–4	–0.371	0.7105	Log_{10} (TN:TP) = 2.52 – 0.119 × Log_{10}(TN)
Log_{10} TP vs. Log_{10} TN:TP	–0.714	–20	–2.351	0.0187	Log_{10} (TN:TP) = 3.70 – 1.108 × Log_{10}(TP)
(b) A-CUD					
A-CUD year (2000–2013)	t_{au}	s	z	p value	Linear model
Log_{10} TN vs. Log_{10} TN:TP	–0.187	–17	–0.876	0.3811	Log_{10} (TN:TP) = 5.90 – 1.111 × Log_{10}(TN)
Log_{10} TP vs. Log_{10} TN:TP	0.868	–79	–4.270	0.0000	Log_{10} (TN:TP) = 4.54 – 1.616 × Log_{10} (TP)

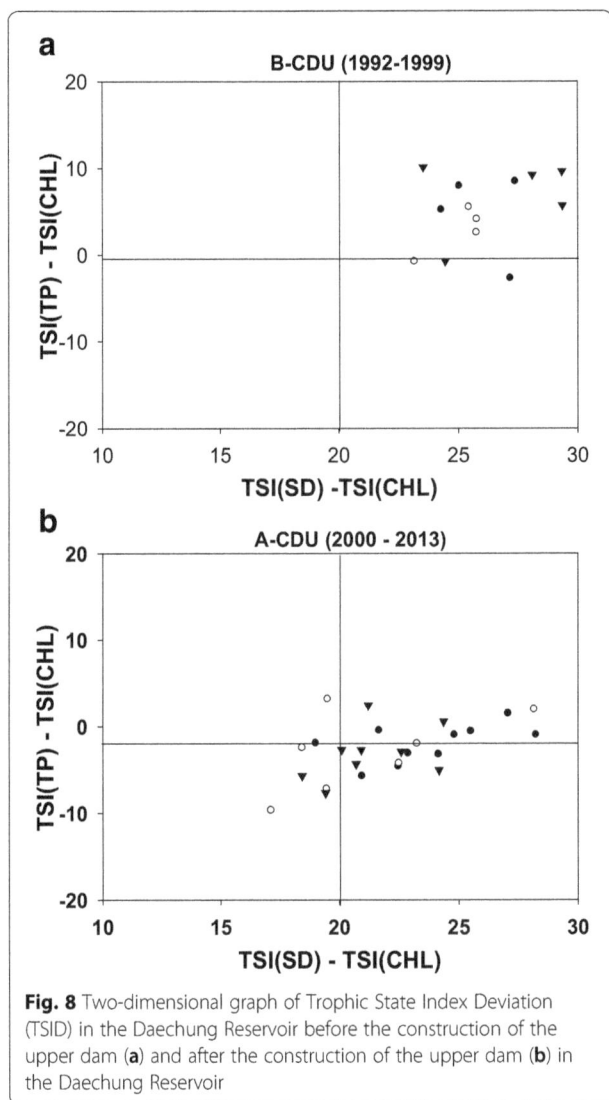

Fig. 8 Two-dimensional graph of Trophic State Index Deviation (TSID) in the Daechung Reservoir before the construction of the upper dam (**a**) and after the construction of the upper dam (**b**) in the Daechung Reservoir

Abbreviations
DR: Daechung Reservoir; TSIDs: Trophic State Index Deviations; MSL: Mean sea level; YD: Yongdam Dam; SDCs: Serial discontinuity concepts; TN: Total nitrogen; TP: Total phosphorus; TSS: Total suspended solids; CHL: Chlorophyll-a; TSI: Trophic State Index; CUD: Construction of the upper dam

Acknowledgements
This research was supported by the fund of "2015' CNU Research Project," Chungnam National University.

Authors' contributions
KG got a project for the topic, and NPI analyzed the data with KG. NPI and KG wrote the manuscript and then edited together. All authors read and approved the final manuscript.

Competing interests
The authors declare that they have no competing interests.

References
An, K. G., & Jones, J. R. (2000). Factors regulating bluegreen dominance in a reservoir directly influenced by the Asian monsoon. *Hydrobiologia, 432*, 37–48.
An, K. G., & Park, S. S. (2002). Indirect influence of summer monsoon on Chlorophyll-total phosphorous models in reservoirs: a case study. *Ecological Modelling, 152*, 192–203.
Canfield, D. J., & Bachmann, R. W. (1981). Prediction of total phosphorus concentration, chlorophyll-a and Secchi depths in natural and artificial lakes. *Canadian Journal of Fisheries and Aquatic Sciences, 38*, 414–423.
Carlson, R. E. (1977). A trophic state index for lake. *Limnology and Oceanography, 22*, 361–369.
Cole, T. M., Hannan, H. H., et al. (1990). Dissolved oxygen dynamics. Chapter 4. In K. W. Thornton (Ed.), *Reservoir Limnology: ecological perspectives*. New York: Wiley.
Correll, D. L. (1999). Phosphorus: A rate limiting nutrient in surface waters. *Poultry Science, 78*, 674–682.
Havens, K. E., James, R. T., East, T. L., & Smith, V. H. (2003). N:P ratios, light limitation and cyanobacterial dominance in a subtropical lake impacted by non-point source nutrient pollution. *Environmental Pollution, 122*, 379–390.
Irigoien, X., & Castel, J. (1997). Light Limitation and Distribution of Chlorophyll Pigments in a Highly Turbid Estuary: the Gironde (SW France). *Estuarine, Coastal and Shelf Science, 44*, 507–517.
Kennedy, R. H., Thornton, K. W., & Gunkel, R. C. (1982). The establishment of water quality gradients in reservoirs. *Canadian Water Resources Journal, 7*, 71–87.
Kennedy, R. H., Thornton, K. W., & Ford, D. (1985). Characterization of the reservoir ecosystem. In D. Gunnison (Ed.), *Microbial processes in reservoirs*. Boston: Dr. W. Junk Publishers.
Kimmel, B.L. and A.W. Groeger. 1984. Factors controlling primary production in lakes and reservoirs; A perspective. In: *Lake and Reservoir Management*. U.S. EPA-440/5-84-001. Pp. 277-281.
Kimmel, B. L., Lind, O. T., & Paulson, L. H. (1990). Reservoir primary production. Chapter 6. In K. W. Thornton et al. (Eds.), *Reservoir Limnology: Ecological Perspectives*. New York: Wiley.
Lehman, P. W., Sommer, T., & Rivard, L. (2007). The influence of floodplain habitat on quantity and quality of riverine phytoplankton carbon produced during the flood season in San Francisco Estuary. *Aquatic Ecology, 42*, 363–378.
Moss, B. R. (1998). *Ionic contents in the lake water were diluted by the rainwater and most pronounced in the riverine zone by high flow. Ecology of Fresh Waters: Man and Medium, Past to Future. By, school of biological sciences, university of Liverpool.* UK: Blackwell Publishing.
Oberholster, P. J., Dabrowski, J., & Botha, A. M. (2013). Using modified multiple phosphorus sensitivity indices for mitigation and management of phosphorus loads on a catchment level. *Fundamental and Applied Limnology/Archiv für Hydrobiologie, 182*, 1–16.
O'Boyle, S., Wilkes, R., McDermott, G., Longphuirt, S. N., & Murray, C. (2015). Factors affecting the accumulation of phytoplankton biomass in Irish estuaries and nearshore coastal waters: A conceptual model. *Estuarine, Coastal and Shelf Science, 155*, 75–88.

became more severe after the upper dam construction and effects of inorganic turbidity in the empirical relations were reduced. Thus, though heavy rainfalls during the monsoon occurred after year 2000, it did not result in light limitation or high inorganic turbidity. Thus, the response of CHL to the phosphorus increased after the upper dam construction. Also, slight increases in pH, water temperature, and conductivity (as an indicator of ionic content) were observed in the reservoir. Overall, our analysis suggests that the upper dam construction modified the response of trophic components (phytoplankton) to the nutrients or nutrient ratios, which is closely associated with flow regime, resulting in modifications of ecological functions and trophic relations in the low trophic levels. Careful reservoir management or flow regime regulation of the upper dam (YD) is required in the future to reduce the eutrophication and algal blooms in DR.

Oh, H. M., Lee, S. J., Kim, J. H., Kim, H. S., & Yoon, B. D. (2001). Seasonal Variation and Indirect Monitoring of Microcystin Concentrations in Daechung Reservoir, Korea. *Applied Environmental Microbiology, 67*, 1484–1489.

Park, J. H., Duan, L., Kim, B., Mitchell, M. J., & Shibata, H. (2010). Potential effects of climate change and variability on watershed biogeochemical processes and water quality in Northeast Asia. *Japan Environment International, 36*, 212–225.

Prepas, E. E., & Rigler, F. A. (1982). Improvements in qualifying the phosphorus concentration in lake water. *Canadian Journal of Fisheries and Aquatic Sciences, 39*, 822–829.

Puckridge, J. T., Walker, K. R., & Costelloe, J. F. (2000). Hydrological persistence and the ecology of dryland rivers. *Regulated Rivers: Research & Management, 16*, 385–402.

Sartory, D. P., & Grobbelaar, J. U. (1984). Extraction of chlorophyll-a from freshwater phytoplankton for spectrophotometric analysis. *Hydrobiologia, 114*, 177–187.

Sedell, J. R., Reeves, G. H., Hauer, F. R., Stanford, J. A., & Hawkins, C. P. (1990). Role of refugia in recovery from disturbances: Modern fragmented and disconnected river systems. Section 4. *Ecosystem and Landscape Constraints on Lotic Community Recovery Environmental Management, 14*, 711–724.

Sterner, R. W. (2008). On the Phosphorus Limitation Paradigm for Lakes. *International Review of Hydrobiology, 93*, 433–445.

Straskraba, M. (1996). Lake and reservoir management. *Verhandlungen des Internationalen Verein Limnologie, 26*, 193–209.

Ward, J.V. and J.A. Stanford (1983). The serial discontinuity concept of lotic ecosystem. . In (eds, T.D. Fontaine and S.M. Bartel), *Dynamics of lotic ecosystems* (pp 29–42). Ann Arbor, Michigan, USA.

Wetzel RG (2001). *Limnology: lake and river ecosystems* 3rd Edition, ISBN: 978-0-12-744760-5.

2

Morphology and taxonomy of the *Aphanizomenon* spp. (Cyanophyceae) and related species in the Nakdong River, South Korea

Hui Seong Ryu, Ra Young Shin and Jung Ho Lee[*]

Abstract

Background: The purpose of this study is to describe the morphological characteristics of the *Aphanizomenon* spp. and related species from the natural samples collected in the Nakdong River of South Korea.

Results: Morphological characteristics in the four species classified into the genera *Aphanizomenon* Morren ex Bornet et Flahault 1888 and *Cuspidothrix* Rajaniemi et al. 2005 were observed by light microscopy. The following four taxa were identified: *Aphanizomenon flos-aquae* Ralfs ex Bornet et Flahault, *Aphanizomenon klebahnii* Elenkin ex Pechar, *Aphanizomenon skujae* Komárková-Legnerová et Cronberg, and *Cuspidothrix issatschenkoi* (Usačev) Rajaniemi et al. *Aph. flos-aquae* and *Aph. klebahnii* always formed in fascicles; the others only occurred in solitary. *Aph. flos-aquae* was similar to *Aph. klebahnii*, whereas these species differed from each other by the size and shape of fascicles, which was macroscopic in *Aph. flos-aquae* and microscopic in the *Aph. klebahnii*. One of their characteristics was that trichomes are easily disintegrating during microscopic examination. *C. issatschenkoi* could be clearly distinguished from other species by hair-shaped terminal cell. Its terminal cell was almost hyaline and markedly pointed. Young populations of the species without heterocytes run a risk of a misidentification. *Aph. skujae* was characterized by akinete. Morphological variability of akinetes from natural samples collected in the Nakdong River was rather smaller than those reported by previous study.

Conclusions: *C. issatschenkoi* are described for the first time in the Nakdong River. In addition, *Aph. klebahnii* and *Aph. skujae* are new to South Korea.

Keywords: Aphanizomenon, Cuspidothrix, Cyanobacteria, Nakdong River, Nostocales

Background

The genus *Aphanizomenon* Morren ex Bornet et Flahault 1888 (type species: *Aph. flos-aquae*) belongs to order Nostocales and family Nostocaceae, which has a world-wide distribution (Rajaniemi et al. 2005a). The species of genus *Aphanizomenon* and several of its members have been described as the cause for harmful bloom (Mcdonald and Lehman 2013; Ma et al. 2015). Some species can produce hepatotoxic and neurotoxic, such as aphantoxin, anatoxin-a, cylindrospermopsin, and saxitoxin, cyanobacterial secondary metabolites which can cause critical problems (Paerl and Huisman 2009; Ballot et al. 2010; Zhang et al. 2015). Therefore, it is very important for the accurate species identification of *Aphanizomenon* because of water bloom with several toxin-producing species (Guzmán-Guillén et al. 2015).

In the Nakdong River, *Microcystis* and *Anabaena* have been considered as the representative bloom-forming cyanobacteria genera (Yu et al. 2014). After the construction of eight weirs, the number of its bloom has been recently growing in mid-upperstream (Ryu et al. 2016). Nevertheless, two *Aphanizomenon* floras (*Aph. flos-aquae* and *Aph. issatschenkoi*) have been described until a recent date (Park 2004) in South Korea; only one *Aphanizomenon* species has been reported in the ecological study of

* Correspondence: jungho@daegu.ac.kr
Department of Biology Education, Daegu University, Gyeongbuk 38453, South Korea

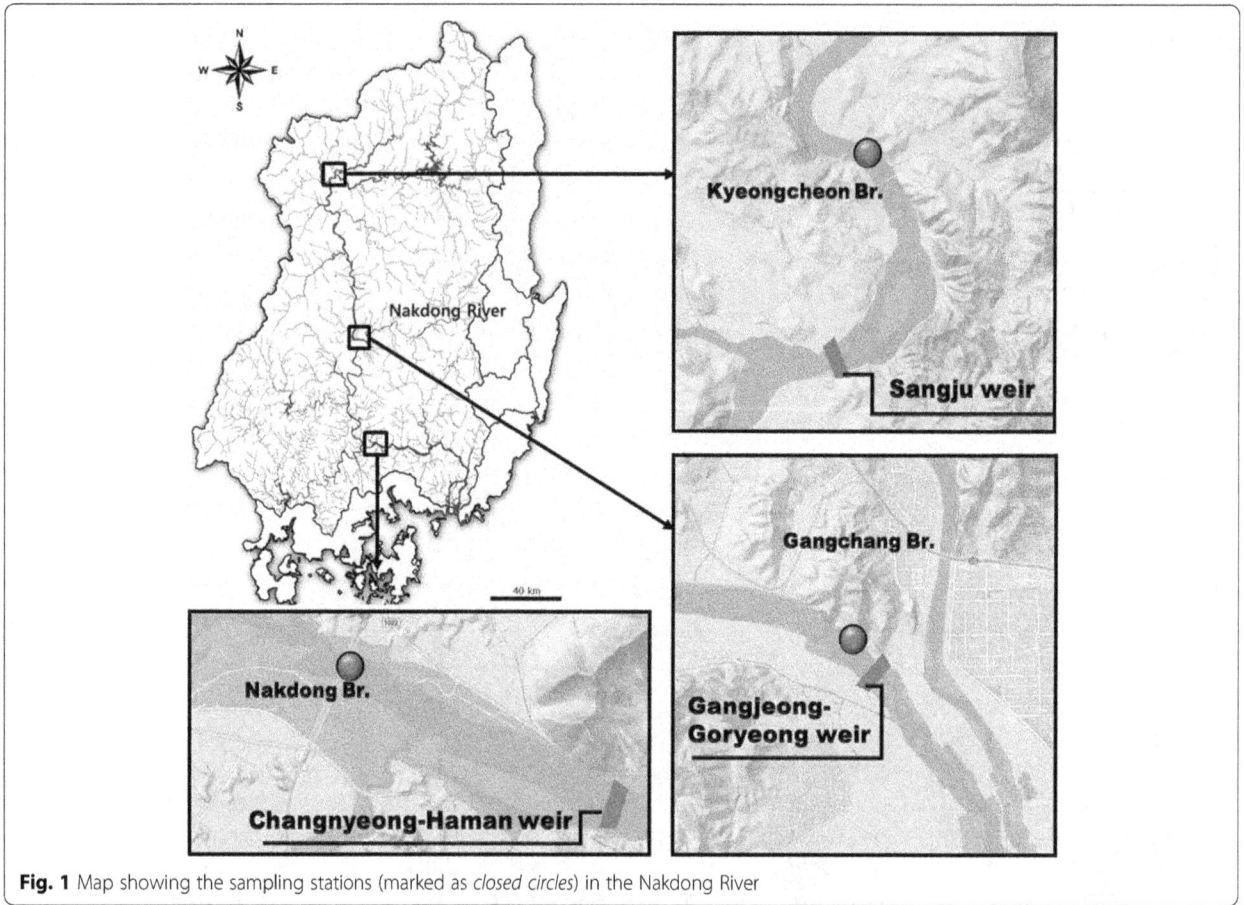

Fig. 1 Map showing the sampling stations (marked as *closed circles*) in the Nakdong River

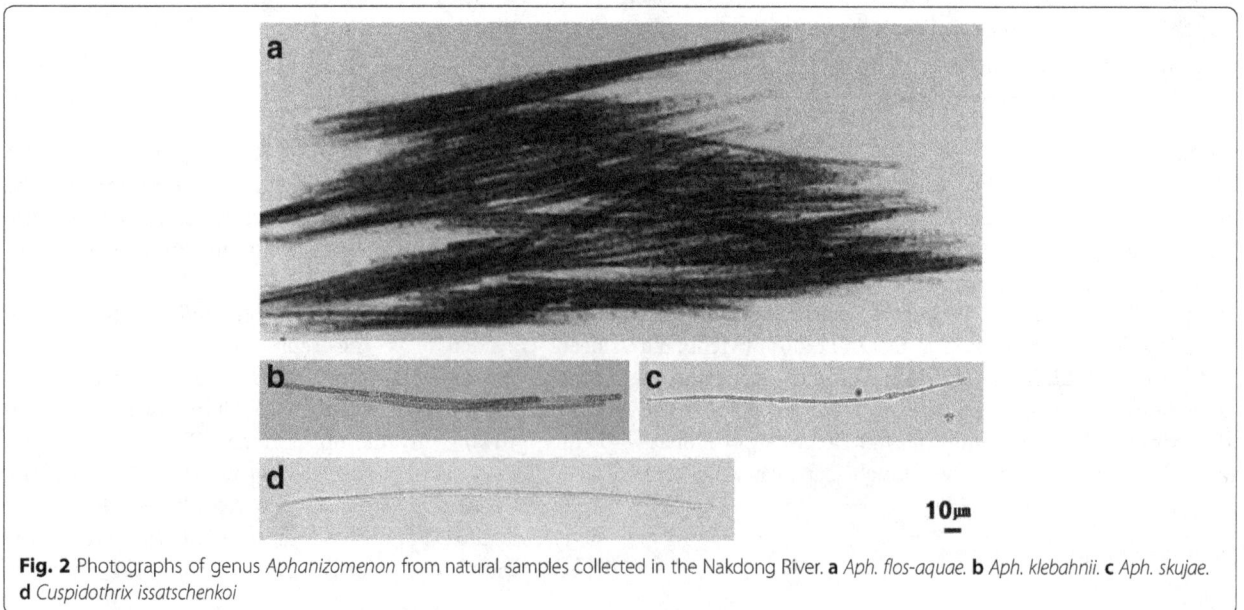

Fig. 2 Photographs of genus *Aphanizomenon* from natural samples collected in the Nakdong River. **a** *Aph. flos-aquae.* **b** *Aph. klebahnii.* **c** *Aph. skujae.* **d** *Cuspidothrix issatschenkoi*

Nakdong River: *Aph. flos-aquae* (Choi et al. 2007; Yu et al. 2014). Recent studies using polyphasic approach, e.g., involving morphology but also ecology and phylogenetics, have revealed that the genus *Apohanizomenon* is in reality very heterogeneous (Cirés and Ballot 2016). According to newly defined approach, 22 taxa identified and described throughout the world have been assigned to the new genera *Aphanizomenon* (e.g., *Aph. flos-aquae* Ralfs ex Bornet et Flahault), *Cuspidothrix* (e.g., former *Aph. issatschenkoi* (Usačev) Proshkina-Lavrenko), *Sphaerospermopsis* (e.g., former *Aphanizomenon aphanizomenoides*

(Forti) Hortobágyi and Komárek), *Chrysosporum* (e.g., former *Aphanizomenon ovalisporum* Forti), *Anabaena/Aphanizomenon* like (e.g., *Aphanizomenon gracile* (Lemmermann) Lemmermann), and *Anabaena*-like group (e.g., *Aphanizomenon volzii* (Lemmermann) Komárek) (Lyra et al. 2001; Gugger et al. 2002; Rajaniemi et al. 2005b; Komárek and Komárková 2006; Zapomělová et al. 2012; Komárek 2013).

The classification of genus *Aphanizomenon* which frequently form blooms is in some cases difficult that is due to lack of the study for morphology and taxonomy

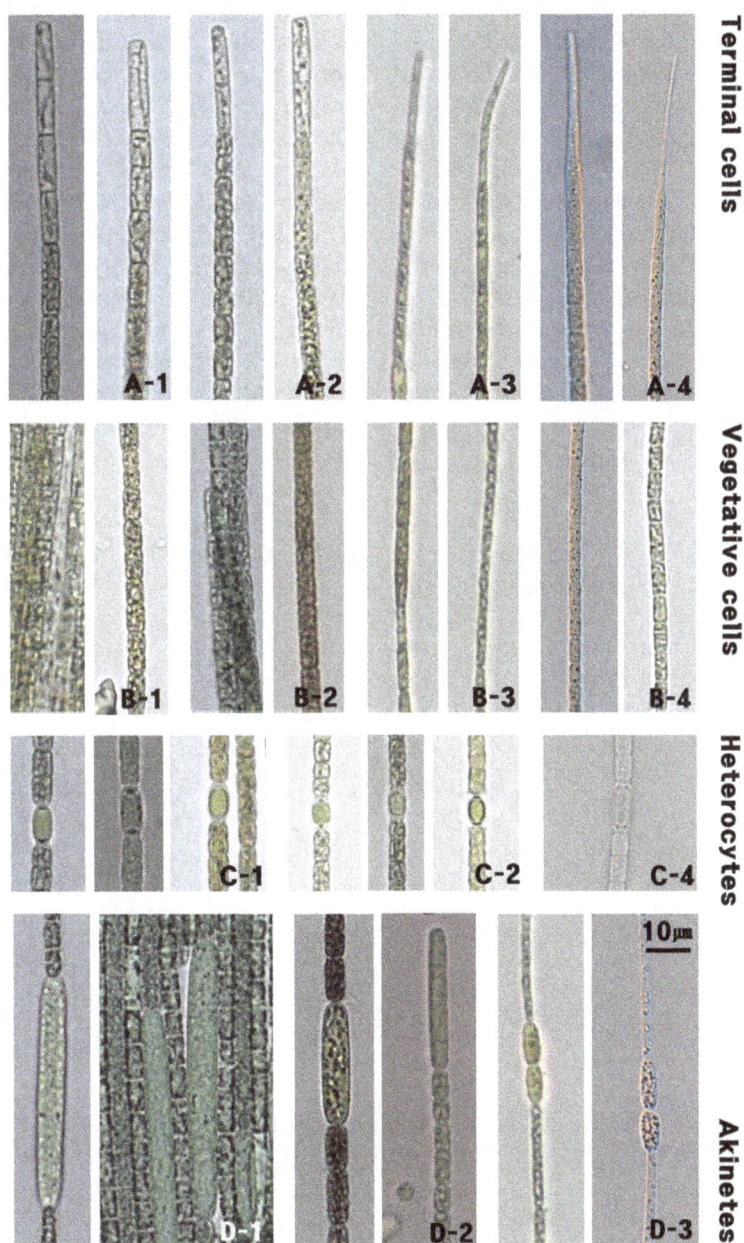

Fig. 3 Terminal cells, vegetative cells, heterocytes, and akinetes of the four *Aphanizomenon* taxa from natural samples collected in the Nakdong River. **A-1**, **B-1**, **C-1**, **D-1** *Aph. flos-aquae*. **A-2**, **B-2**, **C-2**, **D-2** *Aph. klebahnii*. **A-3**, **B-3**, **D-3** *Aph. skujae*. **A-4**, **B-4**, **C-4** *Cuspidothrix issatschenkoi*

in South Korea. The purpose of this study is to describe the morphological characteristics of the *Aphanizomenon* spp. and related species from the natural samples collected in the Nakdong River, South Korea.

Methods

The cyanobacteria samples were collected on three stations of the Nakdong River where the stations located in Sangju (N 35° 27′ 14.69″/E 128° 15′ 27.11″), Daegu (N 35° 50′ 35.58″/E 128° 27′ 33.92″), and Haman (N 35° 23′ 40.89″/ E 128° 31′ 11.84″), respectively (Fig. 1). The samples were collected from June 2015 to May 2016 with 1-month interval using the plankton net (mesh size 32 μm). It was preserved in 4% Lugol's solution or formaldehyde water and was transferred at the laboratory. The morphology of trichomes, vegetative cells, heterocytes, and akinetes were studied using Nikon ECLIPSE 80i light microscope with a digital camera. NIS-Elements F 3.0 software was used for image analysis. The following parameters were selected to describe the morphology of the studied specimens: length and width of vegetative cell, heterocytes, and akinetes; morphology of terminal cell; distance between heterocytes and distance between a heterocyte and the nearest akinete (counted as the number of cells); presence or absence of terminal heterocytes and gas vesicles; and shape of trichomes and its aggregation in colonies. All measurements were obtained with the preserved materials.

Results and discussion

Within the genus *Aphanizomenon*, three clusters were distinguished by Komárek and Komárková (2006) and Komárek (2013) for classification. The first cluster (i) included the type species *Aph. flos-aquae* Ralfs ex Bornet et Flahault 1888 and *Aphanizomenon klebahnii* Elenkin ex Pechar 2008, together with *Aph. yezoense*, *Aph. paraflexuosum*, *Aph. flexuosum*, *Aph. solvenicum*, *Aph. platense*, and *Aph. hungaricum*. Trichomes of *Aph. flos-aquae* and *Aph. klebahnii* taxa always formed macroscopic and microscopic fascicles, and those were able to cause intensive water blooms in eutrophic stagnant water (Hindák 2000). *Aph. flos-aquae* was common species with *Microcystis* spp. and *Anabeana* spp. and the major component of the water bloom in the Nakdong River (Park et al. 2015; Yu et al. 2014). Whereas *Aph. klebahnii* was described for the first time in the South Korea. Cluster (ii) included species with slightly curved or flexuous trichomes. The terminal cells were narrowed, elongated, and hyaline with sharply pointed. Akinetes were distant to heterocytes. This cluster included *Cuspidothrix issatschenkoi* (Usačev) Rajaniemi et al. 2005, together with *C. elenkinii*, *Aph. tropicalis*, *Aph. capricorni*, and *Aph. ussatchevii*. *C. issatchenkoi* was described for the first time in the Nakdong River. Cluster (iii) was comprised of species described as morphotype of *Aphanizomenon gracile* with straight, solitary trichomes and with narrowed ends, which belong into the vicinity of *Dolichospermum* according to molecular sequences. *Aph. skujae* Komárková-Legnerová et Cronberg 1992 belonged to this cluster, together with *Aph. gracile*, *Aph. Schindleri*, *Aph. manguinii*, *Aph. chinense*, and *Aph. sphaericum*. Identification of the species is the first report in South Korea.

Table 1 Diacritical morphological characteristics of four *Aphanizomenon* taxa reviewed from natural samples collected in the Nakdong River

Species	Fascicles trichomes	Terminal cells	Vegetative cells	Heterocytes	Akinetes
Aphanizomenon flos-aquae	Band-like, up to 2 cm long, straight or bent, often grouped in fascicles	Elongated cylindrical, not narrowed, without aerotopes, almost hyaline	Cylindrical to slightly barrel-shaped, 4.0–12.1 × 3.6–5.6 μm[a] (n = 45)	Intercalary, solitary, cylindrical, 6.6–8.5 × 3.3–3.9 μm[a] (n = 19)	Intercalary, long cylindrical, distant from heterocytes, 30–62 × 5.2–7.5 μm[a] (n = 22)
Aphanizomenon klebahnii	Spindle-like, up to 3 mm long, straight or slightly arcuated, often grouped in fascicles	Elongated cylindrical, without aerotopes, and with remaining cytoplasm in the form of fine granulation	Cylindrical or slightly barrel-shaped, 3.9–8.3 × 3.6–4.9 μm[a] (n = 31)	Solitary, intercalary, oval to cylindrical, 5.5–6.7 × 3.2–4.0 μm[a] (n = 6)	Intercalary, solitary, elongated cylindrical, 26–39 × 4.5–5.9 μm[a] (n = 17)
Aphanizomenon skujae	Solitary, straight, bent or irregularly curved	Narrowed and elongated, bluntly pointed, containing a smaller amount of pigment and sporadic aerotopes	4.8–8.4 × 1.2–2.5 μm[a] (n = 19)	Solitary, intercalary, oval to cylindrical, 6–15 × 2–3 μm[b]	Solitary or up to 3 in a row, cylindrical with rounded ends, wider than trichomes, 7.6–11.8 × 3.8–4.6 μm[a] (n = 13)
Cuspidothrix issatschenkoi	Solitary, straight, bent, or slightly coiled	Tapered like hair-shaped, almost hyaline, continually pointed	Cylindrical to long-cylindrical, usually with scarce aerotopes, 4.4–7.0 × 2.5–3.3 μm[a] (n = 36)	Solitary, intercalary, 6.6–8.7 × 3.4–3.7 μm[a] (n = 5)	Solitary or 2–3 in a row, distant from heterocytes, long cylindrical with rounded ends, 8.5–12.5 × 4–4.6 μm[c]

n number of identified samples
[a]This study
[b]Komárek 2013
[c]Rajaniemi et al. 2005b

The four investigated species in the Nakdong River were classified in the genus *Aphanizomenon* (*Aph. flos-aquae*, *Aph. klebahnii*, *Aph. skujae*) and in the genus *Cuspidothrix* (*C. issatschenkoi*). Morphological characteristic of trichomes, heterocytes, and akinetes from natural samples collected in the Nakdong River is shown in Figs. 2 and 3 and Table 1 and that investigated from studies cited is shown in Fig. 4.

Systematics of genus *Aphanizomenon* and genus *Cuspidothrix*

Class Cyanophyceae Sachs 1874

　Order Nostocales Borzi 1914

　　Family Nostocaceae C.A. Agardh 1824 ex Korchner 1898

　　　Genus *Aphanizomenon* Morren ex Bornet et Flahault 1888

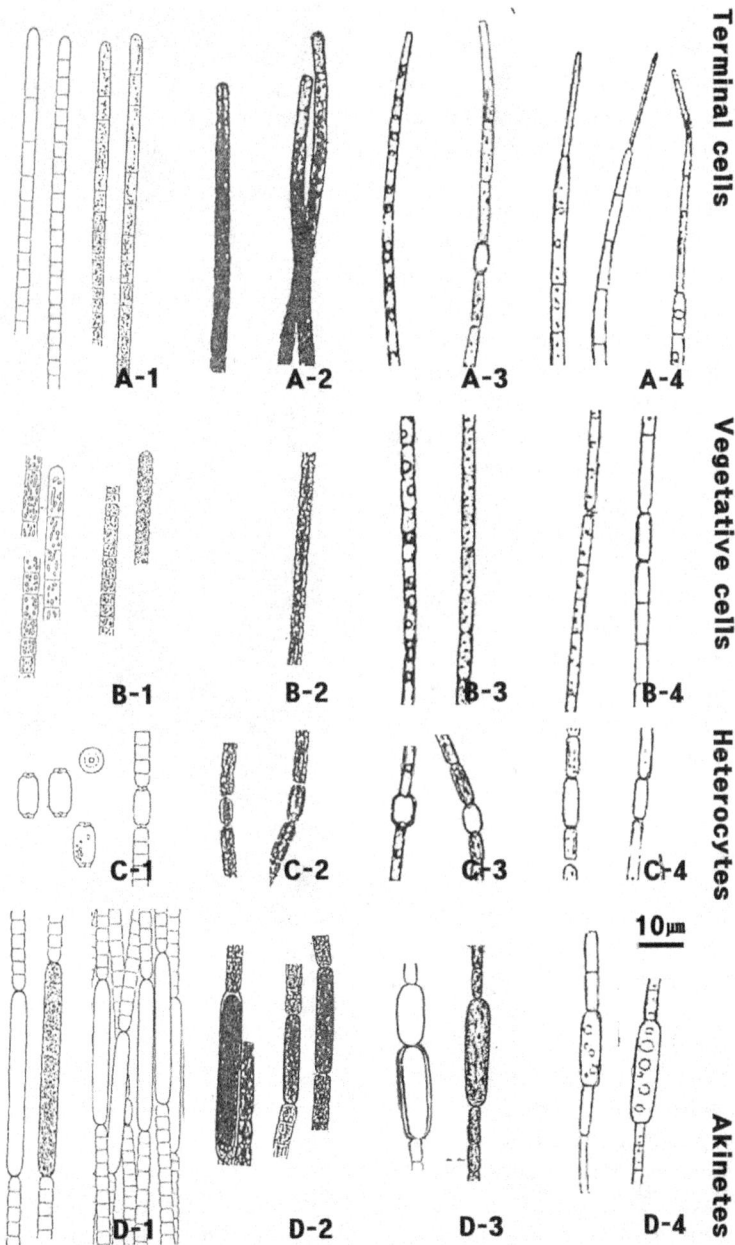

Fig. 4 Terminal cells, vegetative cells, heterocytes, and akinetes of the four *Aphanizomenon* taxa investigated from studies cited. **A-1**, **B-1**, **C-1**, **D-1** *Aph. flos-aquae*—after Komárek (1958). **A-2**, **B-2**, **C-2**, **D-2** *Aph. klebahnii*—after Komárek et Kováčik (1989). **A-3**, **B-3**, **D-3** *Aph. skujae*—after Skuja (1956) and Komárková-Legnerová et Cronberg (1992). **A-4**, **B-4**, **C-4** *Cuspidothrix issatschenkoi*—after Usačev from Kondraeva 1968 and after Hindák et Moustaka (1988)

Aphanizomenon flos-aquae Ralfs ex Bornet et Flahault 1888

Aphanizomenon klebahnii Elenkin ex Pechar 2008

Aphanizomenon skujae Komárková-Legnerová et Cronberg 1992

Genus *Cuspidothrix* Rajaniemi et al. 2005

Cuspidothrix issatschenkoi (Usačev) Rajaniemi et al. 2005

Morphology and taxonomy of individual species

Aphanizomenon flos-aquae Ralf ex Bornet et Flahault (Fig. 2a; Fig. 3A-1~D-1)

(Smith 1950, p. 585, fig. 503; Hirose et al. 1977, p. 85, pl. 36 3a-3d; Komárek and Kováčik 1989, fig. 8; John et al. 2002, p. 96, pl. 18g-j; Rajanieimi et al. 2005b, Fig. 7. a; Komárek and Komárková 2006, Fig. 6; Komárek 2013, p. 688, Fig. 853)

Synonyms: *Aphanizomenon incurvum* Morren 1835; *Aphnizomenon cyaneum* Ralfs 1850; *Aphanizomenon holsaticum* Richter 1896; *Aphanizomenon americanum* Reinhard 1941

This species was common in Asian freshwater not only in South Korea (Park et al. 2015; Ryu et al. 2016) but also in China (Wu et al. 2010; Ma et al. 2015) and in Japan (Takano and Hino 2009; Yamamoto 2009). Studied samples were collected from all stations. The species was characterized by a tendency to aggregate trichomes in parallel fascicles which can reach a macroscopic size of up to 2 cm. Trichomes of the species were straight or bent, cylindrical, and slightly constricted at the cross-walls. Other morphological features include an isopolar and at the ends cylindrical-rounded (Fig. 2a). Trichomes were easily disintegrating by shaking or fixing solution during microscopic examination (Fig. 5). Cells were cylindrica to slightly barrel-shaped, isodiametric with olive-green protoplast and numerous aerotopes, 4–12.1 × 3.6–5.6 μm; terminal cells elongated and up to 19.1 long, without aerotope, almost hyline, usually with characteristic remains of cytoplasm in form of an irregular (Fig. 3A-1). Heterocytes were intercalary, solitary (up to 3) in a trichome, cylindrical, and 6.6–8.5 × 5–8.5 μm. Morphological variability of akinetes from natural samples collected in the Nakdong River was rather smaller. Akinetes were reported by Komárek (2013) as 40–220 × 6–10.8 μm, however mostly 30–62 × 5.2–7.5 μm, intercalary, long cylindrical, and distant from heterocyte (Table 1).

Ecology: This species is planktonic in eutrophic reservoir (Komárek 2013). It has shown positive growth within a wide range of temperatures (16–25 °C) (Preussel et al. 2009) and can grow below 10 °C (Üveges et al. 2012). It has a competitive advantage under situations of low light intensities (Mehnert et al. 2010). We collected this specimen in waterbodies of mesotrophic or eutrophic status (range of total phosphorus 0.017–0.040 mg L^{-1}).

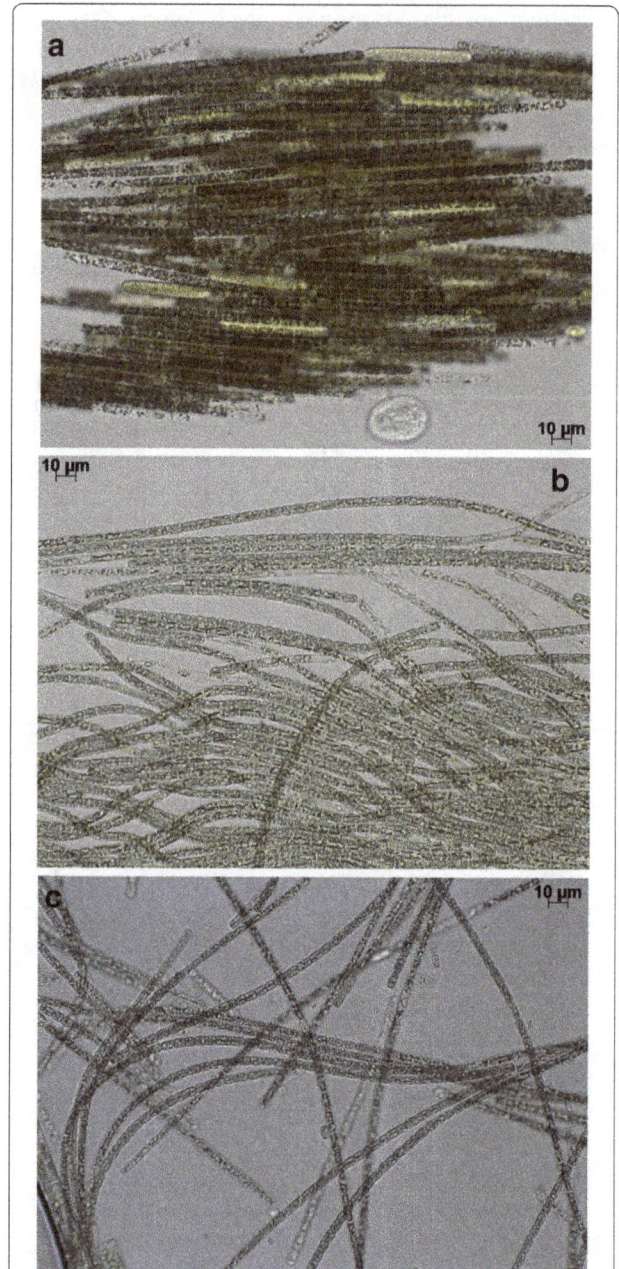

Fig. 5 Serial photos of easily disintegrating characteristics of *Aphanizomenon flos-aquae* from natural samples collected in the Nakdong River. **a** Trichomes were grouped in fascicles on sampling day (no shaking, no fixing). **b** Trichomes were disintergrated by shaking on sampling day (no fixing). **c** Trichome were disintergrated by fixing solution and shaking after 3 days

Material examined: Sangju (Jun. 2015, Oct. 2015, Dec. 2015, Nov. 2015, Jan. 2015, May. 2016), Daegu (Jun. 2015, Oct. 2015, Dec. 2015, Nov. 2015, Jan. 2016, Feb. 2016, Mar. 2016, Apr. 2016, May 2016), Haman (Jun. 2015, Oct. 2015, Dec. 2015, Nov. 2015, Jan. 2016, Feb. 2016, Mar. 2016, Apr. 2016, May 2016)

Aphanizomenon klebahnii Elenkin ex Pechar (Fig. 2b; Fig. 3A-2~D-2)

(Hirose et al. 1977, p. 85, pl. 36 4a-4b; Komárek and Kováčik 1989, fig. 9; Komárek 2013, p. 690, Fig. 855)

Synonyms: *Aphanizomenon flos-aquae* var. *klebahnii* Elenkin 1909; *Aphnizomenon klebahnii* Elenkin 1909 (Nomen alternat.)

Blooms of this species have been frequently observed in the Japanese lakes (Yamamoto 2009); however, the species has not been recorded in South Korea. Single free-floating trichomes of the species were aggregated in parallel in spindle-like fascicles and up to 2 mm long. Trichomes were straight or slightly arcuated cylindrical; almost not or only very slightly constricted at the cross-walls; isopolar, on both ends with elongated cylindrical; and not narrowed cells (Fig. 2b). Cells were cylindrical or slightly barrel-shaped, isodiametric, with olive-green protoplast with numerous aerotopes, and 3.9–8.3 × 3.6–4.9 µm; terminal cells were elongated, up to 12 µm long, without aerotopes, and with remaining cytoplasm in the form of fine granulation (Fig. 3A-2). Heterocytes were solitary, intercalary, 1–2 in the trichome, oval to cylindrical, and 5.5–6.7 × 3.2–4.0 µm. Akinetes were developed asymmetrically on the trichome, intercalary, solitary or rarely in pairs, elongated cylindrical, and 26–39 × 4.5–5.9 µm (Table 1). This species was very similar to the species, *Aph. flos-aquae* Ralfs ex Bornet et Flahault, namely in shape, size, and color of cells, terminal cell, and heterocytes. However, this species differed from *Aph. flos-aquae* only by the size and shape of fascicles, which is macroscopic in *Aph. flos-aquae* and microscopic in *Aph. klebahnii* (Fig. 6, Table 1).

Ecology: This species is planktonic in eutrophic up to hypertrophic reservoir (Komárek 2013). It is adapted to high water temperatures (Yamamoto and Nakahara 2006). We collected this specimen in waterbodies of mesotrophic or eutrophic status (range of total phosphorus 0.028–0.036 mg L^{-1}).

Material examined: Daegu (Oct. 2015, Nov. 2015), Haman (Oct. 2015, Feb. 2016)

Aphanizomenon skujae Komárková-Legnerová et Cronberg (Fig. 2c; Fig. 3A-3, B-3, D-3)

(Komárek and Komárková 2006, Fig. 17)

Synonyms: *Aphanizomenon* cf. *flos-aquae* var. *klebahnii* sensu Skuja 1956

Identification of this species is the first report in South Korea. Trichomes of the species were solitary, fine, straight, bent or irregularly curved, and without mucilage (Fig. 2c). Cells usually were with aerotopes, particularly in the central part, towards the ends sometimes more hyaline, 4.8–8.4 × 1.2–2.5 µm; terminal cells were narrow and elongated, containing a smaller amount of pigment and sporadic aerotopes. Terminal cells were bluntly pointed (Fig. 3A-3). Heterocytes were not found from natural samples collected in the Nakdong River. Akinetes were oval to cylindrical with rounded ends, conspicuously wider than vegetative cells, and with an akinete/trichome width ratio often greater than twofold. Morphological variability of akinetes from natural samples collected in the Nakdong River was rather small. It was reported by Komárek (2013) as 20–34 × 2.7–4.7 µm but mostly 7.6–11.8 × 3.8–4.6 µm. Akinetes range from 1 to 2 in number (rarely up to 3) in a row, with smooth and colorless exospore; the content was greenish and granular (Table 1).

Ecology: This species is planktonic in lakes. It is distributed in northern and colder parts of temperate zone in Eurasia (Komárek 2013). We collected this specimen in waterbodies of oligotrophic status (total phosphorus 0.009 mg L^{-1}).

Material examined: Sangju (Jun. 2015)

Cuspidothrix issatschenkoi (Usačev) Rajaniemi et al. 2005 (Fig. 2d; Fig. 3A-4~C-4)

(John et al. 2002, p. 96, pl. 18k; Rajanieimi et al. 2005b, Fig. 7. c; Komárek and Komárková 2006, Fig. 31; Figueiredo et al. 2011, Fig. 1. a-c; Ballot et al. 2010. Fig. 1; Komárek 2013, p. 668, Figs. 822-823)

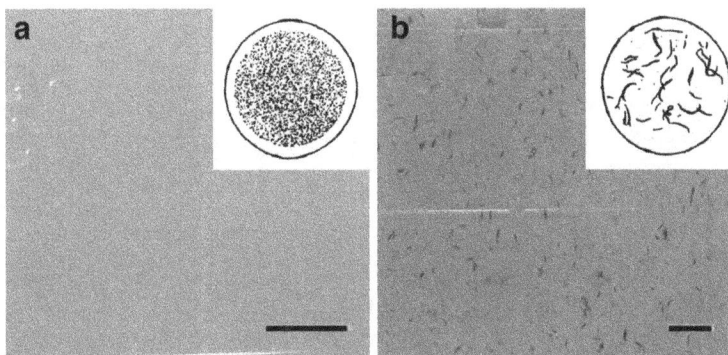

Fig. 6 Comparison of size and shape of fascicles from natural samples collected in the Nakdong River. **a** *Aphanizomenon klebahnii* (microscopic size) and **b** *Aphanizomenon flos-aquae* (macroscopic size). *Scale bar* is 10 mm. Shape of fascicles in *circles* was depicted by Pechar and Kalina from Komárek and Komárková (2006)

Synonyms: *Aphanizomenon issatschenkoi* Usačev 1938

The presence of this species has been reported in freshwaters from many European countries (Kastovsky et al. 2010) and in Asia including China (Wu et al. 2010; Ballot et al. 2010), Japan (Watanabe 1985), and Singapore (Pham et al. 2011). In South Korea, this species has been described only one time until a recent date in the Han River (Park 2004) and it is described for the first time in the Nakdong River. The species was characterized by solitary, bent, or slightly coiled trichomes. The trichomes were isopolar, cylindrical in central part, and continually narrowed or pointed towards ends (developed trichomes). Other morphological features include a not or slightly constricted at the cross-walls and subsymmetric (Fig. 2d). Cells were cylindrical to long cylindrical, usually with scarce aerotopes, 4.4–7.0 × 2.5–3.3 μm; terminal cell was almost hyaline and markedly pointed. The hair-shaped terminal cell was in general narrower than the vegetative cells and continually elongated (Fig. 3A-4). Heterocytes were solitary, intercalary, 1–2 (rarely 3) on a trichome, cylindrical, and 6.6–8.7 × 3.4–3.7 μm. Akinetes were not found from natural samples collected in the Nakdong River. This species was easily recognized by trichomes with hair-shaped terminal cells. However, young field populations of this species without heterocytes can be easily misidentified as the very similar *Rapidiopsis mediterranea* Skuja or as the non-heterocystous life stages of *Cylindrospermopsis raciborskii* (Woloszynska) Seenayya and Subba Raju (Moustaka-Gouni et al. 2010).

Ecology: This species is sporadically planktonic in mesotrophic and eutrophic reservoirs (Komárek 2013). It has shown positive growth within a moderate range of temperatures (22–28 °C) (Dias et al. 2002). It has also been observed thriving in freshwater, as well as in oligohaline and brackish waters (Marshall et al. 2005). We collected this specimen in waterbodies of eutrophic status (range of total phosphorus 0.032–0.043 mg L^{-1}).

Material examined: Haman (Oct. 2015, Nov. 2015, Apr. 2016)

Conclusions

The four investigated species in the Nakdong River were classified in the genus *Aphanizomenon* (*Aph. flos-aquae, Aph. klebahnii, Aph. skujae*) and in the genus *Cuspidothrix* (*C. issatschenkoi*). *C. issatschenkoi* are described for the first time in the Nakdong River. In addition, *Aph. klebahnii* and *Aph. skujae* are new to South Korea.

Acknowledgements
Not applicable.

Funding
This research was supported by the Daegu University Research Grant 2013.

Authors' contributions
RHS carried out the design of the study, performed the fieldwork, and drafted the manuscript. SRY participated in the microscopic analysis. LJH participated in the design and coordination of manuscript and helped draft the manuscript. All authors read and approved the final manuscript.

Competing interests
The authors declare that they have no competing interests.

References
Ballot, A., Fastner, J., Lentz, M., & Wiedner, C. (2010). First report of anatoxin-a-producing cyanobacterium *Aphanizomenon issatschenkoi* in northeastern Germany. *Toxicon, 56*, 964–971.

Choi, C. M., Kim, J. H., Lee, J. S., Jung, G. B., Lee, J. T., & Moon, S. G. (2007). Phytoplankton flora and community structure in the lower Nakdong River. *Korean Journal of Environmental Agriculture, 26*(2), 159–170.

Cirés, S., & Ballot, A. (2016). A review of the phylogeny, ecology and toxin production of bloom-forming *Aphanizomenon* spp. and related species within the Nostocales (cyanobacteria). *Harmful Algae, 54*, 21–43.

Dias, E., Pereira, P., & Franca, S. (2002). Production of paralytic shellfish toxins by *Aphanizomenon* sp. LMECYA 31 (cyanobacteria). *Journal of Phycology, 38*(4), 705–712.

Figueiredo, D. R., Ana, M. M., Gonçalves, B. B., Castro, F., Gonçalves, M., Pereira, J., & Correia, A. (2011). Differential inter- and intra-specific responses of *Aphanizomenon* strains to nutrient limitation and algal growth inhibition. *Journalof Plankton Research, 33*, 1606–1616.

Gugger, M., Lyra, C., Henriksen, P., Coute, A., Humbert, J. F., & Sivonen, K. (2002). Phylogenetic comparison of the cyanobacterial genera *Anabaena* and *Aphanzomenon*. *International Journal of Systematic and Evolutionary Microbiology, 52*, 1867–1880.

Guzmán-Guillén, R., Manzano, I. L., Moreno, I. M., Ortega, A. I. P., Moyano, R., Blanco, A., & Cameán, A. M. (2015). Cylindrospermopsin induces neurotoxicity in tilapia fish (*Oreochromis niloticus*) exposed to *Aphanizomenon ovalisporum*. *Aquatic Toxicology, 161*, 17–24.

Hindák, F. (2000). Morphological variation of four planktic nostocalean cyanophytes—members of the genus *Aphanizomenon* or *Anabaena*? *Hydrobiologia, 438*, 107–116.

Hindák, F., & Moustaka, M. T. (1988). Planktic cyanophytes of Lake Volvi, Greece. *Archiv für Hydrobiologie/Algological Studies, 50–53*, 497–528.

Hirose, H. M., Akiyama, T., Imahori, H., Kasaki, H., Juamo, S., Kobayasi, H., Takahashi, E., Tsumura, T., Hirano, M., & Yamagishi, T. (1977). *Illustrations of the Japanese freshwater algae* (p. 933 pp). Tokyo: Uchidarokakugo Publishing Co., Ltd.

John D.M., Whitton B.A. and Brook A.J. 2002. The freshwater algal flora of the British isles: An identification guide to freshwater and terrestrial algae. Cambridge University Press, 702 pp.

Kastovsky, J., Hauer, T., Mares, J., Krautova, M., Besta, T., Komarek, J., Desortova, B., Hetesa, J., Hindakova, A., Houk, V., Janecek, E., Kopp, R., Marvan, P., Pumann, P., Skacelova, O., & Zapomělová, E. (2010). A review of the alien and expansive species of freshwater cyanobacteria and algae in the Czech Republic. *Biological Invasions, 12*(10), 3599–3625.

Komárek, J. (1958). Die taxonomische revision der planktishen blaualgen der Tschechoslowakei. In *Algologische Studien*, P. 10-206, Academia, Praha.

Komárek, J. (2013). Cyanoprokaryota 3. Teil/3rd part: Heterocytous Genera. In B. Bübel, G. Gärtner, L. Krienitz, & M. Schagerl (Eds.), *SüBwasserflora von Mutteleuropa, 19/3*. Springer Spektrum (p. 1131 pp).

Komárek, J., & Komárková, J. (2006). Diversity of *Aphanizomenon*-like Cyanobacteria. *Czech Phycology, Olomouc, 6*, 1–32.

Komárek, J., & Kováčik, L. (1989). Trichome structure of four *Aphanizomenon* taxa (Cyanophyceae) from Czechoslovakia, with notes on the taxonomy and delimitation of the genus. *Plant Systematics and Evolution, 164*, 47–64.

Komárková-Lengnerová, J., & Cronberg, G. (1992). New and recombined filamentous cyanophytes from lakes in South Scania, Sweden. *Archiv für Hydrobiologie/Algological Studies, 67*, 21–37.

Kondrateva, N. V. (1968). Sin'o-zeleni vodorosti-Cyanophyta.-[Blue-green algae-Cyanophyta.]. In *Viznač. Prosnov. Vodorost. Ukr. RSR 1(2): 1-524*, Vidav. "Naukova Dumka", Kiev.

Lyra, C., Soumalainen, S., Gugger, M., Vezie, C., Sundman, P., Paulin, L., & Sivonen, K. (2001). Molecular characterization of planktic cyanobacteria of *Anabaena*, *Aphanizomenon*, *Microcystis* and *Planktothrix* genera. *International Journal of Systematic and Evolutionary Microbiology, 51*, 513–526.

Ma, H., Wu, Y., Gan, N., Zheng, L., Li, T., & Song, L. (2015). Growth inhibitory effect of *Microcystis* on *Aphanizomenon flos-aquae* isolated from cyanobacteria bloom in Lake Dianchi. *Harmful Algae, 42*, 43–51.

Marshall, H. G., Burchardt, L., & Lacouture, R. (2005). A review of phytoplankton composition within Chesapeake Bay and its tidal estuaries. *Journal of Plankton Research, 27*(11), 1083–1102.

McDonald, K. E., & Lehman, J. T. (2013). Dynamics of *Aphanizomenon* and *Microcystis* (cyanobacteria) during experimental manipulation of an urban impoundment. *Lake and Reservoir Management, 29*(2), 272–276.

Mehnert, G., Leunert, F., Cirés, S., Jöhnk, K. D., Rücker, J., Nixdorf, B., & Wiedner, C. (2010). Competitiveness of invasive and native cyanobacteria from temperate freshwaters under various light and temperature conditions. *Journal of Plankton Research, 32*(7), 1009–1021.

Moustaka-Gouni, M., Kormas, K. A., Polykarpou, P., Gkelis, S., Bobori, D. C., & Vardaka, E. (2010). Polyphasic evaluation of *Aphanizomenon issatschenkoi* and *Raphidiopsis mediterranea* in a Mediterranean lake. *Journal of Plankton Research, 32*(6), 927–936.

Paerl, H. W., & Huisman, J. (2009). Climate change: a catalyst for global expansion of harmful cyanobacterial blooms. *Environmental Microbiology Reports, 1*(1), 27–37.

Park, H. K. (2004). *Phytoplankton of Lake Paldang, Han River Environment Research Center* (p. 131 pp).

Park, H. K., Shin, R. Y., Lee, H. J., Lee, K. L., & Cheon, S. U. (2015). Spatio-temporal characteristics of cyanobacterial communities in the middle-downstream of Nakdong River and Lake Dukdong. *Journal of Korean Society on Water Environment, 31*(3), 286–294.

Pharm, M. N., Onodera, H., Andrinolo, D., Franca, S., Araujo, F., Lagos, N., & Oshima, Y. (2011). *A checklist of the algae of Singapore. Singpore: Raffles Museum of Biodiversity Research* (pp. 1–100). Singapore: National University of Singapore.

Preussel, K., Wessel, G., Fastner, J., & Chorus, I. (2009). Responde of cylindrospermopsin production and release in Aphanizomenon flos-aquae (Cyanobacteria) to varying light and temperature conditions. *Harmful Algae, 8*(5), 645–650.

Rajaniemi, P., Hrouzek, P., Kaštovska, K., Willame, R., Rantala, A., Hoffmann, L., Komárek, J., & Sivonen, K. (2005). Phylogenetic and morphological evaluation of the genera *Anabaena*, *Aphanizomenon*, *Trichormus* and *Nostoc* (Nostocales, Cyanobacteria). *International Journal of Systematic and Evolutionary Microbiology, 55*, 11–26.

Rajaniemi, P., Komárek, J., Willame, R., Hrouzek, P., Kaštovska, K., Hoffmann, L., & Sivonen, K. (2005). Taxonomic consequences from the combined molecular and phenotype evaluation of selected *Anabaena* and *Aphanizomenon* strains. *Algologicals Studies, 117*, 371–391.

Ryu, H. S., Park, H. K., Lee, H. J., Shin, R. Y., & Cheon, S. U. (2016). Occurrence and succession pattern of cyanobacteria in the upper region of the Nakdong River: factors influencing *Aphanizomenon* bloom. *Journal of Korean Society on Water Environment, 32*(1), 52–59.

Skuja, H. (1956). Taxonomische und biologische studien üder das phytoplankton schwedischer Binnengewässer. *Nova acta Regiae Societatis Scientiarum Upsaliensis, Serie, 16*(3), 1–104.

Smith G.M. 1950. The fresh-water algae of the United States, Mcgraw-Hill Book Company, Inc., 719pp.

Takano, K., & Hino, S. (2009). Phylogenic analysis of Aphanizomenon flos-aquae distributed in Japan on partial sequence of rbcLX. *Japanese Journal of Limnology, 69*(3), 247–253.

Üveges, V., Tapolczai, K., Krienitz, L., & Padisák, J. (2012). Photosynthtic characteristics and physiological plasticity of an *Aphanizomenon flos-aquae* (Canobacteria, Nostocaceae) winter bloom in a deep oligo-mesotrophic lake(Lake Stechlin, Germanay). *Hydrobiologia, 698*, 263–272.

Watanabe, M., (1985). Phytoplankton studies of Lake Kasumigaura. (2). On some rare or interesting algae. *Bulletin National Science Museum Tokyo, Series, B11*(4), 137–142.

Wu, Z. X., Shi, J. Q., Lin, S., & Li, R. H. (2010). Unraveling molecular diversity and phylogeny of *Aphanizomenon* (Nostocales, Cyanobacteria) strains isolated from China. *Journal of Phycology, 46*(5), 1048–1058.

Yamamoto, Y. (2009). Environmental factors that determine the occurrence and seasonal dynamics of *Aphanizomenon flos-aquae*. *Journal of Limnology, 68*(1), 122–132.

Yamamoto, Y., & Nakahara, H. (2006). Importance of interspecific competition in the abundance of *Aphanizomenon flos-aquae* (Cyanophyceae). *Limnology, 7*, 163–170.

Yu, J. J., Lee, H. J., Lee, K. L., Lyu, H. S., Hwang, J. H., Shin, R. Y., & Chen, S. U. (2014). Relations between distribution of the dominant phytoplankton species and water temperature in the Nakdong River, Korea. *Korean Journal of Ecology and Environment, 47*(4), 247–257.

Zapomělová, E., Skácelová, O., Pumann, P., Kopp, R., & Janeček, E. (2012). Biogeographically interesting planktonic Nostocales (Cyanobacteria) in the Czech Republic and their polyphasic evaluation resulting in taxonomic revisions of *Anabaena bergii* Ostenfeld 1908 (*Chrysosporum* gen. nov.) and *A. tenericaulis* Nygaard 1949 (*Dolochospermum tenericaule* comb. nova). *Hydrobiologia, 698*(1), 353–365.

Zhang, D. L., Liu, S. Y., Zhang, J., Hu, C. X., Li, D. H., & Liu, Y. D. (2015). Antioxidative responses in Zebrafish Liver exposed to sublethal doses *Aphanizomenon flos-aquae* DC-1 aphatoxins. *Ecotoxicology and Environmental Safety, 113*, 425–432.

3

Zonation and soil factors of salt marsh halophyte communities

Jeom-Sook Lee[1], Jong-Wook Kim[4*], Seung Ho Lee[2], Hyeon-Ho Myeong[3], Jung-Yun Lee[4] and Jang Sam Cho[5]

Abstract

Background: The structures and soil factors of *Suaeda glauca-Suaeda japonica* zonal communities and *Phragmites australis-S. japonica* zonal communities were studied in salt marshes of west and south coasts of South Korea to provide basic data for coastal wetland conservation and restoration.

Results: *S. glauca* community mean length was 67 m and *S. japonica* community mean length was 567 m in zonal communities, and *P. australis* and *S. japonica* community mean length were 57 m and 191 m in zonal communities. Regarding the electrical conductivity, sodium content, and clay contents in Upnae-ri, Shinan-gun, there were significant differences among zonal communities at significance level of 0.05 for two-sided *t* test. However, other factors were not significantly different.

Conclusions: The results indicate that multiple factors such as electronic conductivity, total nitrogen level, clay, and sodium might play important roles in the formation of zonal plant communities of salt marshes.

Keywords: Zonation, Salt marsh plant, Soil factor, *Suaeda glauca*, *S. japonica*, *Phragmites australis*

Background

Zonal distribution of higher plants in salt marshes has been studied extensively for over a century. However, mechanisms of generating the segregation of salt marsh plant species are poorly understood (Caçador et al. 2007; Emery et al. 2001). In order to explain plant zonation, shore height is frequently used as an indicator of abiotic gradient in intertidal ecosystems. This is based on the implicit assumption that shore height is directly correlated with inundation frequency and/or duration (Bockelmann et al. 2002; Sánchez et al. 1996). The objective of this study was to determine structures of zonal communities and factors that might control salt marsh plant patterns and zonations.

Methods

The structures and soil factors of two zonal community types of South Korea were monitored and can be used as basic data for conservation and restoration of coastal wetland ecosystems (Fig. 1). Six *Suaeda glauca-Suaeda japonica* zonal communities (Table 1, No. 1–6 in 1998) and five *Phragmites australis-S.*

* Correspondence: keco@mokpo.ac.kr
[4]Department of Biological Science, Mokpo National University, Muan-gun 58554, South Korea
Full list of author information is available at the end of the article

japonica zonal communities (Table 1, No. 1 in 1998, No. 2–3 in 2005, No. 5–6, 2015) were sampled from the west coast to the south coast of South Korea.

Results and discussion

S. glauca community mean length was 67 m and *S. japonica* community mean length was 567 m in zonal communities, and *P. australis* community mean length was mean 57 m and *S. japonica* community mean length was 191 m in zonal communities (Table 1). *S. glauca* community was found in Unpo-ri, Songhyun-ri, and Chulpo-ri. The community height was 70–80 cm. Its coverage in study areas was 70–80 %. The *S. japonica* community was found in both *S. glauca-S. japonica* and *P. australis-S. japonica* zonal communities. Its community height was 35–45 cm. Its coverage in study areas was 85–100 %. The area of salt marshes in Chulpo-ri and Sinduk-ri was 4–5 km². *P. australis* communities were found in Woopo-ri, Nongjoo-ri, and Dongkeom-ri. Community height was 64–125 cm with coverage of 85–100 %.

Soil factors in *S. glauca*, *S. japonica*, and *P. australis* communities of Upnae-ri, Shinan-gun, are shown in Fig. 2. Electrical conductivity ± SE in *S. glauca*, *S. japonica*, and *P. australis* communities were 1.38 ± 0.0015,

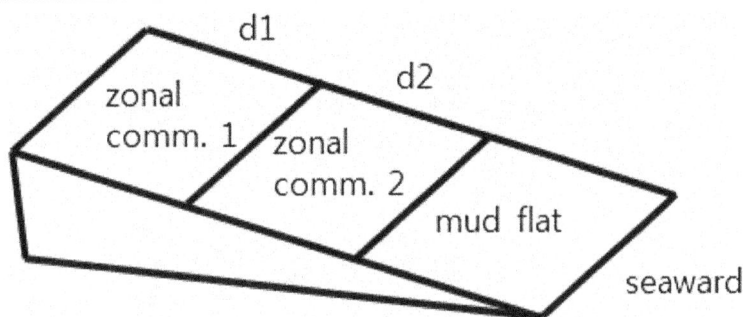

Fig. 1 The first community length of occupation ($d1$, m) and second community length of occupation ($d2$, m) in salt marsh

1.28 ± 0.0045, and 1.01 ± 0.0055 mS/cm, respectively ($n = 10$). Total nitrogen $\pm$ SE in *S. glauca*, *S. japonica*, and *P. australis* communities were 0.21 ± 0.0026, 0.55 ± 0.0026, and 0.69 ± 0.0025 mg/g, respectively ($n = 10$). Higher total nitrogen level in *S. japonica* community than that in *S. glauca* community might be due to higher density in *S. japonica* community. Higher total nitrogen level in *P. australis* community might be due to higher biomass in *P. australis* community. Total phosphate $\pm$ SE in *S. glauca*, *S. japonica*, and *P. australis* communities were 0.05 ± 0.0008, 0.04 ± 0.0009, and 0.04 ± 0.0005 mg/g, respectively ($n = 10$). Such slight difference might be due to dilution of inland and coastal wastewater by tide. Sodium contents $\pm$ SE in *S. glauca*, *S. japonica*, and *P. australis* communities were 15.3 ± 0.0137, 12.3 ± 0.0052, and 5.8 ± 0.0104 mg/g, respectively ($n = 10$). Clay content $\pm$ SE in *S. glauca*, *S. japonica*, and *P. australis* communities were 26.0 ± 0.0344, 25.0 ± 0.0446, and 8.0 ± 0.0274 mg/g, respectively ($n = 10$). Regarding the electrical conductivity, sodium content, and clay contents in both *S.* *glauca-S. japonica* and *P. australis-S. japonica* communities and total phosphate in *S. glauca-S. japonica* community in Upnae-ri, Shinan-gun, there were significant differences among zonal communities at significance level of 0.05 for two-sided *t* test. However, there were little differences in total phosphate levels.

Conclusions

Halophyte distributions are related to multiple reactions of flooding and salinity concentrations (Benito et al. 1990; Caçador et al. 2007; Mert and Varder 1977). In South Korea, halophyte distributions have been determined for soil-water relation and soil texture (Ihm et al. 2007; Rogel et al. 2001) as well as flooding frequency (Lee 1990). A combination of multiple factors such as flooding, soil salinity, and competition have been suggested to play important roles in the formation of zonal plant communities in salt marshes (Pennings and Callaway 1992; Silvestri et al. 2005).

Table 1 Zonal community name, first community length of occupation, second community length of occupation (d1 and d2, m), locations of six *Suaeda glauca-S. japonica*, and five *Phragmites australis-S. japonica* zonal communities

Zonal community name	First community length d1 (m)	Second community length d2 (m)	Locations
Suaeda glauca-S. japonica1	67	990	Kimje-gun Unpo-ri
Suaeda glauca-S. japonica2	100	100	Kimje-gun Sopo-ri
Suaeda glauca-S. japonica3	33	462	Buan-gun Songhyun-ri
Suaeda glauca-S. japonica4	100	627	Buan-gun Chulpo-ri
Suaeda glauca-S. japonica5	67	924	Kochang-gun Sinduk-ri
Suaeda glauca-S. japonica6	33	297	Kochang-gun Wolsan-ri
Mean $\pm$ SD	67 ± 30	567 ± 350	
Phragmites australis-S. japonica1	33	231	Yonggwang-gun Hasa-ri
Phragmites australis-S. japonica2	55	210	Gangwha-gun Dongkeom-ri
Phragmites australis-S. japonica3	17	118	Bosung-gun Jeonil-ri
Phragmites australis-S. japonica4	144	216	Suncheon-gun Nongjoo-ri
Phragmites australis-S. japonica5	36	180	Buan-gun Woopo-ri
Mean $\pm$ SD	57 ± 51	191 ± 45	

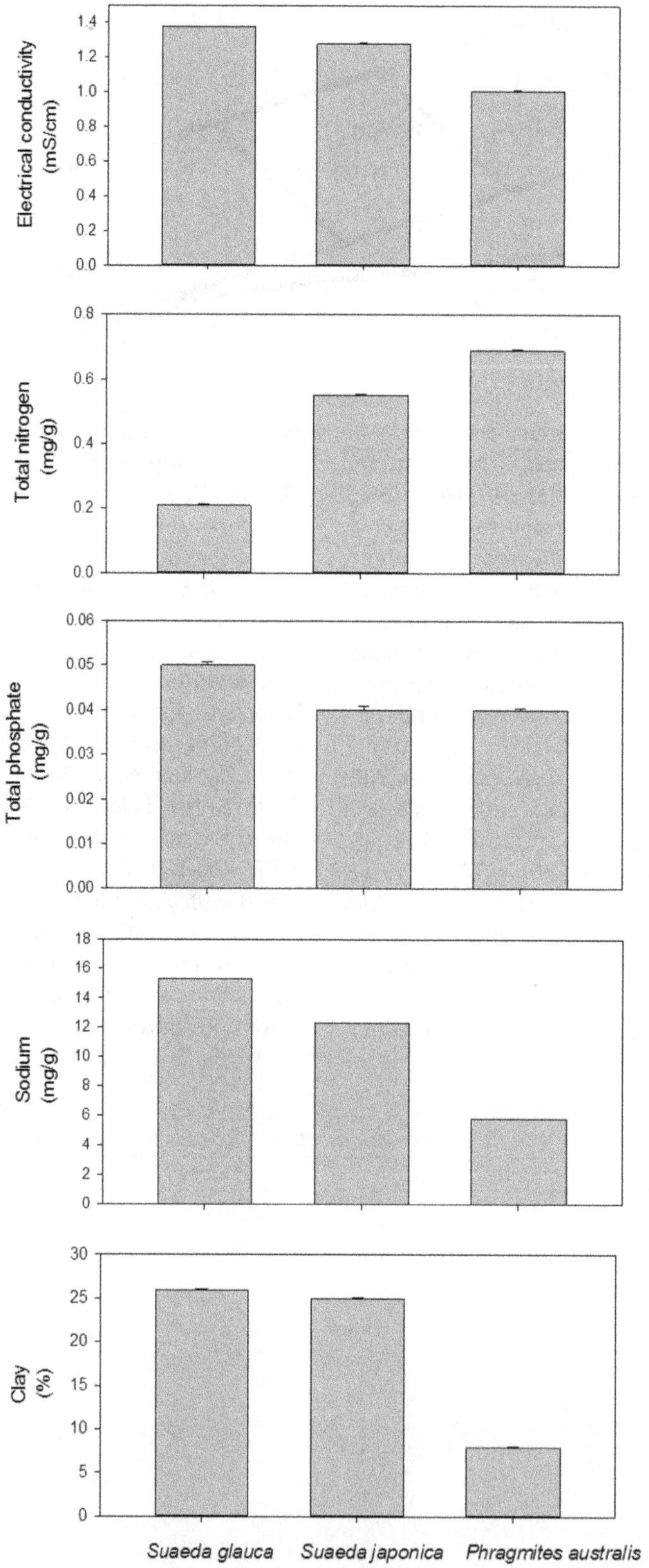

Fig. 2 Electrical conductivity, total nitrogen, total phosphate, sodium, and clay contents (mean ± SE) in *S. glauca*, *S. japonica*, and *P. australis* communities in salt marshes of Upnae-ri, Shinan-gun

Acknowledgements

Emeritus Professor Byung-Sun Ihm deserves to be acknowledged for his ideas and comments on the manuscript at Mokpo National University, South Korea.

Funding

This research was funded with resources from Mokpo National University.

Authors' contributions

The study was designed by JSL and JWK. SHL, HHM, and JYL collected and analyzed the data. SHL has helped in the statistical analysis of the data. JWK and SHL has drafted the manuscript. All authors approved the final manuscript.

Competing interests

The authors declare that they have no competing interests.

Author details

[1]Department of Biology, Kunsan National University, Gunsan 54150, South Korea. [2]Marine & Environmental Research Laboratory, Ansan 15486, South Korea. [3]Division of Ecosystem Research, National Park Research Institute, Wonju 26441, South Korea. [4]Department of Biological Science, Mokpo National University, Muan-gun 58554, South Korea. [5]Division of Ecological Assessment, National Institute of Ecology, Seocheon 33657, South Korea.

References

Benito, I, Agirre, A, & Onaindia, M (1990). Zonation of halophytic vegetation along a tide exposure gradient and associated processes. *Anales de Biologia, 16,* 163–175.

Bockelmann, AC, Bakker, JP, Neuhaus, R, & Lage, J (2002). The relation between vegetation zonation, elevation and inundation frequency in a Wadden Sea salt marsh. *Aquatic Botany, 73,* 211–221.

Caçador, I, Tibério, S, & Cabral, HN (2007). Species zonation in Corroios salt marsh in the Tagus estuary (Portugal) and its dynamics in the past fifty years. *Hydrobiologia, 587,* 205–211.

Emery, NC, Ewanchuk, PJ, & Bertness, MD (2001). Competition and salt-marsh plant zonation: stress tolerators may be dominant competitors. *Ecology, 82,* 2471–2485.

Ihm, B-S, Lee, J-S, Kim, J-W, & Kim, J-H (2007). Coastal plant and soil relationships in the southwestern coast of South Korea. *J Plant Biol, 50,* 331–335.

Lee, J-S (1990). *On establishment of halophytes along tidal level gradient at salt marshes of Manhyong and Dongjin river estuaries.* South Korea, Seoul National University.

Mert, HH, & Varder, Y (1977). Salinity, osmotic pressure, and transpiration relationships of *Salicornia herbaceae* in its natural habitats. *Phyton, 18,* 71–78.

Pennings, SC, & Callaway, RM (1992). Salt marsh plant zonation: the relative importance of competition and physical factors. *Ecology, 73,* 681–690.

Rogel, JÁ, Silla, RO, & Ariza, FA (2001). Edaphic characterization and soil ionic composition influencing plant zonation in a semiarid Mediterranean salt marsh. *Geoderma, 99,* 81–98.

Sánchez, JM, Izco, J, & Medrano, M. (1996). Relationships between vegetation zonation and altitude in a salt-marsh system in northwest Spain. *Journal of Vegetation Science, 7,* 695–702.

Silvestri, S, Defina, A, & Marani, M (2005). Tidal regime, salinity and salt marsh plant zonation. *Estuar Coast Shelf S, 62,* 119–130.

4

Physiological effects of biocide on marine bivalve blue mussels in context prevent macrofouling

Md Niamul Haque[1] and Sung-Hyun Kwon[2*]

Abstract

Background: Mussels are stubborn organisms attached to solid substrata by means of byssus threads. The abundance of marine mussel *Mytilus edulis* in marine facilities like power stations was reason to select among fouling animals.

Methods: Mortality patterns as well as physiological behavior (oxygen consumption, foot activity, and byssus thread production) of two different size groups (14- and 25-mm shell length) of *M. edulis* were studied at different hydrogen peroxide concentrations (1–4 mg l^{-1}).

Results: Studied mussels showed progressive reduction in physiological activities as the hydrogen peroxide concentration increased. Mussel mortality was tested in 30 days exposure, and 14 mm mussels reached the highest percentage of 90% while 25 mm mussels reached 81%. Produced data was echoed by Chick-Watson model extracted equation.

Conclusions: This study points that, while it could affect the mussel mortality moderately in its low concentrations, hydrogen peroxide has a strong influence on mussels' physiological activities related to colonization. Therefore, hydrogen peroxide can be an alternative for preventing mussel colonization on facilities of marine environment.

Keywords: Bivalve blue mussel, Biofouling, Hydrogen peroxide, Physiological behavior, Mortality

Background

Using seawater in cooling systems is a common practice in many parts of the world where there is a shortage of fresh water (Freese and Nozaic 2007), and that practice in coastal power plants is well-known (Mattice and Zittel 1976, Khalanski and Bordet 1980, Rajagopal. 1991). Fouling is the accumulation of unwanted material on solid surfaces to the detriment of function. The fouling nature is different according with the processes involved in its genesis. Usually, fouling is categorized into five types: biological (biofouling), corrosion, particulate, chemical reaction, and crystallization fouling (Epstein 1981). However, biofouling is one of the major operational problems associated with the usage of seawater in cooling systems, beside other problems like corrosion and scaling (Jenner et al. 1998). There are two broad categories of biofouling: macroscopic and microscopic. In macrofouling or macroinvertebrate fouling, clams, barnacles, and mussels block the seawater from properly flowing through the heat exchangers.

On the other hand, microbiologic fouling or microfouling is caused by the growth of slime and algae (Vaccaro et al. 1977). Controlling of macrofouling is the goal of this study. Consequently, mussels are bivalve molluscs belonging to the family *Mytilidae*. The common or blue mussel, *Mytilus edulis*, is among the most abundant and widely distributed invertebrate species inhabiting intertidal and shallow sub-tidal waters in the North Atlantic (Stewart 1994). It is also found in Arctic waters, Greenland, Atlantic coast, and Pacific coast, and as well as in European waters as far south as the Mediterranean and North Africa (Seed 1976). Sessile mussels such as *M. edulis* are often a major fouling species when their settlement and growth result in blockage of free flow of water in the conduits (Rajagopal 1991) and clogging of condenser tubes (Holmes 1970).

Though much work has been done in the area of the relative sensitivity (lethal and sub lethal) of the respective species to chlorine, there is lack of literature on

* Correspondence: shkwon@gnu.ac.kr
[2]Department of Marine Environmental Engineering, College of Marine Science, Engineering Research Institute (ERI), Gyeongsang National University, Cheondaegukchi-Gil 38, Tongyeong, Gyeongnam 650-160, South Korea
Full list of author information is available at the end of the article

hydrogen peroxide applications against marine mussel *M. edulis*. There are several published reports available on the response of other common tropical fouling mussels such as *P. viridis*, *P. perna*, *B. striatulus*, and *M. philippinarum* to chlorine (Rajagopal et al. 1997, 1995; Rajagopal et al. 2003a, b).

Since no information exists on the lethal and sub lethal effects of hydrogen peroxide on this mussel species, it is considered worthwhile to generate this data by exposing the mussels to a range of hydrogen peroxide concentrations. How does *M. edulis* respond physiologically under hydrogen peroxide conditions? Development of an antifouling strategy for *Mytilus edulis* would require that these questions be answered by way of careful experimentation. In the present study, we attempt to find answers in the laboratory using mussels collected from the site by subjecting different size groups of them to a range of hydrogen peroxide concentrations.

The objectives of this study, therefore, were to (1) understand the physiological response of *M. edulis* to hydrogen peroxide environments and (2) study lethality in 30 day's duration. It is presumed that hydrogen peroxide environment might be effective against settling and growing of blue mussels in cooling water system, although this biocide was a little successful in the case of zebra mussels.

Methods
Mussel assortment and preservation
Mussels for the experiments were collected from Jinhae-gu, Changwon-si, South Korea (35° 07′ 39.5″ N and 128° 44′ 19.8″ E). Mussels were grown for commercial purpose. The experimental mussels were collected as attached with growing rope, and mussels attached with ropes were preserved in continuous seawater flowing glass aquarium. The mussels were gently removed from the rope by cutting their byssus threads using a pair of scissors and immediately transferred to an ice box. The ice box had conveyed in laboratory within minimum time and minimum hassle. Seawater collected from the Gyeongsangnam-do Fisheries Resources Institute, South Korea, was used for accustoming *M. edulis* under standard laboratory conditions. Mussels acclimated for at least 48 h in the laboratory were used for each experiment.

Peroxide sample preparation and laboratory study method
Seawater collected from Gyeongsangnam-do Fisheries Resources Research Institute, South Korea, was used for the experiment, after a day's storage. Factors that may change the response of the mussels such as salinity (mean ± SD; 33.47 ± 0.2‰ salinity, 20.0 ± 0.4 °C temperature, 6.2 ± 0.5 mg l^{-1} dissolved oxygen, and 7.8 ± 0.1 pH) did not show any considerable variation during the course of the experiments. The experiments were conducted in

continuous static peroxide system, following the slight adjusted procedure outlined by Rajagopal et al. (1997). Seawater was stored in a 150-l aquarium tank, and peroxide stock solution (1000 mg l^{-1}) prepared from 30% solution (MERK, Germany), was stored in a 1-l volumetric flask. Using a micropipette, an appropriate mix of the two was employed to maintain the desired peroxide concentration in a 2-l glass beaker, with the outside at the 2-l mark. Mixing of the water was speed up by the glass stirrer. After 2 days of acclimation, 20 for 14 mm, and 20 for 25 mm size randomly picked mussels were introduced into the experimental glass beaker containing seawater of known peroxide (1.0, 2.0, 3.0, and 4.0 mg l^{-1}) concentrations. Hydrogen peroxide concentration was determined by thiosulfate titration, using a Hach hydrogen peroxide test kit, model HYP-1. The levels of the total residual H_2O_2 were monitored at 30-min intervals.

Sub lethal responses
Oxygen consumption, foot activity index, and byssus thread production of two size groups of *Mytilus edulis* were also studied at five different peroxide concentrations (0, 1, 2, 3, and 4 mg/L). Experiment were run as detailed as below.

Oxygen consumption
The oxygen consumption was determined via the method of Bruijs et al. (2001). A closed glass respiratory chamber (750 ml), placed inside a double-walled glass beaker (to minimize temperature changes), was filled with Millipore (0.45 μm) filtered seawater (500 ml) previously aerated to 100% oxygen saturation. Five mussel of a particular size group were placed together in the chamber for each measurement. In each experiment, 8 replicate measurements were taken (5 mussels in each experiment × 5 peroxide doses including control × 2 size groups × 8 replicates = 400 mussels). Control measurements were performed using the same setup but without mussels. The oxygen content of the water was determined at the start and end of each run (1 h) by Winkler method (Strickland and Parsons 1972). The amounts of oxygen used by the animals were taken as the average differences in oxygen concentration between the measurements with animals and the controls. Oxygen consumption was expressed in ml O_2/mussels/h. The rate of oxygen consumption (ROC) was calculated as

$$ROC = \frac{\text{oxygen in control (ml/l)} - \text{oxygen in experimental set(ml/l)}}{\text{duration (h)}}.$$

(1)

Foot activity indexing
For foot activity index, six mussels were kept in 2 l of seawater and left undisturbed for 24 h. Every 10 min,

the number of mussels with foot extended outside their shell was noted (Holmes 1970). No attempt was made to follow the foot activity of individual mussels. For each experiment, the foot activity of all mussels was analyzed and was expressed as times/mussels/hour (5 mussels per experiment × 5 peroxide concentrations (including control) × 2 size groups × 8 replicate) = 400 mussels).

Byssus thread production
Byssus thread production was determined following procedures outlined by Van Winkle (1970) and Rajagopal et al. (1995). After 48 h of acclimation, one mussel was placed in a 1 l glass beaker containing 0.75 L of seawater of known peroxide concentration (1 mussel per experiment × 5 peroxide concentrations (including control) × 2 size groups × 8 replicate) = 80 mussels). By using only one mussel per container, there was no need to code the mussels, and any problems of counting threads (when mussels clump, which they invariably did) were prevented (Rajagopal et al. 1997). The byssus threads produced by mussels were counted after 24 h and expressed as threads mussel^{-1} day^{-1}(Van Winkle 1970).

Mortality experiment
Two size groups of *Mytilus edulis* (shell length in mm ± SD; 14.0 ± 0.24. and 25.2 ± 0.34) were tested at six different peroxide concentrations including control (control, 1.0, 2.0, 3.0, 4.0, and 5 mg L^{-1}).

In the earliest experiments, comparable mortality responses were observed between fed (mixed algal culture) and non-fed *Brachidontes varoabilis*, when exposed to chlorination. Similar observations were also reported earlier for *P. viridis* (Rajagopal et al. 1995), *B. striatulus* (Rajagopal et al. 1997), and *P. perna* (Rajagopal et al. 2003a, b). Therefore, *Mytilus edulis* used in the present study were not fed during the course of the experiment. Mortality was assessed at 6-h intervals. The criterion for mortality of mussels was a shell valve gape with no response of exposed mantle tissue to external stimuli (Rajagopal et al. 1997). Dead mussels were immediately removed from the flask. The number of dead animals in each experiment was recorded, along with their shell lengths and total weights for each observation event. The same experiment was repeated three times for each size group and peroxide concentration (14 and 25 mm (20 mussels in each experiment × 5 peroxide concentration (including control) × 2 size groups × 3 replicates = 720 mussels)).

Kinetic model (disinfection) for mussel mortality
An important feature of kinetic modeling is not only to simplify but also to idealize a complex phenomenon of cleansing systems. With the data

from these experiments designated above, we attempted to determine the coefficients of selected models. The major principles of disinfection kinetics were articulated by Chick and recognized the close similarity between microbial inactivation by chemical disinfectants and chemical reactions (Chick, 1908). From Chick's law, if N_0 is the number of organisms when t equal 0, it can be expressed as

$$\log(N/N_0) = -k * t, \tag{2}$$

where

N = a number of microorganism at contact time t
N_0 = a number of initial microorganism at contact time, $t = 0$
k^* = inactivation rate constant
t = contact time

Watson (1908) proposed an empirical logarithmic function to relate the rate constant of inactivation, "k^*" to the disinfectant concentration "C" (). In general, disinfection systems are designed by the "Ct" values derived from Chick-Watson kinetics based on the data obtained from laboratory inactivation studies.

$$k* = k C^n \tag{3}$$

$$\log(N/N_0) = -k C^n t, \tag{4}$$

where

k = constant for a specific microorganism and set of conditions
C = disinfectant concentration
n = coefficient of dilution

The Watson function, Eq. (4), is based on the assumption that microorganisms are genetically similar and of a single strain of synchronous development and the killing action would be a single-hit and single-site type. The assumptions are necessary in order to derive the Chick-Watson model based on a chemical reaction mechanism. In many cases, the n value for Chick-Watson law is close to 1.0, and hence, a fixed value of the product of concentration and time (Ct product) results in a fixed degree of inactivation (AWWA 1999).

Results
Sub lethal responses
Oxygen consumption
In control experiments, mussel in the 25 and 14 mm size group showed a maximum oxygen consumption rate of 3.13 ml O$_2$ mussel^{-1}h^{-1}. In control, the oxygen consumption did not show potential difference with respect to the size of mussels. Oxygen consumption of blue mussel species at different peroxide levels showed a progressive decline as the peroxide increased from 1.0 to 4.0 mg l^{-1} (Fig. 1). For example, oxygen consumption of 14 mm mussel showed a

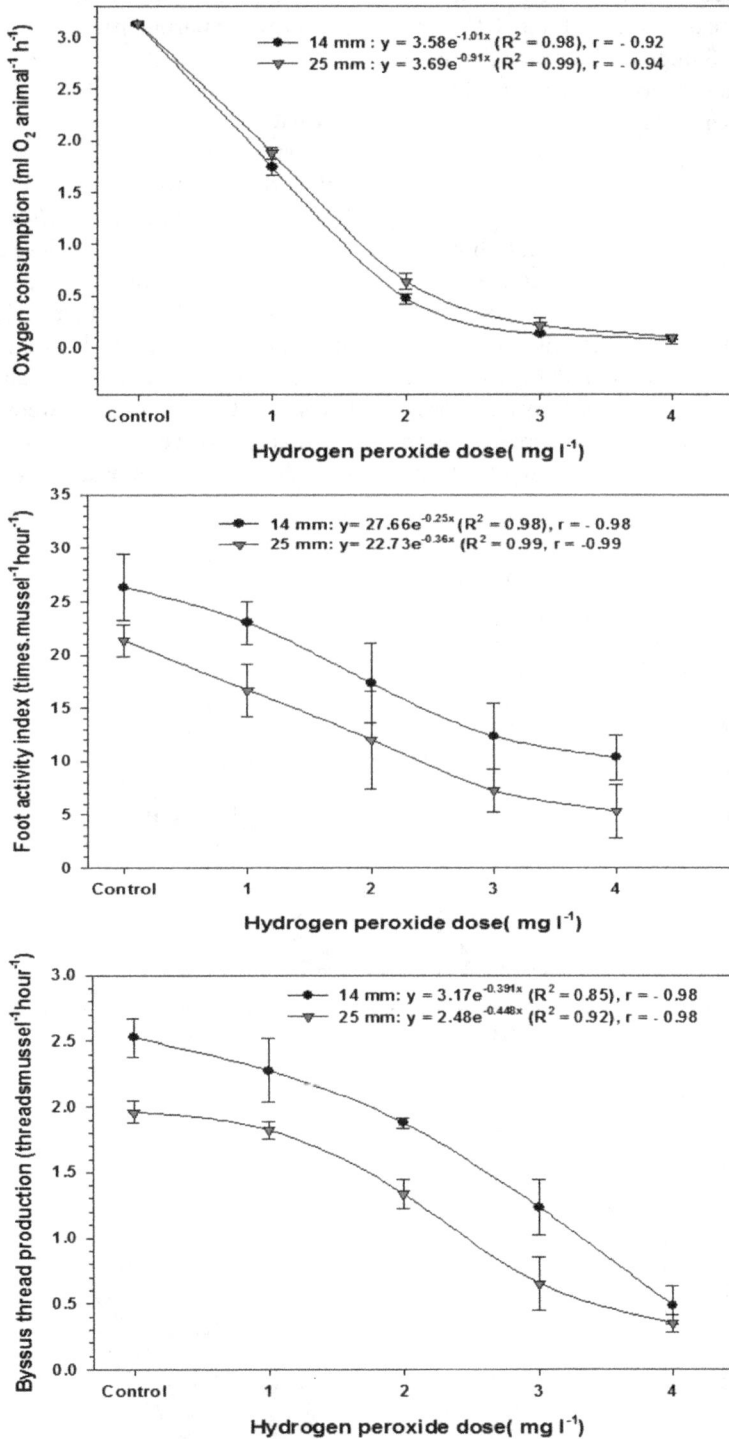

Fig 1 Relationship between H_2O_2 concentration and physiological activities of *Mytilus edulis*

reduction of 98% with 4 mg l^{-1} peroxide concentration and 44% with 1 mg l^{-1} peroxide concentration compare with control. Data also show a clear size dependent variation in oxygen consumption. As the size increased, progressive increase in oxygen consumption was observed.

Foot activity index

The highest foot activity index decrease was measured 60% in 4 mg l^{-1} peroxide experiments with 14 mm mussel (Fig. 1) and 75% with 25 mm mussel when compared with control. As peroxide concentration decreased, the foot activity index also decreased accordingly. For example, foot activity index of 14 mm mussel was decrease 12% at 1 mg l^{-1} peroxide when it was 53% at 3 mg l^{-1}. Foot activity index decreasing with peroxide concentration for 14 and 25 mm mussel was shown in Fig. 1.

Byssus thread production

The byssus thread production of *M. edulis* showed a progressive decline as the peroxide concentration inceased. The smaller mussels showed higher byssus production. For example, 14 mm size mussel showed 2% higher byssus thread production than 25 mm at 4 mg l^{-1} peroxide (Fig. 1). However, byssus thread production of subjected to continuous peroxide application showed a reduction of 10 to 80% of 14 mm

and 7% to 82% of 25 mussels group when compared to control experiments.

Mortality

Two size group *M. edulis* mussels were tested at the standard laboratory condition. The time required for 100% mortality of *Mytilus edulis* mussel exposed to different hydrogen peroxide levels was measured. With 30 days exposure to peroxide, the lethal percentage has not reached 100% mortality. However, the cumulative mean mortality of mussel within studied period was represented by Fig. 2. No mortality observed in control tanks. Smaller size mussels were dead at higher percentage than the bigger size ones. For example, 14 mm mussels showed 90% mortality whereas 25 mm mussels

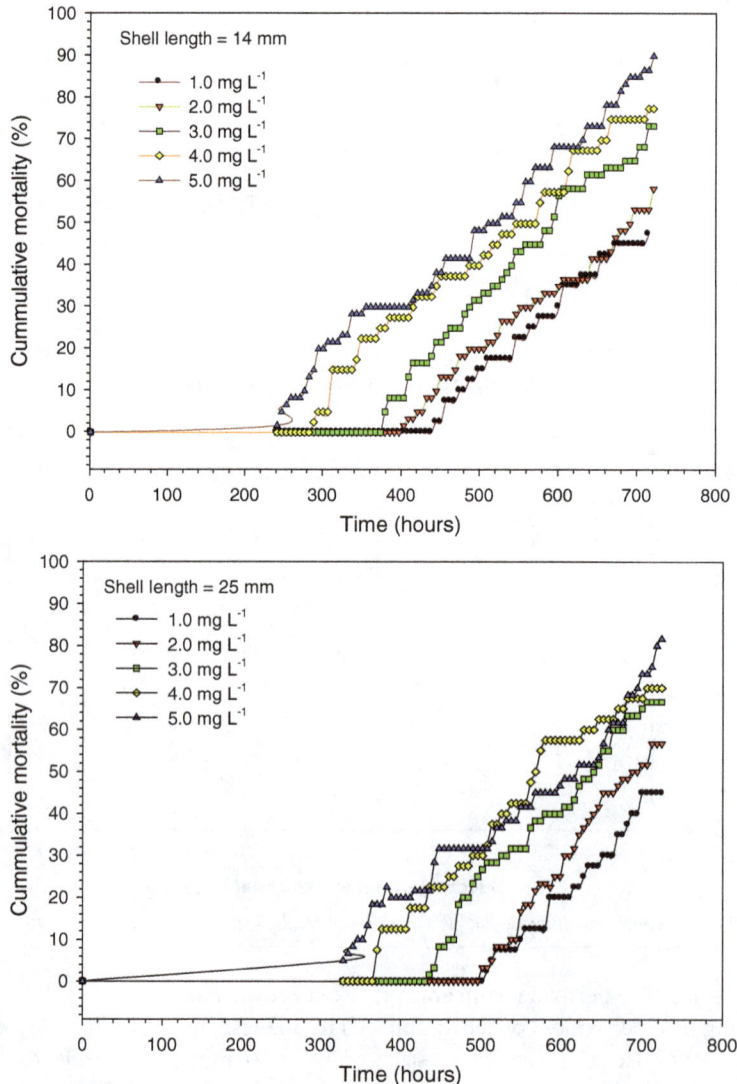

Fig. 2 Time course variation of mussel mortality [14 mm (*top*) and 25 mm (*bottom*)] of blue mussels at five different hydrogen peroxide concentrations (1–5 mg l^{-1})

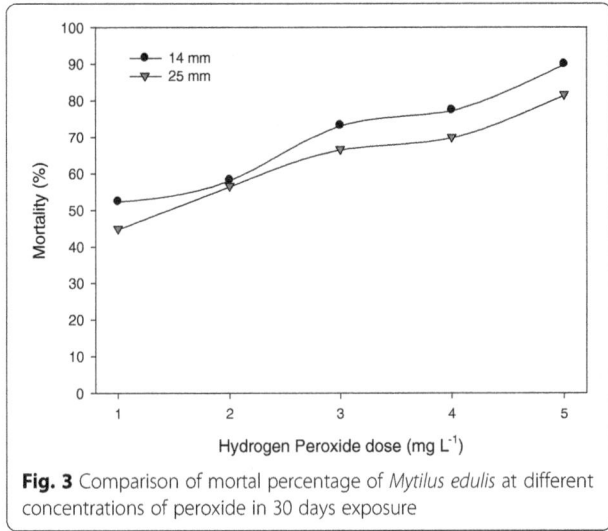

Fig. 3 Comparison of mortal percentage of *Mytilus edulis* at different concentrations of peroxide in 30 days exposure

reached 81% mortality in 5 mg l^{-1} peroxide concentration (Fig. 3).

Kinetic-based analysis of mortality

Based on the Chick-Watson model, the inactivation degrees on Ct value were presented in Fig. 4. As seen in Eq. (4), the estimated k values were found to be "0.0005" and "0.0004" for 14 and 25 mm size mussels, respectively.

$$\log(N/N_0) = -0.0005\ Ct \qquad (5)$$

$$\log(N/N_0) = -0.0004\ Ct \qquad (6)$$

The log inactivation value was greater for 14 mm size mussel than 25 mm mussel. At the same, Ct

value for 5 mg l^{-1}, the maximum log inactivation values were 2-log and 1.5-log for 14- and 25-mm size mussels, respectively (Table 1).

Discussion

In the present study, peroxide concentration as low as 1 mg l^{-1} to higher 4 mg l^{-1} was used to assess the physiological processes of mussel. Even at the lower level, peroxide tends to interfere with the physiological processes at 1 mg l^{-1} peroxide, for instance. The experiments conducted in the present study were aimed to ascertain the effects of peroxide administered at low levels on the physiological processes such as oxygen consumption, foot activity, and byssus thread production of test mussels. It may be noted that all these three activities are directly correlated to valve opening of mussels. The data in general show decreased activity with increasing peroxide concentrations. Valve opening was affected even at the lowest concentration (1.0 mg l^{-1}), indicating that the entire mussel could sense the presence of chlorine at such low levels. By monitoring valve opening in *M. edulis* species, a 14-mm mussel was known to open its valve more frequently than any 25 mm size mussels.

Laboratory tests of the biocide using model organisms are necessary to find its optimal concentrations for adequate fouling control at different sites (Claudi and Mackie 1994, Rajagopal et al. 2002). The most obvious feature of bivalve is the protective shell. During peroxide application, mussels shut their valves (Khalanski and Bordet 1980) and halt byssus thread production (Rajagopal et al. 1997). By doing it, bivalves isolate their body tissues from any changes caused by the external environment. At a high toxic external

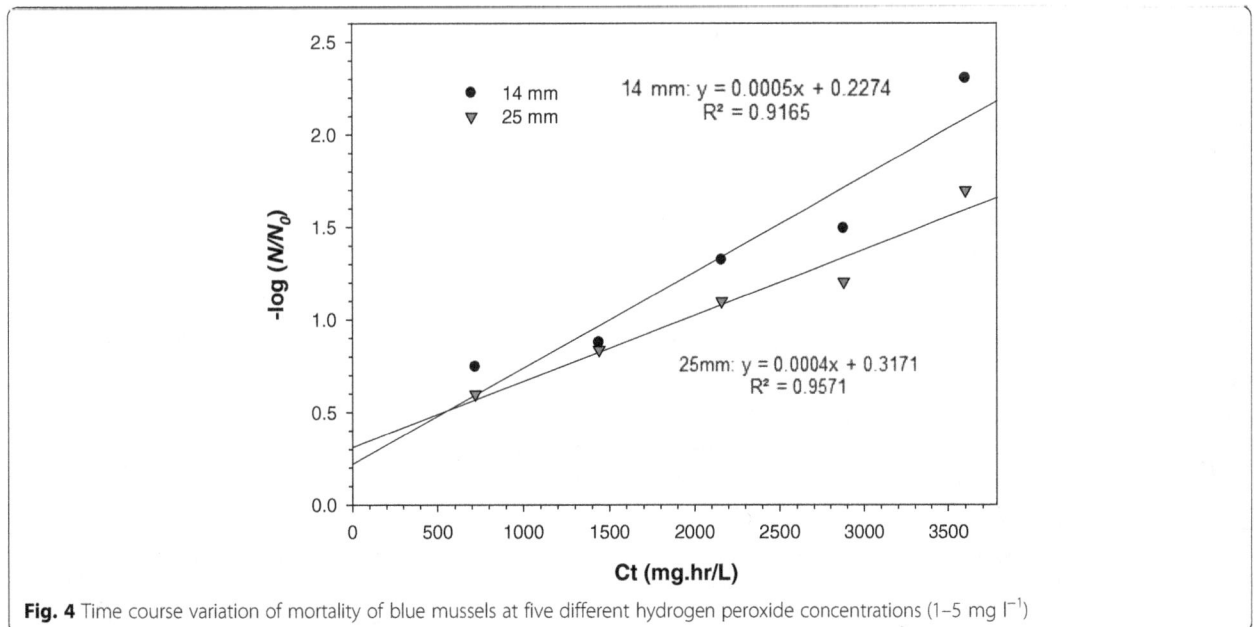

Fig. 4 Time course variation of mortality of blue mussels at five different hydrogen peroxide concentrations (1–5 mg l^{-1})

Table 1 Estimated Ct, log inactivation value, and rate constant of inactivation for death of *Mytilus edulis* with residual hydrogen peroxide

H_2O_2 (mg l^{-1})	$[Ct]$ (mg h l^{-1})	Log inactivation $[-\log (N/N_0)]$		Rate constant of inactivation, [K]	
C	$t = 30$ day (720 h)	14 mm	25 mm	14 mm	25 mm
1.0	720	0.74444	0.59784	0.0005	0.0004
2.0	1440	0.87547	0.83625		
3.0	2160	1.32176	1.09861		
4.0	2880	1.49165	1.20397		
5.0	3600	2.30259	1.69645		

environment, mussels are forced to shut their valves and rest in stored food reserves and anaerobic respiration until energy resources are depleted or metabolic wastes reach toxic levels. The lag period of Fig. 2 might portray those reserved energy resources of test mussels. However, influence of other factors such as sex (Martin et al. 1993a), season (Jenner 1985), and reproductive differences (Martin et al. 1993b)—that may cause greater biocide resistance—cannot be discounted.

H_2O_2 can not only act as an oxidizing agent on the skin of living organism but also produce hydroxyl radicals, which are known to be harmful to living cells. Especially, the radical formation could be accelerated in the presence of metallic ions like in seawater mood. In that context, we assume that hydroxyl radicals along with oxidizing factor of H_2O_2 may terminate lives of blue mussels through a complicated, internal breakdown mechanism. The Chick-Watson model with $n = 1$ reflected fairly the data shown in Fig. 4.

Conclusions

From the present study, it is clear that marine bivalve *Mytilus edulis* is sensitive even to lower peroxide environment. Physiological behavior (oxygen consumption, foot activity, and byssus thread production) showed decreasing pattern unanimously with increasing peroxide concentrations. Young mussel showed higher physiological activity than older mussel. The higher activity seems to be more vulnerable or susceptible to any external disturbances like hydrogen peroxide, though. Therefore, early stage in mussel cultivation is always better in order to prevent its colonization, followed by macrofouling. As a result, hydrogen peroxide is found to be a prospective biocide for preventing mussel colonization on marine environment.

Acknowledgements

This research was a part of the project entitled as "Improvement Technologies of Hypoxic Water Mass in Semi-Enclosed Coastal Bay," funded by the Ministry of Oceans and Fisheries, and BK21 plus program, South Korea.

Authors' contributions
Both authors read and approved the final manuscript.

Competing interests
The authors declare that they have no competing interest.

Author details
[1]Department of Ocean System Engineering, College of Marine Science, Gyeongsang National University, Cheondaegukchi-Gil 38, Tongyeong, Gyeongnam 650-160, South Korea. [2]Department of Marine Environmental Engineering, College of Marine Science, Engineering Research Institute (ERI), Gyeongsang National University, Cheondaegukchi-Gil 38, Tongyeong, Gyeongnam 650-160, South Korea.

References

AWWA. (1999). *Water quality and treatment* (5th ed., pp. 22–32). New York: McGraw-Hill.

Bruijs, M. C. M., Kelleher, B., van der Velde, G., & bij de Vaate, A. (2001). Oxygen consumption, temperature and salinity tolerance of the invasive amphipod *Dike rogammarus villosus*: indicators of further dispersal via Ballast water transport. *Archiv für Hydrobiologie, 152*, 633–646.

Chick, H. (1908). An investigation into the laws of disinfection. *Journal of Hygiene (Cambridge), 8*, 92–158.

Claudi, R., & Mackie, G. L. (1994). *Practical manual for zebra mussel monitoring and control.* London: Lewis Publishers.

Epstein, N. (1981). Thinking about heat transfer fouling: a 5x5 matrix. *Heat Transf. Eng, 4*, 43–56.

Freese, S. D., & Nozaic, D. J. (2007). Chlorine based disinfectants: how do they compare? *Water Science and Technology, 55*(10), 403–411.

Holmes, NJ, (1970). The effects of chlorination on mussels. Report No. RD/L/R 1672, Central Electricity Research Laboratories, Leatherhead, Surrey, pp. I-20.

Jenner, HA (1985). Chlorine minimization in macrofouling control in the Netherlans. In: Jolly RL, Bull RJ, Davies WP, Katz S, Roberts MH, Jacobs VA(Eds.), Water chlorination: Chemistry, Environmental Impact and Health Effects. Vol 5 Lewis publishers, London, pp.1425-1433.

Jenner, HA, Whitehouse, JW, Taylor, CJL and Khalanski, M (1998). Cooling water management in European power stations: biology and control, pp.1–225. Hydroe'cologieAppliquee 1–2, Chatou, Paris: Electricite' de France.

Khalanski, M, Bordet, F (1980). Effects of chlorination on marine mussels. In: Jolly RL, Brungs WA, Cumming RB(Eds.), Water Chlorination: Chemistry, Environmental Impact and Health Effects, vol. 3. Ann Arbor, Michigan, pp. 5577567.

Martin, ID, Mackie, GL, Baker, MA, (1993a). Control of the biofouling mollusc, *Dreissena polymorpha* (Bivalvia: Dreissenidae), with sodium hypochlorite and with polyquaternary ammonia and benzothiazole compounds. *Archives of Environmental Contamination and Toxicology, 24*, 381-388.

Martin, ID, Mackie, GL, Baker, MA. (1993b). Acute toxicity tests and pulsed-dose delayed mortality at 12 and 22°C in the zebra mussel (*Dreissena polymorpha*). *Archives of Environmental Contamination and Toxicology, 24*, 389-398.

Mattice, J. S., & Zittel, H. E. (1976). Site-specific evaluation of power plant chlorination. *Journal of the Water Pollution Control Federation, 48*, 2284–2308.

Rajagopal, S (1991). Biofouling problems in the condenser cooling circuit of a coastal power station with special reference to green mussel, *Pernavividis* (L.). Ph. D. thesis. University of Madras, India, pp. I-113.

Rajagopal, S., Venugopalan, V. P., Nair, K. V. K., & Azariah, J. (1995). Response of green mussel, *Perna viridis* (L.) to chlorine in the context of power plant biofouling control. *Marine and Freshwater Behaviour and Physiology, 25*, 261–274.

Rajagopal, S., Nair, K. V. K., Van der Velde, G., & Jenner, H. A. (1997). Response of mussel, *Brachidontes striatulus*, to chlorination: an experimental study. *Aquatic Toxicology, 39*, 135–149.

Rajagopal, S., Van der Velde, G., & Jenner, H. A. (2002). Does status of attachment influence chlorine toxicity in zebra mussel, *Dreissena polymorpha*. *Environmental Toxicology and Chemistry, 21*, 342–346.

Rajagopal, S, Venugopalan, VP, Van der Velde, G, Jenner, HA. (2003a). Tolerance of five species of tropical marine mussels to continuous chlorination. *Marine Environment Research, 55*, 277–291.

Rajagopal, S, Venugopalan, VP, Van der Velde, G, Jenner, HA. (2003b). Response of fouling brown mussel, *Pernaperna* (L.), to chlorine. *Archives of Environmental Contamination and Toxicology, 44,* 269–276

Seed, R. (1976). Ecology. In B. L. Bayne (Ed.), *Marine mussels: their ecology and physiology* (pp. 13–66). Cambridge, UK: Cambridge University Press.

Stewart, P. L. (1994). Environmental requirements of the blue mussel (*Mytilus edulis*) in eastern Canada and its response to human impacts. *Canadian Technical Report of Fisheries and Aquatic Sciences, 2004.* 41.

Strickland, T. D. H., & Parsons, T. R. (1972). *A practical hand book of seawater analysis.* Ottawa: Fisheries Research Board of Canada.

Vaccaro, R. F., Azam, F., & Hadson, R. E. (1977). Response of natural marine bacterial populations to copper: controlled ecosystem population experiment. *Bulletine of Marine Science, 27*(1), 17–22.

Van Winkle, W. V. (1970). Effects of Environmental Factors on Byssal Thread Formation. *Marine Biology, 7,* 143–148.

Watson, H. E. (1908). A note on the variation of the rate of disinfection with change in the concentration of the disinfectant. *Journal of Hygiene (Cambridge), 8,* 536–542.

5

Newly recorded species of the genus *Synura* (Synurophyceae) from Korea

Bok Yeon Jo and Han Soon Kim[*]

Abstract

Background: Species in the heterokont genus *Synura* are colonial and have silica scales whose ultrastructural characteristics are used for classification. We examined the ultrastructure of silica scales and molecular data (nuclear SSU rDNA and LSU rDNA, and plastid *rbc*L sequences) to better understand the taxonomy and phylogeny within the section *Petersenianae* of genus *Synura*. In addition, we report the first finding of newly recorded *Synura* species from Korea.

Results: We identified all species by examination of scale ultrastructure using scanning and transmission electron microscopy (SEM and TEM). Three newly recorded species from Korea, *Synura americana*, *Synura conopea*, and *Synura truttae* were described based on morphological characters, such as cell size, scale shape, scale size, keel shape, number of struts, distance between struts, degree of interconnections between struts, size of base plate pores, keel pores, base plate hole, and posterior rim. The scales of the newly recorded species, which belong to the section *Petersenianae,* have a well-developed keel and a characteristic number of struts on the base plate. We performed molecular phylogenetic analyses based on sequence data from three genes in 32 strains (including three outgroup species). The results provided strong statistical support that the section *Petersenianae* was monophyletic, and that all taxa within this section had well-developed keels and a defined number of struts on the base plate.

Conclusions: The phylogenetic tree based on sequence data of three genes was congruent with the data on scale ultrastructure. The resulting phylogenetic tree strongly supported the existence of the section *Petersenianae*. In addition, we propose newly recorded *Synura* species from Korea based on phylogenetic analyses and morphological characters: *S. americana*, *S. conopea*, and *S. truttae*.

Keywords: *Synura americana*, *S. conopea*, *S. truttae*, Morphology, Ultrastructure, Scale, Molecular phylogeny, Taxonomy

Background

Ehrenberg established the genus *Synura* in 1834 (Ehrenber 1834), with *S. uvella* as the type species. *Synura* is the most common and widespread genus in many phytoplankton floras (Kristiansen & Preisig 2007). The species in this genus are colonial flagellates with two visible flagella and two chloroplasts, and are covered by imbricate silica scales. Several scale morphologies (apical scales, body scales, transition scales, and caudal scales) occur at different locations on the surface of the same cell. These body scales are the most important character for species identification (Kristiansen & Preisig 2007).

Early classification of *Synura* species using light microscopy (LM) was based largely on features such as cell size and shape, general outline of scales, and the spine or keel (Ehrenber 1834). Previous taxonomical studies of *Synura* have traditionally stressed the distinguishing features of these scales.

The classification of *Synura* species using electron microscopy (EM) is based on scale ultrastructure (Korshikov 1929; Petersen & Hansen 1956; Petersen & Hansen 1958; Fott & Ludvík 1957; Asmund 1968; Balonov & Kuzmin 1974; Péterfi & Momeu 1977; Takahashi 1967; Takahashi 1972; Takahashi 1973; Takahashi 1978; Cronberg 1989; Škaloud et al. 2012; Škaloud et al. 2013; Škaloud et al. 2014). In fact, examination of the ultrastructural features of the silica scales has revolutionized *Synura* taxonomy. The first classification scheme to consider scale ultrastructure

* Correspondence: kimhsu@knu.ac.kr
Department of Biology, Kyungpook National University, Daegu 41566, South Korea

suggested that the genus *Synura* is divided into two sections: *Petersenianae* and *Uvellae* (Petersen & Hansen 1956). Subsequent classification schemes have made additional subgeneric distinctions (Balonov & Kuzmin 1974; Péterfi & Momeu 1977; Takahashi 1967; Takahashi 1972; Takahashi 1973; Takahashi 1978; Cronberg 1989).

The first molecular analyses investigated the genetic variability in 15 individuals of *Synura petersenii* by comparison of nuclear internal transcribed spacer (ITS) sequences (Wee et al. 2001). Subsequent molecular analyses examined ITS sequences from 21 other individuals (Kynčlová et al. 2010). Also, phylogenetic analyses investigated about 100 *S. petersenii* using seven-protein gene and confirmed the high degree of cryptic, species-level diversity within this nominal species (Boo et al. 2010). A recent taxonomic assessment of observed cryptic diversity redefined the species concept within the *S. petersenii* morphotype and recognized six cryptic lineages as separate species: *Synura americana, Synura conopea, Synura glabra, Synura macropora, Synura petersenii,* and *Synura truttae* (Škaloud et al. 2012). Most recently, the classification of *Synura* described an additional four new species within the *S petersenii* species complex based on scale morphology and sequence data (ITS, *rbc*L, and *cox*1) (Škaloud et al. 2014).

Several researchers have studied the genus *Synura* from different regions in Korea by the use of EM (Kim 1997). These studies described nine species and provided very short descriptions based on scale ultrastructure (Kim 1997; Kristiansen 1990). Most recently, the first molecular multigene phylogeny of a large number of *S. petersenii* confirmed the high degree of cryptic, species-level diversity (Boo et al. 2010).

The purpose of the present study was to provide a better understanding of the taxonomy and molecular phylogeny within the section *Petersenianae* of genus *Synura* by analysis of the ultrastructure of the silica scales and molecular data (nuclear SSU rDNA and LSU rDNA, and plastid *rbc*L sequences) and to describe three species of *Synura* that are new to Korea.

Methods
Strains and cultures
The information and accession numbers for the 32 strains (including three outgroup species) examined in this study are in Table 1. Strains were either obtained from culture collections or collected with a 20-µm mesh plankton net (Bokyeong Co., Pusan, Korea) from small ponds in Korea. The details of the culture methods were previously published (Jo et al. 2011; Jo et al. 2013).

Morphological investigations
For field emission scanning electron microscopy (SEM), cells were filtered using nylon membrane filters (Whatman Ltd., Maidstone, UK), rinsed in distilled water, fixed in 1%

OsO$_4$, dehydrated, and then prepared and viewed as described previously (Jo et al. 2011). Voucher specimens were stored at the Kyungpook National University Herbarium. For field emission transmission electron microscopy (TEM), cells were prepared by air drying onto formvar coated copper grids. The grids were viewed in a JEM 1010 TEM (JEOL Ltd., Tokyo, Japan) at 80 kV. Images were recorded on Kodak EM Film 4489 (Eastman Kodak Co., Rochester, NY, USA) and scanned to digital format using an Epson Perfection V700 Photo scanner (Epson Korea Co., Ltd, Seoul, Korea). The terminology used to describe scale ultrastructure follows a previous method (Škaloud et al. 2012).

DNA extraction, amplification, sequence alignment, and phylogenetic analyses
DNA extraction, PCR amplification, PCR product purification, and sequence alignment were conducted as previously described (Jo et al. 2011; Jo et al. 2013). Phylogenetic analyses were performed using a combined dataset of 5011 characters (nr SSU rDNA = 1638, nr LSU rDNA = 2548, and pt *rbc*L = 825) by maximum likelihood (ML) and Bayesian inference (BI). Although nuclear ITS1 and ITS2 sequences were also determined, these sequences were used to examine groups of genetically identical strains and as a barcode to identify species. The sequences of three species of Chrysophyceae (*Chromulina* sp., *Ochromonas danica,* and *Ochromonas* sp.) were used as outgroups to root the tree. Primer regions and ambiguously aligned regions were removed prior to phylogenetic analyses. Prior to ML analysis, the best-fit model for individual and concatenated data sets was traced under Bayesian information criterion (BIC) using Modeltest 3.7 (Posada & Crandall 1998). GTR + I + G model for all the individual and concatenated data sets was selected. We used the GTR + I + G nucleotide model as implemented in RAxML v8 (Stamatakis 2014). Bayesian analyses were run using MrBayes 3.2 (Ronquist et al. 2012) with a random starting tree and ran for 2×10^6 generations, keeping on tree every 1000 generations. The burn-in point was identified graphically by tracking the likelihoods in Tracer v.1.6 (Rambaut et al. 2013). Trees were visualized using the FigTree v.1.4.2 program (Rambaut A. FigTree v1.4.2 2014). Each analysis was conducted as previously described (Jo et al. 2011; Jo et al. 2013).

Results and discussion
Morphological characteristics
We identified all species based on scale ultrastructure from SEM and TEM. This analysis led to identification of three species that are new to Korea: *S. americana, S. conopea,* and *S. truttae* (Figs. 1, 2 and 3 and Table 2). The scales of the newly recorded species, all in the section

Table 1 List of strains used in the molecular study and GenBank accession number

Taxa/strain	GenBank accession			
	Nuclear ITS	Nuclear SSU	Nuclear LSU	Plastid *rbcL*
S. americana Kynčlová and Škaloud				
Chimu112407C	KP268712	KM590551	KM590617	KM590838
Johae010508F	KP268711	JX455151	JX455155	JX455147
CCMP862	GU338124	GU325583	—	GU325485
CCMP863	GU338125	GU325584	—	GU325486
KNUJO-CM2015I226	KX610938	KX610941	KX610944	KX610947
S. asmundiae (Cronberg and Kristiansen) Škaloud, Kristiansen and Škaloudová				
S90D10	KP268729	KM590553	KM590619	KM590840
S90D11	KP268730	HF549069	—	HF549079
S. bjoerkii (Cronberg and Kristiansen) Škaloud, Kristiansen and Škaloudová				
SC57A6	KP268731	HF549070	—	HF549080
S. conopea Kynčlová and Škaloud				
Sugyeji041808B	KP268690	KM590557	KM590623	KM590844
Yeonseong120807E	KP268689	KM590558	KM590624	KM590845
CCMP859	GU338121	GU325580	—	GU325482
NIES1007	GU338119	GU325578	—	GU325479
KNUJO-YG2016O117	KX610939	KX610942	KX610945	KX610948
S. glabra Korshikov emend. Kynčlová and Škaloud				
Bonggye101407K	KP268722	KM590564	KM590630	KM590851
Cheonma041908B	KP268716	KM590565	KM590631	KM590852
Dohak111107C	KP268721	JX455149	JX455153	JX455145
Geumma020610B	KP268718	KM590568	KM590634	KM590855
Hwangsan012508A	KP268724	KM590571	KM590637	KM590858
S. macracantha (Petersen and Hansen) Asmund				
S90B5	KP268732	HF549064	KM590648	HF549075
S. petersenii Korshikov emend. Škaloud and Kynčlová				
Buje100307A	KP268710	KM590586	KM590657	KM590873
Gamgok111107C	KP268707	KM590587	KM590658	KM590874
Swaeji103109I	KP268705	KM590589	KM590660	KM590876
Yongseong112407A	KP268706	KM590590	KM590661	KM590877

Table 1 List of strains used in the molecular study and GenBank accession number (*Continued*)

Youngji101407A	KP268708	JX455150	JX455154	JX455146
S. truttae (Siver) Škaloud and Kynčlová				
Hanjeong080611J	KP268702	KM590609	KM590680	KM590896
Jangjuk032611J	KP268703	KM590610	KM590681	KM590897
CAUP2	GU338138	GU325598	—	GU325500
CAUPD5	GU338140	GU325600	—	GU325502
KNUJO-HJ2015122	*KX610940*	*KX610943*	*KX610946*	*KX610949*
Chromulina sp.				
SAG 17.97	—	EF165103	GU935638	EF165151
Ochromonas danica Pringsheim				
SAG 933.7	—	JQ281514	GU935636	GU935657
Ochromonas sp.				
SAG 933.10	—	EF165109	GU935637	GU935658

New sequences are indicated in italic type

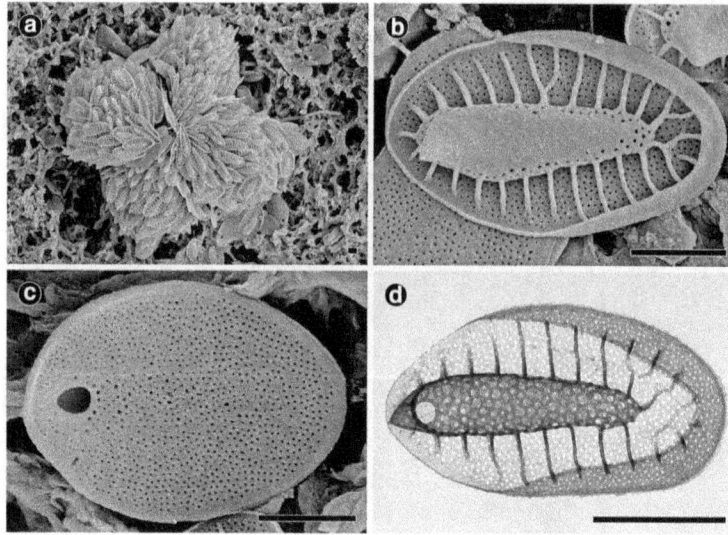

Fig. 1 Morphology of the colony and scales of *Synura americana* (**a–c**: SEM, **d**: TEM). All *scale bars*, 1 μm. **a** SEM image of colony forming cells. **b** Top surface of a body scale. **c** Bottom surface of a body scale. **d** TEM image of body scale

Petersenianae, have well-developed keels and a number of struts on the base plate. The terminology used to describe the ultrastructure of these scales follows a previous method (Škaloud et al. 2012). Other studies have described newly recorded species of *Synura* from Korea based on morphological characters, such as cell size, scale shape, scale size, keel shape, number of struts, distance between struts, degree of interconnections between struts, size of the base plate pores, keel pores, base plate hole, and posterior rim (Škaloud et al. 2012). Two of our species (*S. americana* and *S. conopea*) are morphologically similar to *S. petersenii*, suggesting a close relationship. *S. conopea* was most similar to

S. petersenii in terms of cell shape and transverse folds, although these species differ in keel reticulation. *S. conopea* is distinguished by its smaller scales and its large and closely arranged keel pores. *S. americana* is characterized by rounded scales, a near absence of transverse folds, an occasionally triangular keel, and long rear scales. *S. truttae* is characterized by small scale size, keel tips, large base plate hole, and short distance between struts.

Taxonomic description
S. americana **Kynčlová and Škaloud 2012 (Fig. 1)**
Reference: Škaloud et al. 2012, p. 320, Figs. 62–69.

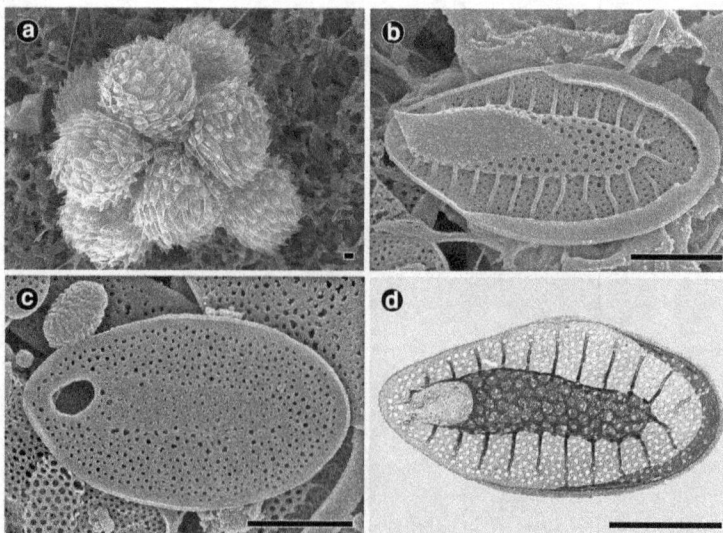

Fig. 2 Morphology of the colony and scales of *Synura conopea* (**a–c**: SEM, **d**: TEM). All *scale bars*, 1 μm. **a** SEM image of colony forming cells. **b** Top surface of a body scale. **c** Bottom surface of a body scale. **d** TEM image of a body scale

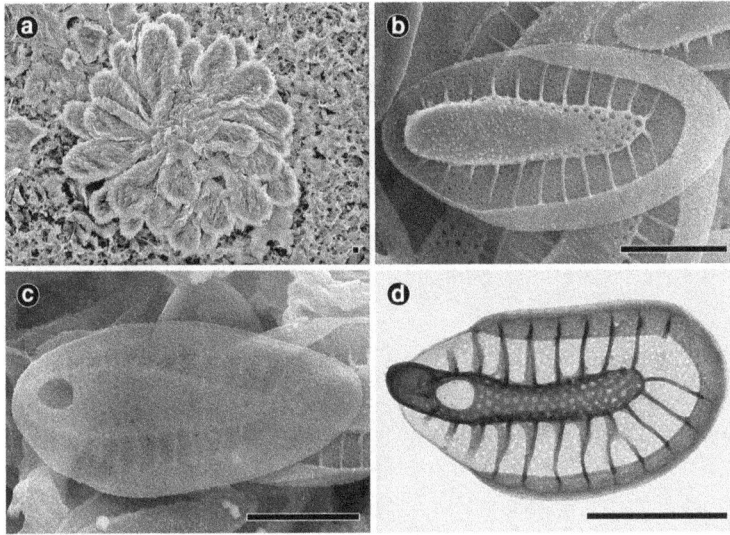

Fig. 3 Morphology of the colony and scales of *Synura truttae* (**a–c**: SEM, **d**: TEM). All *scale bars*, 1 μm. **a** SEM image of colony forming cells. **b** Top surface of a body scale. **c** Bottom surface of a body scale. **d** TEM image of a body scale

Specimens examined: KNUJO-CM20151226.

Description: Colonies globular and 22–51 μm in diameter (Fig. 1a). Cells pyriform (22–28 × 8–12 μm) and entirely covered by rounded scales (Fig. 1a). Body scales 3.0–4.2 × 1.7–2.3 μm (Fig. 1b–d). The keel often terminates at an acute tip (Fig. 1b) and is ornamented by medium-sized pores (Fig. 1d). In some cases, the keel is wider in the anterior region, giving it a triangular shape (Fig. 1b). The basal plate, ornamented by numerous small pores, is anteriorly perforated by a rounded base plate hole that is 0.08–0.27 μm in diameter (Fig. 1b–d). Numerous struts (21–24) extend regularly from the keel to the edge of the scale but almost never interconnect the transverse folds (Fig. 1b and d). The spacing between struts is 0.27–0.30 μm (Fig. 1b and d).

Site of collection: Chimu, Daesan-myeon, Haman-gun, Gyeongsangnam-do, Korea (35°20′21″N, 128°25′47″E).

Date of collection: 26 Dec 2015.

Distribution: Widely distributed. Canada (Wee et al. 2001), Colombia (Cronberg 1989), Czech Republic (Škaloud et al. 2012; Kynčlová et al. 2010), Denmark (Kristiansen 1988), Germany (Kies & Berndt 1984), Korea (Boo et al. 2010, this study), North America (Kling & Kristiansen 1983; Kristiansen 1975; Wee 1981), and USA (Wee et al. 2001; Boo et al. 2010).

S. conopea Kynčlová and Škaloud 2012 (Fig. 2)

Reference: Škaloud et al. 2012, p. 324, Figs. 78–85.

Specimens examined: KNUJO-YG20160117, NIBRFL0000131748, and NIBRFL0000131749.

Description: Colonies globular and 25–47 μm in diameter (Fig. 2a). Cells pyriform (20–28 × 8–12 μm) and entirely covered by lanceolate scales (Fig. 2a). Body scales 3.3–4.1 ×

1.4–1.9 μm (Fig. 2b–d). The keel terminates at an acute tip (Fig. 2b) and is usually broadened apically and ornamented by medium to large-sized pores (Fig. 2d). The basal plate, ornamented by numerous medium-sized pores, is anteriorly perforated by a round to oblong base plate hole that is 0.19–0.32 μm in diameter (Fig. 2b–d). Numerous struts (24–30) extend regularly from the keel to the edge of the scale but are usually not interconnected by transverse folds (Fig. 2b and d). The spacing between struts is 0.23–0.26 μm (Fig. 2b and d).

Site of collection: Yongji, Yongchon-ri, Toseong-myeon, Goseong-gun, Gangwon-do, Korea (38°13′43″N, 128°33′49″E).

Date of collection: 17 Jan 2016.

Distribution: Widely distributed. Argentina (Vigna & Munari 2001), Brazil (Couté & Franceschini 1988), Czech Republic (Škaloud et al. 2012; Kynčlová et al. 2010), Greenland (Jacobsen 1985), Ireland (Řezáčová & Škaloud 2005), Japan (Boo et al. 2010), and Korea (Boo et al. 2010, this study).

S. truttae (Siver 1987) Škaloud and Kynčlová 2012 (Fig. 3)

Basionym: *S. petersenii* f. *truttae* (Siver 1987), p. 111, Figs. 12–14.

Reference: Škaloud et al. 2012, p. 318, Figs. 52–61.

Specimens examined: KNUJO-HJ20151222.

Description: Colonies globular and 35–48 μm in diameter (Fig. 3a). Cells pyriform (22–31 × 11–13 μm) and entirely covered by lanceolate scales (Fig. 3a). Body scales elongated and 3.3–3.8 × 1.5–1.8 μm (Fig. 3a–d). The keel of the body scales has no apparent tip or a much reduced tip and is ornamented by small pores (Fig. 3b). The keel tip frequently has several (two to

Table 2 Summary of the major characteristic features observable with EM used in this study to distinguish between taxa of the section *Petersenianae*

Taxon	Cell size (μm)	Scale size (μm)	Base plate hole size (μm)	Number of struts	Distance of struts (μm)	Interconnection of struts by transverse folds	Other
S. americana Kynčlová & Škaloud	22–28 × 8–12	*3.0–4.2 × 1.7–2.3	0.08–0.27	21–24	0.27–0.30	Almost never interconnected	Occasional triangular shape of the keel
S. conopea Kynčlová & Škaloud	20–28 × 8–12	*3.3–4.1 × 1.4–1.9	0.19–0.32	24–30	0.23–0.26	Usually not interconnected	Large and closely arranged keel pores
S. glabra Korshikov emend. Kynčlová & Škaloud	19–28 × 10–14	*2.4–3.4 × 1.5–2.4	0.14–0.32	17–22	0.25–0.29	Never interconnected	With a small, narrow and sometimes bent keel
S. petersenii Korshikov emend. Škaloud & Kynčlová	20–31 × 8–12	*3.6–4.6 × 1.8–2.3	0.24–0.36	26–34	0.24–0.28	Mainly interconnected	Large scale dimensions, common presence of transverse folds
S. truttae (Siver) Škaloud & Kynčlová	22–31 × 11–13	*3.3–3.8 × 1.5–1.8	0.32–0.56	27–33	0.19–0.24	Mainly interconnected	Keel tip has several (two to four) very short teeth on its top end and large base plate hole

*The dimensions of body scales

four) very short teeth on its top (Fig. 3d) and is covered by a number of small bumps. The basal plate, ornamented by numerous small pores, is anteriorly perforated by a large, round to oblong base plate hole that is 0.32–0.56 μm in diameter (Fig. 3b–d). Numerous struts (27–33), which are often interconnected, regularly extend from the keel to the edge of the scale (Fig. 3b and d). Scales with nearly absent transverse folds (Fig. 3b–d). The spacing between struts is 0.19–0.24 μm (Fig. 3b and d).

Site of collection: Hanjeong, Girin-ri, Soseong-myeon, Jeongeup-si, Jeollabuk-do, Korea (35°33′55″N, 126°46′30″E).

Date of collection: 22 Dec 2015.

Distribution: Widely distributed. Czech Republic (Škaloud et al. 2012; Kynčlová et al. 2010), Korea (This study), and USA (Siver 1987; Siver & Wujek 1993; Siver & Lott 2004).

Molecular data

The 5011 nucleotides of the combined data set (nuclear SSU and LSU rDNA, and plastid *rbc*L) were determined for 32 strains (Table 1). Although the nuclear ITS1, 5.8S, and ITS2 sequences were also determined, these sequences were only used for to confirm identification, not to assess phylogenetic relationships. The combined sequences had 5011 nucleotides, 4039 variable sites, and 725 parsimoniously informative sites. The molecular data contained 12 new sequences (3 new nr SSU rDNA sequences, 3 new nr LSU rDNA sequences, 3 new nr ITS sequences, and 3 new pt *rbc*L sequences) and 102 published sequences (29 nr SSU rDNA sequences, 20 nr LSU rDNA sequences, 25 nr ITS sequences, and 28 pt *rbc*L sequences).

Phylogenetic analyses

We analyzed nr SSU and LSU rDNA, and pt *rbc*L sequences from 32 strains (including three outgroup species). The phylogenetic tree based on the Bayesian analysis was rooted with three species of Chromulinaceae serving as outgroups. The Bayesian and ML analyses recovered a tree with identical topologies (Fig. 4). The phylogenetic tree consisted of species of the section *Petsenianae*, each of which has a well-developed keel and a number of struts on

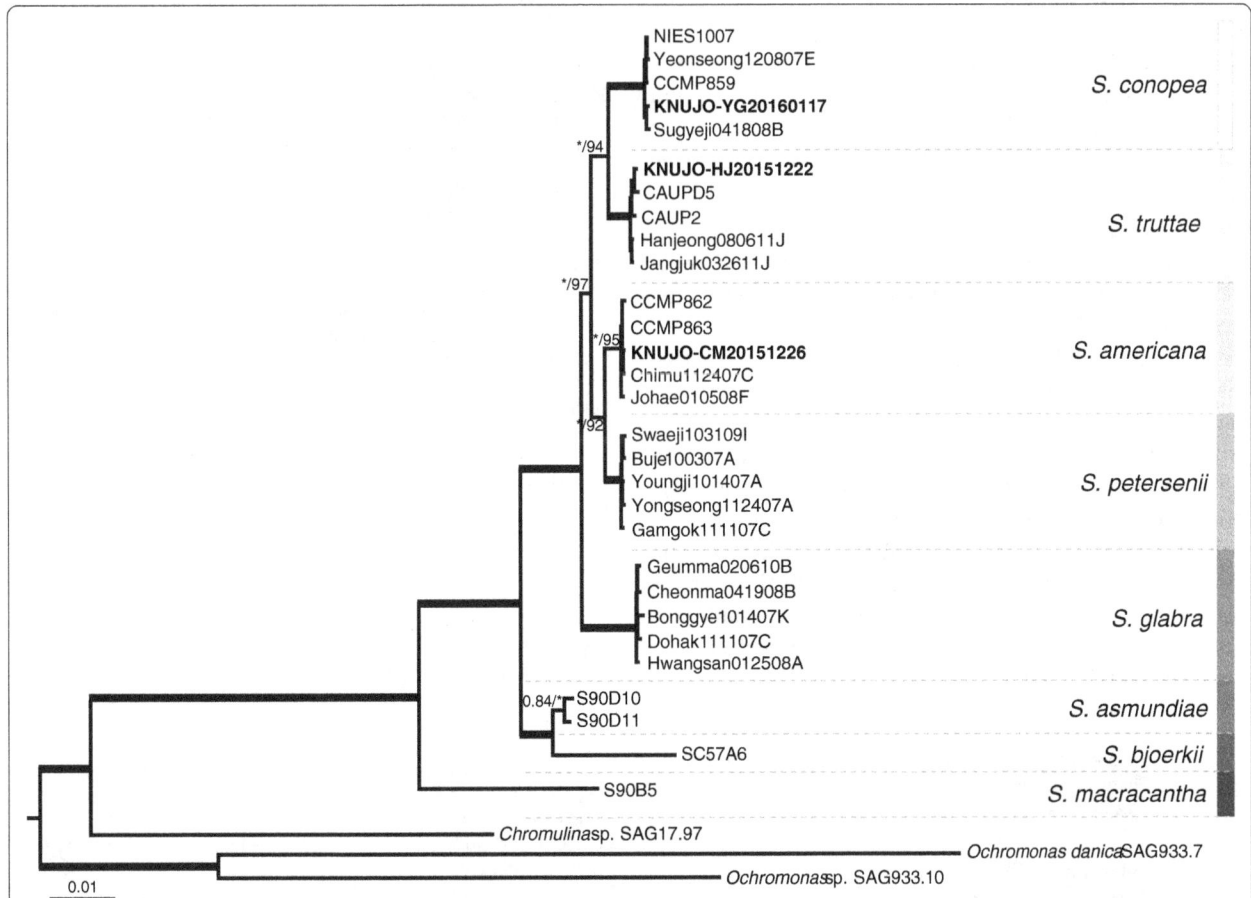

Fig. 4 Consensus Bayesian tree of the genus *Synura* based on a combined nuclear SSU and LSU rDNA, and plastid *rbc*L sequences data. Bayesian posterior probability (pp) and maximum-likelihood (ML) bootstrap values are shown above or below the branches. The bold branches indicate strongly supported values (pp = 1.00 and ML = 100). *Scale bar*, 0.01 substitutions/site

the base plate. The section *Petersenianae* formed a strongly supported monophyletic lineage (pp = 1.00 and ML = 100). The single strain of *Synura macracantha* diverged at the base of the tree, followed by *Synura bjoerkii* and *Synura asmundiae*. The single strain of *S. bjoerkii* was closely related to *S. asmundiae*, which included two strains (pp = 1.00 and ML = 100). *Synura glabra* formed a sister group with *S. americana*, *S. conopea*, *S. petersenii*, and *S. truttae* (pp = 1.00 and ML = 100), and *S. americana* and *S. petersenii* diverged at the next *S. glabra*. The five strains of *S. petersenii* formed a strongly supported monophyletic lineage (pp = 1.00 and ML = 100) and was a sister group to the five strains of *S. americana*, which included KNUJO-CM20151226 (pp = 1.00 and ML = 92). The five strains of *S. americana* were monophyletic group (pp = 1.00 and ML = 95), and the intraspecific similarity based on nuclear ITS rDNA sequence data ranged from 99.9% to 100.0%. The five strains of *S. truttae* (including KNUJO-HJ20151222) were a sister group to the five strains of *S. conopea*, which included KNUJO-YG20160117 (pp = 1.00 and ML = 94). The five strains of *S. conopea* formed a monophyletic lineage with strong support values (pp = 1.00 and ML = 100), and the intraspecific similarity based on nuclear ITS rDNA sequence data ranged from 98.5% to 100.0%. The five strains of *S. truttae* formed a monophyletic lineage with strong support values (pp = 1.00 and ML = 100), and the intraspecific similarity based on nuclear ITS rDNA sequence data was 100.0%.

Conclusions

In summary, we used molecular analysis of three genes and data on the scale ultrastructure to investigate the phylogenetic relationships within *Synura*, with a focus on the section *Petersenianae*. The phylogenetic tree based on a combined dataset was well congruent with the ultrastructural characteristics of scales. The phylogenetic tree was comprised of members of the section *Petersenianae*. The section *Petersenianae* was monophyletic with strong support values and characterized by a well-developed keel and a number of struts on the base plate. In addition, our morphological observations and molecular analyses confirmed unambiguously that this is the first report of *S. americana*, *S. conopea*, and *S. truttae* in Korea.

Funding
This work was supported by a grant from the National Institute of Biological Resources (NIBR) funded by the Ministry of Environment (MOE) of the Republic of Korea (NIBR201501209).

Authors' contributions
Both authors read and approved the final manuscript.

Competing interests
The authors declare that they have no competing interests.

References
Asmund, B. (1968). Studies on Chrysophyceae from some ponds and lakes in Alaska. VI. Occurrence of *Synura* species. *Hydrobiologia, 31*, 497–515.
Balonov, I. M., & Kuzmin, G. V. (1974). Vidy roda *Synura* Ehrenberg (Chrysophyta) v vodokhranilischchakh Volzhskogo Kaskada. *Botanicheskii Zhurnal, 59*, 1675–1686.
Boo, S. M., Kim, H. S., Shin, W., Boo, G. H., Cho, S. M., Jo, B. Y., Kim, J. H., Kim, J. H., Yang, E. C., Siver, P. A., Wolfe, A. P., Bhattacharya, D., Andersen, R. A., & Yoon, H. S. (2010). Complex phylogeographic patterns in the freshwater alga *Synura* provide new insights into ubiquity vs. endemism in microbial eukaryotes. *Molecular Ecology, 19*, 4328–4338. doi:10.1111/j.1365-294X.2010.04813.x.
Couté, A., & Franceschini, I. M. (1988). Scale-bearing chrysophytes from acid waters of Florianópolis, Santa Catarina Island, South Brazil. *Algological Studies, 88*(Suppl 123), 37–66.
Cronberg, G. (1989). Scaled chrysophytes from the tropics. *Nova Hedwigia Beiheft, 95*, 191–232.
Ehrenber, C. G. (1834). Dritter Beitrag zur Erkenntnis grosser Organisation in der Richtung des kleinsten Raumes. *Abhandlungen der Königlichen Akademie der Wissenschaften Berlin, 1833*, 145–336.
Fott, B., & Ludvík, J. (1957). Die submikroskopische Struktur der Kieselschuppen bei *Synura* und ihre Bedeutung fur die Taxonomie der Gattung. *Preslia, 29*, 5–16.
Jacobsen, B. A. (1985). Scale-bearing Chrysophyceae (Mallomonadaceae and Paraphysomonadaceae) from West Greenland. *Nordic Journal of Botany, 5*, 381–398.
Jo, B. Y., Shin, W., Boo, S. M., Kim, H. S., & Siver, P. A. (2011). Studies on ultrastructure and three-gene phylogeny of the genus *Mallomonas* (Synurophyceae). *Journal of Phycology, 47*, 415–425. doi:10.1111/j.1529-8817.2010.00953.x.
Jo, B. Y., Shin, W., Kim, H. S., Siver, P. A., & Andersen, R. A. (2013). Phylogeny of the genus *Mallomonas* (Synurophyceae) and descriptions if five new species on the basis of morphological evidence. *Phycologia, 52*, 266–278. doi:10.2216/12-107.1.
Kies, L., & Berndt, H. (1984). Die *Synura*-Arten (Chrysophyceae) Hamburgs und seiner nordöstlichen Umgebung. *Mitteilungen aus dem Institut für Allgemeine Botanik Hamburg, 19*, 99–122.
Kim, H. S. (1997). Silica-scaled chrysophytes (Synurophyceae) in several reservoirs, swamps, and a highland pond from Changnyong County, Korea. *Algae, 12*, 1–10.
Kling, H. J., & Kristiansen, J. (1983). Scale-bearing Chrysophyceae (Mallomonadaceae) from Central and Northern Canada. *Nordic Journal of Botany, 3*, 269–290.
Korshikov, A. (1929). Studies on the chrysomonads I. *Archiv für Protistenkunde, 67*, 253–290.
Kristiansen, J. (1975). Chrysophyceae from Alberta and British Columbia. *Syesis, 8*, 97–108.
Kristiansen, J. (1988). Seasonal occurrence of silica-scaled chrysophytesunder eutrophic conditions. *Hydrobiologia, 161*, 171–184.
Kristiansen, J. (1990). Studies on silica-scaled chrysophytes from Central Asia. *Archiv für Protistenkunde, 138*, 298–303.
Kristiansen, J., & Preisig, H. R. (2007). Chrysophyte and Haptophyte algae. 2. Teil/Part 2: synurophyceae. In B. Büdel, G. Gärtner, L. Krienitz, H. R. Preisig, & M. Schagerl (Eds.), *Süsswasserflora von Mitteleuropa* (p. 252). Berlin, Heidelberg: Spektrum Akademischer Verlag.
Kynčlová, A., Škaloud, P., & Škaloudová, M. (2010). Unveiling hidden diversity in the *Synura petersenii* species complex (Synurophyceae, Heterokontophyta). *Nova Hedwigia Beiheft, 136*, 283–298. doi:10.1127/1438-9134/2010/0136-0283.
Péterfi, L. S., & Momeu, L. (1977). Remarks on the taxonomy of some *Synura* species based on the fine structure of scales. *Muzeul Brukenthal Studii si Comuicări-Stiinte Naturale, 21*, 15–23.
Petersen, J. B., & Hansen, J. B. (1956). On the scales of some *Synura* species. *Biol Medd Kgl. Danske Videnskabernes Selskab., 23*(2), 3–27.
Petersen, J. B., & Hansen, J. B. (1958). On the scales of some *Synura* species. II. *Biol Medd Kgl Danske Videnskabernes Selskab, 23*(7), 1–13.
Posada, D., & Crandall, K. A. (1998). MODELTEST: testing the model of DNA substitution. *Bioinformatics, 14*(9), 817–8. doi:10.1093/bioinformatics/14.9.817.
Rambaut A. FigTree v1.4.2. 2014. Available online at: http://tree.bio.ed.ac.uk/software/figtree/
Rambaut A, Suchard MA, Drummond AJ. Tracer v.1.6. 2013. Available online at: http://tree.bio.ed.ac.uk/software/tracer/.
Řezáčová, M., & Škaloud, P. (2005). Silica-scaled chrysophytes of Ireland. With an appendix: geographic variation of scale shape of *Mallomonas caudata*. *Nova Hedwigia Beiheft, 128*, 101–124.

Ronquist, F., Teslenko, M., Van Der Mark, P., Ayres, D. L., Darling, A., Höhna, S., Larget, B., Liu, L., Suchard, M. A., & Huelsenbeck, J. P. (2012). MrBayes 3.2: efficient Bayesian phylogenetic inference and model choice a cross a large model space. *Systematic Biology, 61*(3), 539–42. doi:10.1093/sysbio/sys029.

Siver, P. A. (1987). The distribution and variation of *Synura* species (Chrysophyceae) in Connecticut, USA. *Nordic Journal of Botany, 7,* 107–116. doi:10.1111/j.1756-1051.1987.tb00922.x.

Siver, P. A., & Lott, A. M. (2004). Further observations on the scaled Chrysophycean and Synurophycean flora of the Ocala National Forest, Florida, USA. *Nordic Journal of Botany, 24,* 211–233. doi:10.1111/j.1756-1051.2004.tb00835.x.

Siver, P. A., & Wujek, D. E. (1993). Scaled Chrysophyceae and Synurophyceae from Florida, USA. IV. The flora of Lower Lake Myakka and Lake Tarpon. *Florida scientist (USA), 56,* 109–117.

Škaloud, P., Kristiansen, J., & Škaloudová, M. (2013). Developments in the taxonomy of silica-scaled chrysophytes from morphological and ultrastructural to molecular approaches. *Nordic Journal of Botany, 31,* 385–402. doi:10.1111/j.1756-1051.2013.00119.x.

Škaloud, P., Kynčlová, A., Benada, O., Kofroňová, O., & Škaloudová, M. (2012). Toward a revision of the genus *Synura*, section Petersenianae (Synurophyceae, Heterokontophyta): morphological characterization of six pseudo-cryptic species. *Phycologia, 51,* 303–329. http://dx.doi.org/10.2216/11-20.1.

Škaloud, P., Škaloudová, M., Procházková, A., & Němcová, Y. (2014). Morphological delineation and distribution patterns of four newly described species within the *Synura petersenii* species complex (Chrysophyceae, Stramenopiles). *European Journal of Phycology, 49,* 213–229. doi:10.1080/09670262.2014.905710.

Stamatakis, A. (2014). RAxML version 8: a tool for phylogenetic analysis and post-analysis of large phylogenies. *Bioinformatics, 30*(9), 1312–3. doi:10.1093/bioinformatics/btu033.

Takahashi, E. (1967). Studies on genera *Mallomonas, Synura* and other plankton in fresh-water with the electron microscope. VI. Morphological and ecological observations on genus *Synura* in ponds and lakes in Yamagata Prefecture. *Bulletin Yamagata University Agricultural Science, 5,* 99–118.

Takahashi, E. (1972). Studies on genera *Mallomonas* and *Synura*, and other plankton in freshwater with electron microscope. VIII. On three new species of Chrysophyceae. *The botanical magazine= Shokubutsu-gaku-zasshi, 85,* 293–302.

Takahashi, E. (1973). Studies on genera *Mallomonas* and *Synura*, and other plankton in fresh water with the electron microscope VII. New genus *Spiniferomonas* of the Synuraceae (Chrysophyceae). *The botanical magazine= Shokubutsu-gaku-zasshi, 86,* 75–88.

Takahashi, E. (1978). *Electron microscopical studies of the Synuraceae (Chrysophyceae) in Japan: taxonomy and ecology.* Tokyo: Tokai University Press.

Vigna, M. S., & Munari, C. (2001). Seasonal occurrence of silica scales chrysophytes in a Buenos Aires lake. *Nova Hedwigia Beiheft, 122,* 195–209.

Wee, J. L. (1981). Studies on silica-scaled chrysophytes from Iowa. II. Common *Synura* species. *Proceedings of the Iowa Academy of Sciences, 88,* 70–73.

Wee, J. L., Fasone, L. D., Sattler, A., Starks, W. W., & Hurley, D. L. (2001). ITS/5.8S DNA sequence variation in 15 isolates of *Synura petersenii* Korshikov (Synurophyceae). *Nova Hedwigia Beiheft, 122,* 245–258.

6

The effect of phosphorus removal from sewage on the plankton community in a hypertrophic reservoir

Sungmin Jung[1], Kiyong Kim[2], Yunkyoung Lee[1], Jaeyong Lee[1], Yukyong Cheong[1], Arif Reza[1], Jaiku Kim[1], Jeffrey S. Owen[3] and Bomchul Kim[1*]

Abstract

Background: When developing water quality improvement strategies for eutrophic lakes, questions may arise about the relative importance of point sources and nonpoint sources of phosphorus. For example, there is some skepticism regarding the effectiveness of partial reductions in phosphorus loading; because phosphorus concentrations are too high in hypertrophic lakes, in-lake phosphorus concentrations might still remain within typical range for eutrophic lakes even after the reduction of phosphorus loading. For this study, water quality and the phytoplankton and zooplankton communities were monitored in a hypertrophic reservoir (Lake Wangsong) before and after the reduction of phosphorus loading from a point source (a sewage treatment plant) by the installation of a chemical phosphorus-removal process.

Results: Before phosphorus removal, Lake Wangsong was classified as hypertrophic with a median phosphorus concentration of 0.232 mg L^{-1} and a median chlorophyll-a concentration of 112 mg L^{-1}. The dominant phytoplankton were filamentous cyanobacteria for the most of the ice-free season. Following the installation of the advanced treatment process, phosphorus concentrations were reduced to 81 mg L^{-1}, and the N/P atomic ratio increased from 42 to 102. Chlorophyll-a concentrations decreased to 42 μg L^{-1}, and the duration of cyanobacterial dominance was confined to the summer season. Cyanobacteria in spring and autumn were replaced by diatoms and cryptomonads. Filamentous cyanobacteria in summer were replaced by colony-forming unicellular *Microcystis* spp. It was remarkable that zooplankton biomass increased despite the decrease in phytoplankton biomass, and especially cladoceran zooplankton which increased drastically. These responses to the reduction of point source P loading to Lake Wangsong imply that reducing the point source P loading can have a big impact even when nonpoint sources account for a large fraction of the total annual phosphorus loading.

Conclusions: Our results also show that the phytoplankton community can shift to decreased cyanobacterial dominance and the zooplankton community can shift to higher cladoceran dominance, even when phosphorus concentrations remain within the typical range for eutrophic lakes following the reduction of phosphorus loading.

Keywords: Eutrophication, Phosphorus removal, Phytoplankton, Point source, Zooplankton

* Correspondence: bkim@kangwon.ac.kr
[1]Department of Environmental Science, Kangwon National University, Chuncheon 24341, South Korea
Full list of author information is available at the end of the article

Background

Eutrophication is the most common water quality problem in lakes and reservoirs (Azevedo et al. 2015; Reed-Andersen et al. 2000), and phosphorus is the major limiting factor of eutrophication (Schindler 1977; Withers and Jarvie 2008). Phosphorus (P) comes mostly from animal excretion and fertilizer, and the sources are classified as point sources or nonpoint sources (Kundu et al. 2015; Carson et al. 2015). In most rural watersheds, nonpoint P sources such as agricultural fields and forest are the major sources of the total P load. However, if there is significant urban development within the watershed, sewage discharge can contribute a considerable portion of P loading (Neal et al. 2000; Jarvie et al. 2006). In watersheds consisting of complex terrain with urban and rural land use, P loading from agricultural fields is usually larger than the P loading from sewage. However, the export of P from nonpoint sources is exported mostly on rainy days in the form of storm runoffs (Li et al. 2015). On the contrary, P loading from point sources have uniform flow rates, and on dry days, their contribution can be larger than those of nonpoint sources.

Hydraulic residence time can be another factor that determines the relative importance of storm runoff in lakes and reservoirs (Dillon 1975; Brett and Benjamin 2008). Shock loading from agricultural nonpoint sources can be very large during a storm event, but a large portion of the P loading can be flushed out of reservoirs with short residence times. Therefore, there are still many important questions remaining concerning the relative importance of point and nonpoint P loading. There might be some skepticism regarding the effectiveness of reductions in point source P loading which result in only a partial reduction of the annual P loading (Mainstone and Parr 2002).

Point sources of P include raw sewage discharge or effluent from sewage treatment plants which do not have chemical phosphorus-removal processes. The sewage treatment process can be divided into three phases: primary treatment, secondary treatment, and tertiary treatment (or advanced treatment). Commonly, in developed countries, sewage is treated using secondary treatment processes, biological treatment processes designed mainly for biological oxygen demand (BOD) removal (Cullen and Forsberg 1988). Because P concentrations in the effluent from secondary treatment ($1–2$ mg P L^{-1}) is $30–60$ times higher than the typical criterion for eutrophication (typically 0.03 mg P L^{-1}), advanced P treatment using chemical precipitation has been incorporated in some sewage treatment plants in Korea that discharge effluent into sensitive surface waters. The legal standards for P concentrations in effluent from treatment plants are between 0.2 and 0.5 mg P L^{-1}, still substantially higher than the criterion for eutrophication.

In hypertrophic lakes where P concentrations are too high even after a reduction of point source P loading, the in-lake P concentrations may still remain in the typical range for eutrophic lakes (Schindler 2006). As part of efforts to provide a scientific basis for decision-making and developing strategies for water quality management, detailed information on the consequences of a reduction in P loading and especially the effects on water quality and the phytoplankton and zooplankton communities is required.

Lake Wangsong is a hypertrophic reservoir located in a watershed with mixed urban and rural land use (Table 1). The lake has been hypertrophic since a sewage treatment plant (STP) was constructed on the shore of the lake. Because the STP effluent contained high P concentrations, the discharge from the STP was identified as the major cause of eutrophication, and the advanced treatment process was added to the plant to reduce P concentration in the STP effluent. In this study, water quality and the phytoplankton and zooplankton communities were surveyed before and after the installation of the advanced sewage treatment process in order to examine the effectiveness of the P removal process for the water quality improvement in a hypertrophic lake. We also examined the effect of the reduction in P loading on the phytoplankton and zooplankton communities in this lake.

Methods

Lake Wangsong was constructed on the Hwangkujichon Stream (Uiwang city) for providing irrigation water for agriculture (Fig. 1). Rapid urbanization within the watershed resulted in an increase in sewage discharge, thus discharging a large amount of P into the lake. In 1999, the STP was constructed at the shore of the lake, and the advanced treatment process using chemical precipitation was added in August of 2007 to reduce P concentrations in the effluent.

Table 1 Morphological characteristics of Lake Wangsong

Watershed area (km^2)		15.5
Mean depth (m)		2.5
Total water storage capacity (10^3 m^3)		2079
Maximum surface area (km^2)		0.86
Hydraulic residence time (days)		49
Land use (ha)	Agriculture (upland field)	290 (18.6 %)
	Agriculture (paddy field)	320 (20.6 %)
	Forest	685 (44.1 %)
	Residential area	182 (11.7 %)
	Others	78 (5.0 %)
Total population		27,994
Population density (pop. km^{-2})		1806

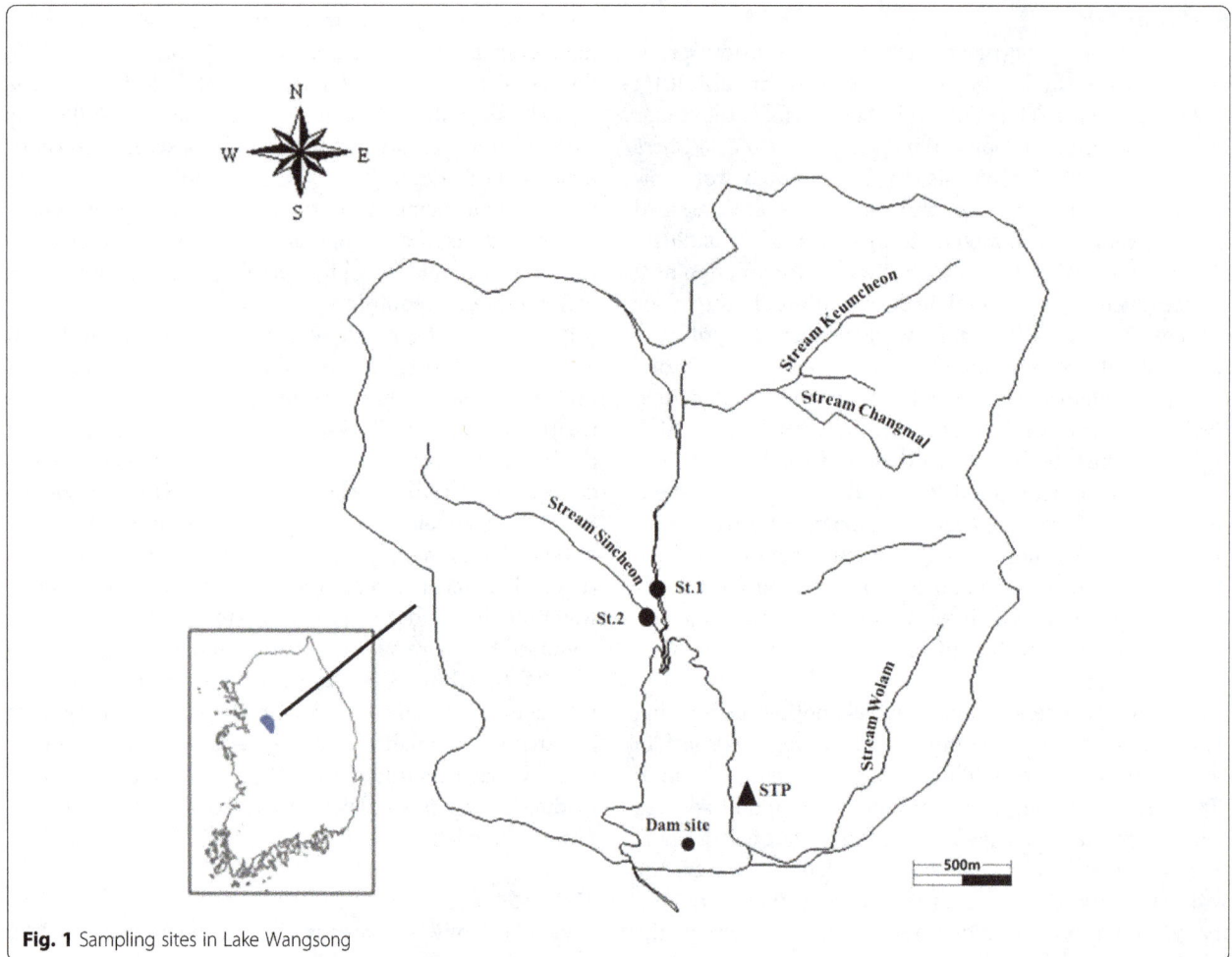

Fig. 1 Sampling sites in Lake Wangsong

The phytoplankton and zooplankton communities and water quality were surveyed before the advanced treatment operation (2003 and 2004) and after the advanced treatment (2008 and 2009). Water samples were collected from the center of the lake at 0, 2, and 5-m depths. Two main inflowing streams were surveyed in order to assess P loading from the watershed in the dry season. In this study, water quality measurements of the STP effluent were measured and, also, the data from the STP management office were employed together for calculating the P loading from the STP effluent.

Water samples were collected using an inflatable boat and a horizontal Van Dorn water sampler. The water samples were transported in a cooler and stored in a refrigerator until analysis. Dissolved oxygen was measured in situ with a DO meter (YSI, USA). All water quality measurements were conducted according to standard methods (American Public Health Association 2005) except chemical oxygen demand (COD). Total phosphorus (TP) concentration was measured using the ascorbic acid method following persulfate digestion. Chlorophyll-a concentration was measured by the trichromatic spectrophotometric method. Total nitrogen (TN) was measured by the cadmium reduction method following persulfate digestion. Suspended solids (SS) was measured gravimetrically after filtration by GF/F filter. Biological oxygen demand (BOD) was measured by using a DO meter. COD measurements used the permanganate method, the official standard method in Korea (MOE 2009).

Water quality data collected at the surface of the lake by the Gyeonggido local province was also employed in addition to the data collected by this study in order to increase statistical significance in the comparison of water quality before and after the advanced treatment. Stream discharge of inflowing streams was measured using a magnetic flow velocity meter and the current cross-section method. P loading from nonpoint sources in the watershed was estimated by multiplying the unit export coefficients of phosphorus by the area of each land use type as suggested by the Korean Ministry of the Environment (MOE 2014).

Phytoplankton samples were collected in 500-mL polyethylene bottles at 0.5-m depth and preserved with

Lugol's solution. Cell densities were measured using a Sedgewick-Rafter counting chamber and an X300 microscope (Olympus BX50). Zooplankton samples were collected with a plankton net (63-μm mesh) using a slow vertical tow from the bottom (typically 5–6-m depth) to the surface of the lake. Zooplankton samples were preserved with 4 % sucrose formalin (Steedman 1976). The volume of water filtered by a zooplankton net was calculated by multiplying the aperture area of the net by the towing distance, assuming there is no significant loss of filtering efficiency through a 5-m towing.

Results

Nutrient loading and water quality

Water quality parameters in the two main inflowing streams and the STP effluent are shown in Table 2. In general, the nutrient status in the inflowing streams indicates eutrophic conditions. The median TP concentration was 0.131 and 0.190 mg P L^{-1} at St.1 and St.2, respectively. TP concentrations in the STP effluent were much higher than either St.1 or St.2 in 2003–2007 (median TP 0.633 mg P L^{-1}). Following the phosphorus removal (in 2008), TP concentrations in the effluent were lower (median 0.310 mg P L^{-1}).

The annual average daily P loading from nonpoint sources in the Lake Wangsong watershed was 7.5 kg P day^{-1}, which was higher than the P loading from the major point source (STP effluent), 3.7 kg P day^{-1} (Gyeonggi Research Institute 2011). But most of the annual P loading from agricultural nonpoint sources is concentrated in a few rain events during the summer monsoon season in Korea, and only limited amounts of P export occur on dry days. When the daily P loading was measured in dry seasons, the P discharge from the STP effluent accounted for a larger portion (56 and 60 %) than the P loading from the two main inflowing streams (Table 3).

In August of 2007, the advanced P removal treatment started operation. With the reduction of P loading, TP concentrations in the surface have decreased from 0.232 to 0.081 mg P L^{-1} (median), a reduction of 65 % (Table 2). But even after the advanced treatment, TP concentrations in the STP effluent were ten times higher than the typical criterion for eutrophic conditions (0.03 mg P L^{-1}). Consequently, TP in Lake Wangsong also exceeded the threshold of eutrophication by a factor of 2.5.

Because the advanced treatment was focused on the chemical removal of P, which has a lower removal efficiency than nitrogen, TN concentrations did not decrease as much as TP concentrations. With the higher removal efficiency of P than N, the atomic N/P ratio in the lake increased from 68 to 154 following the start of the advanced treatment. Thus, the atomic N/P was much higher than the Redfield ratio (16) (Redfield 1958) implying P limitation of algal growth in Lake Wangsong.

Suspended solids (SS) in the lake did not change after the advanced treatment. The median SS was 19 mg L^{-1}, suggesting that the seston is composed of mostly inorganic particles, possibly arising from the bottom sediment in this shallow reservoir. Assuming that the chlorophyll-a content of algal cells is commonly 1 % of dry weight and the median chlorophyll-a concentration of 42.2 ug L m^{-1} in Lake Wangsong, algal cells would account for an algal biomass of

Table 3 Comparison of daily phosphorus loading on dry days from the watershed (St.1 and St.2) and sewage treatment plant effluent (STP)

Date		St.1	St.2	STP
25 Sep. 2008	Flow rate (m^3 day^{-1})	15,314	7733	11,584
	P loading (kg day^{-1})	1.75 (26.3 %)	0.91 (13.7 %)	3.98 (60.0 %)
31 Oct. 2008	Flow rate (m^3 day^{-1})	10,152	6229	11,910
	P loading (kg day^{-1})	1.71 (28.7 %)	0.9 (15.1 %)	3.37 (56.2 %)

Table 2 Water quality of inflowing streams and lake surface (median, mg L^{-1}; Chl.a, ug L^{-1}, (25th–75th percentile))

	BOD	COD	SS	TN	TP	Chl.a	BOD/COD	N/P (atomic)	Chl.a/TP	Number
Inflow St.1 (2008–2009)	4.6 (3.1–5.0)	6.4 (5.9–6.4)	45.7 (32.0–71.5)	3.73 (3.71–4.89)	0.190 (0.124–0.244)		0.16 (0.09–0.23)	51 (38–54)		10
Inflow St.2 (2008–2009)	2.7 (1.3–5.0)	6.4 (5.0–8.1)	11.9 (5.7–44.1)	3.76 (3.30–4.37)	0.131 (0.113–0.222)		0.42 (0.23–0.79)	48 (38–68)		10
STP effluent (2003–2007)	10.4 (8.3–12.7)	14.9 (12.3–17.7)	12.5 (7.1–23.3)	7.9 (5.9–13.6)	0.633 (0.270–1.299)		0.65 (0.50–0.77)	28 (21–44)		24
STP effluent (2008–2009)	1.6 (1.0–2.2)	6.8 (6.0–7.8)	1.9 (1.4–3.6)	10.6 (8.9–12.5)	0.310 (0.185–0.355)		0.25 (0.15–0.32)	78 (61–88)		29
Lake surface (May 1999–Jun. 2007)	11.4 (8.7–13.7)	15.8 (13.2–18.4)	20.4 (16.0–30.7)	4.75 (3.47–5.87)	0.232 (0.181–0.284)	112.4 (53.9–163.7)	0.70 (0.51–0.87)	42 (27–70)	0.54 (0.22–0.84)	42
Lake surface (Aug. 2007–Nov. 2009)	5.2 (4.1–7.1)	10.2 (8.7–11.3)	19.1 (13.9–30.8)	3.56 (2.31–4.93)	0.081 (0.054–0.137)	42.2 (24.9–110.8)	0.59 (0.47–0.69)	102 (68–154)	0.61 (0.28–1.15)	20

approximately 4 mg L^{-1}, much lower than the SS. Therefore, inorganic particles would account for 15 mg L^{-1} in the SS. In Lake Wangsong, resuspension of sediment is likely common on windy days. The decrease in BOD following the reduction in P loading affected the hypolimnetic DO concentrations, a common criterion for eutrophication (Horne and Goldman 1994). Anoxic conditions in the hypolimnion clearly developed below 4-m depth in September 2003 and May to August 2004, whereas the hypolimnion was oxic in 2009.

Phytoplankton and zooplankton

With decreased in-lake TP, chlorophyll-a concentrations decreased drastically from a median of 112.4 mg m^{-3} before the advanced treatment to 42.2 mg m^{-3} after the advanced treatment; the decrease in chlorophyll-a was 62 %, similar to the decrease in TP (65 %, Table 2). The maximum cell density for cyanobacteria also decreased drastically from 25×10^5 cells mL^{-1} to 9×10^5 cells mL^{-1} (Table 4). The phytoplankton species composition changed together with a decrease in algal standing crop. In 2003 and 2004, cyanobacteria were the dominant phytoplankton species during most of the ice-free season from April to December. It was remarkable that cyanobacteria were dominant even in winter. By contrast, after the advanced treatment, cyanobacteria were dominant in only 4 samples out of 12 monthly samples (Table 4). In the summer months, cyanobacteria were still dominant, but in spring and autumn, cryptomonads replaced cyanobacteria. In the winter samples, diatoms were the dominant phytoplankton. In a statistical comparison, the density of diatoms and cyanobacteria showed significant difference between before and after the advanced treatment ($p < 0.05$).

With the change in phytoplankton species composition, the dominant zooplankton also changed (Table 5). The most remarkable change was that large-sized cladoceran zooplankton (*Daphnia galeata*) reached the maximum density in spring and autumn when cyanobacteria were not dominant. Before the advanced treatment started, the maximum cladoceran density was 250 ind. L^{-1} in July and August, whereas in April 2008, the density of cladocerans was 1242 ind. L^{-1}. In the statistical analysis, the density and dominance of cladocerans between before and after the advanced treatment showed significant differences.

Table 4 Cell densities of three major phytoplankton taxa and dominant species before and after advanced sewage treatment

Year	Month	Cell density (cells mL^{-1})					Dominant species (by biomass, µgC L^{-1})
		Bacillariophyceae	Cyanophyceae	Chlorophyceae	Others	Total	
2003	Sep.	6408	497,304	534	971	505,217	*Oscillatoria* sp.
	Nov.	4269	213,537	5817	123	223,746	*Microcystis* sp.
	Dec.	115	59,486	3326	1840	64,767	*Oscillatoria* sp.
2004	Mar.	381	757	545	25	1708	*Closterium* sp.
	Apr.	185	130,010	122,333	3933	256,461	*Anabaena* sp.
	May	43	1,474,200	54,871	1207	1,530,321	*Lyngbya* sp.
	Jun.	58	2,515,477	1234	13,247	2,530,016	*Microcystis* sp.
	Jul.	3385	508,721	1362	34	513,502	*Lyngbya* sp.
	Aug.	299	1,378,199	59	595	1,379,152	*Microcystis* sp.
Advanced treatment start							
2007	Nov.	36	7088	89	1148	8361	*Cryptomonas ovata*
	Dec..	2196	673	414	61	3344	*Stephanodiscus hantzschii*
2008	Apr.			1055	1570	2625	*Cryptomonas ovata*
	May	2	796	39,744	389	40,931	*Pediastrum duplex*
	Jun.	2	897,039	276	22	897,339	*Microcystis aeruginosa*
	Jul.	230	33,155	469	533	34,386	*Microcystis aeruginosa*
	Aug.	20	160,053	111	2215	162,399	*Microcystis wesenbergii*
	Sep.	210	30,197	2806	393	33,606	*Oscillatoria sancta*
	Dec.	5581	13	1148	47	6789	*Stephanodiscus hantzschii*
2009	Mar.	24		190	315	528	*Cryptomonas* sp.
	May	8		3524	1168	4701	*Cryptomonas* sp.
	Sep.	12	2349	24	133	2518	*Cryptomonas* sp.

Table 5 Standing crop and dominant zooplankton species

Year	Month	Standing crop (ind. L^{-1})				Dominant species
		Cladocera	Copepoda	Rotifera	Total	
2003	Sep.	57	19	13	89	*Bosmina longirostris*
	Nov.	2	18	21	41	*Cyclops vicinus*
	Dec.	21	14	122	157	*Asplanchna priodonta*
2004	Mar.	1	13	85	98	*Cyclops vicinus*
	Apr.	44	29	2731	2804	*Cyclops vicinus*
	May	0	1	10	11	*Nauplius*
	Jun.	48	298	1805	2151	*Thermocyclops taihokuensis*
	Jul.	254	368	1555	2177	*Thermocyclops taihokuensis*
	Aug.	243	271	989	1504	*Nauplius*
Advanced treatment start						
2007	Nov.	5	18	20	43	*Bosmina longirostris*
	Dec.	12	58	12	88	*Cyclops vicinus*
2008	Apr.	1242	1089	1531	3862	*Daphnia galeata*
	May	550	342	626	1518	*Cyclops vicinus*
	Jun.	49	162	54	266	*Daphnia galeata*
	Jul.	77	212	1384	1674	*Copepodite*
	Aug.	22	235	1561	1818	*Brachionus forficula*
	Sep.	97	306	2704	3105	*Keratella cochlearis*
	Oct.	510	1729	1758	3997	*Daphnia galeata*
	Dec.	217	466	3616	4299	*Cyclops vicinus*
2009	Mar.	21	120	154	295	*Cyclops vicinus*
	May	0	0	2275	2275	*Daphnia galeata*
	Sep.	9	82	493	574	*Cyclops vicinus*

Discussion

The annual P loading to lakes from point sources and nonpoint sources is commonly comprised mostly of agricultural nonpoint P sources in watersheds with mixed land uses. However, in reservoirs with short hydraulic residence time, storm runoff from nonpoint sources is not stored for a long time within the reservoir, especially in the rainy season. Because Korea is located in the summer monsoon region, most of the annual rainfall occurs in summer. Short hydraulic residence times in reservoirs can be a critical factor affecting nutrient concentrations in reservoirs. In the dry season, the relative importance of point sources is larger than nonpoint sources, because nonpoint sources do not export nutrients during periods with minimum stream flow. In this study, the importance of P removal in the STP effluent was obviously manifested in the water quality improvement after the advanced treatment.

Even after the advanced treatment, Lake Wangsong remained eutrophic. TP and chlorophyll-a concentrations were higher than the criteria for eutrophic lakes (Wetzel 2001), mainly because TP in the STP effluent was still much higher than the typical criterion for eutrophication. TP can be reduced to as low as 0.01 mg P L^{-1} in STP effluent with advanced chemical treatment. In Korea, the phosphorus standards for STP effluent are in the range of 0.2 to 2.0 mg L^{-1}, still much higher than the eutrophication criterion. Therefore, even if sewage is treated according to the government standard for P concentration in the effluent, sewage can be a main cause of eutrophication and further reductions in P in STP effluent are strongly needed for the control of eutrophication.

Importantly, we observed that the dominance of cyanobacteria decreased with P concentrations, even though P concentrations remained in the eutrophic level. The N/P ratio increased due to the P removal in the STP effluent, which might have provided favorable conditions for algae other than cyanobacteria. A low N/P ratio is known to be favorable for cyanobacteria, because N can be a temporary limiting factor and N-fixing cyanobacteria can take advantage of this (Gu and Alexander 1993). Temporary nitrogen depletion during algal blooms can be a controlling factor in the competition among phytoplankton species; that is, decreases in P can provide favorable conditions for other algal species and inhibit the dominance of filamentous cyanobacteria (Fulton 1988). Before the start of the advanced treatment, most of the dominant phytoplankton were filamentous cyanobacteria with many species having the potential to carry out N-fixation, whereas filamentous cyanobacteria were dominant only in 1 sample out of 12 monthly samples after the advanced treatment.

The change of the zooplankton community in Lake Wangsong was also obvious. In 2003 and 2004, rotifers were the dominant zooplankton, whereas cladoceran species were the dominant zooplankton in 2008 and 2009 (Table 5). A remarkable result was that zooplankton biomass increased after the advanced treatment even though the phytoplankton biomass decreased. Generally, zooplankton standing crop is associated with the availability of phytoplankton, the so-called bottom-up effect (Sinistro 2009). In addition, copepod and cladoceran densities increased drastically and rotifer density increased slightly. The increase in cladoceran density is usually regarded as a positive change for water quality due to their large filtering capacity which can result in an increase in water clarity (Sommer et al. 2001; Schrage and Downing 2004). Especially, the maximum cladoceran densities in April 2008 and October 2009 may have caused clear water phases due to the high density of *D. galeata* (Fig. 2). In 2003 and 2004, the

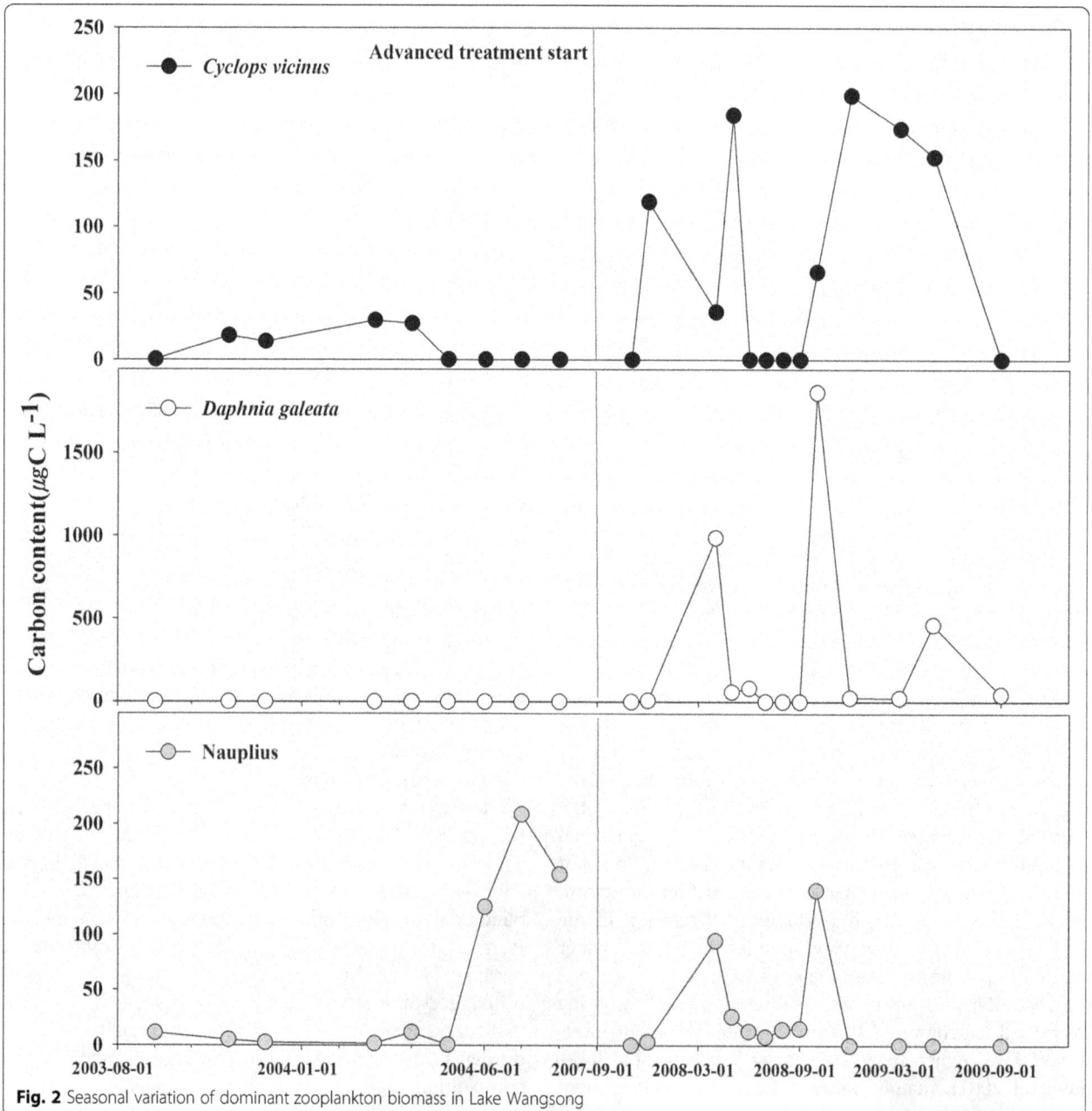

Fig. 2 Seasonal variation of dominant zooplankton biomass in Lake Wangsong

dominant phytoplankton were filamentous cyanobacteria during all seasons, and these species are generally known as inedible prey for zooplankton (Krevš et al. 2010). The change in zooplankton species composition can be explained as the result of change of phytoplankton community from inedible filamentous cyanobacteria to edible diatoms and cryptomonads which were dominant in spring and autumn of 2009 (Bomi et al. 2013).

Conclusions

The reduction of phosphorus from STP effluent resulted in a significant reduction of in-lake P concentrations in Lake Wangsong, which in turn effected a shift in the phytoplankton community. Even though P concentrations remained within the typical range for eutrophic conditions following the reduction in P loading, all the indicators of water quality and aquatic ecosystem health showed improvements: decreased phytoplankton density, decreased hypoxia in the hypolimnion, a shift from cyanobacteria to diatoms in cold seasons, a shift from filamentous cyanobacteria to colony-forming unicellular cyanobacteria, and increased cladoceran zooplankton populations which can improve water clarity and facilitate the transfer of energy through the grazing food chain. This implies that P removal from STP effluent can be important for improving water quality in hypertrophic reservoirs even if the annual P loading from nonpoint sources is larger than the P loading from sewage effluent.

Funding
This study was supported by a 2013 Research Grant from Kangwon National University (no. 120131194). This study was supported by the Center for Aquatic Ecosystem Restoration (CAER) of the Eco-STAR Project from the Ministry of Environment, Republic of Korea (MOE: EW 42-08-10). Support from the Environmental Research Institute at Kangwon National University is also acknowledged.

Authors' contributions
All authors contributed extensively to the work presented in this paper. BK and SJ designed the study and wrote the main paper. YL and JL collected the zooplankton data and analyzed the results. SJ, JK, and YC collected and analyzed the phytoplankton data and conducted the data analysis. KK, JO, and AR analyzed the water quality data. All authors discussed the results and implications and commented on the manuscript at all stages.

Competing interests
The authors declare that they have no competing interests.

Author details
[1]Department of Environmental Science, Kangwon National University, Chuncheon 24341, South Korea. [2]Department of Hydrology, University of Bayreuth, 95447 Bayreuth, Germany. [3]Department of Environmental Science, Hankuk University of Foreign Studies, Yongin 17053, South Korea.

References

American Public Health Association, American Water Works Association, Water Environment Federation. (2005). *Standard methods for the examination of water and wastewater* (21st ed.). Washington, D.C.: American Public Health Association.

Azevedo, L. B., van Zelm, R., Leuven, R. S., Hendriks, A. J., & Huijbregts, M. A. (2015). Combined ecological risks of nitrogen and phosphorus in European freshwaters. *Environmental Pollution, 200*, 85–92.

Bomi, C., Misun, S., Jong, I. K., & Woongghi, S. (2013). Taxonomy and phylogeny of the genus Cryptomonas (Cryptophyceae, Cryptophyta) from Korea. *Algae, 28*, 307–330.

Brett, M. T., & Benjamin, M. M. (2008). A review and reassessment of lake phosphorus retention and the nutrient loading concept. *Freshwater Biology, 53*, 194–211.

Carson, A., Jennings, E., Linnane, S., & Jordan, S. N. (2015). Clearing the muddy waters: using lake sediment records to inform agricultural management. *Journal of Paleolimnology, 53*, 1–15.

Cullen, P., & Forsberg, C. (1988). Experiences with reducing point sources of phosphorus to lakes. *Hydrobiologia, 170*, 321–336.

Dillon, P. (1975). The phosphorus budget of Cameron Lake, Ontario: the importance of flushing rate to the degree of eutrophy of lakes. *Limnology and Oceanography, 20*, 28–39.

Fulton, R. S. (1988). Grazing on filamentous algae by herbivorous zooplankton. *Freshwater Biology, 20*, 263–271.

Gu, B., & Alexander, V. (1993). Estimation of N_2 fixation based on differences in the natural abundance of ^{15}N among freshwater N_2-fixing and non-N_2-fixing algae. *Oecologica, 96*, 43–48.

Gyeonggi Research Institute (2011). *Water quality management and implementation: alternatives for Wangsong Reservoir.*

Horne, A. J., & Goldman, C. R. (1994). *Limnology* (2nd ed.). New York: McGraw-Hill Co.

Jarvie, H. P., Neal, C., & Withers, P. J. (2006). Sewage-effluent phosphorus: a greater risk to river eutrophication than agricultural phosphorus? *Science of the Total Environment, 360*, 246–253.

Krevš, A., Koreivienė, J., & Mažeikaitė, S. (2010). Plankton food web structure during cyanobacteria bloom in the highly eutrophic Lake Gineitiškės. *Ekologija, 56*, 47–54.

Kundu, S., Coumar, M. V., Rajendiran, S., & Rao, A. S. (2015). Phosphates from detergents and eutrophication of surface water ecosystem in India. *Current Science, 108*, 1320–1325.

Li, D., Wan, J., Ma, Y., Wang, Y., Huang, M., & Chen, Y. (2015). Stormwater runoff pollutant loading distributions and their correlation with rainfall and catchment characteristics in a rapidly industrialized city. *PloS One.* doi:10.1371/journal.pone.0118776.

Mainstone, C. P., & Parr, W. (2002). Phosphorus in rivers-ecology and management. *Science of the Total Environment, 282*, 25–47.

Ministry of Environment. (2009). *Standard methods of water sampling and analysis.* Korea: Ministry of Environment.

Ministry of Environment. (2014). *Total maximum daily load program.* Korea: Ministry of Environment.

Neal, C., Jarvie, H. P., Howarth, S. M., Whitehead, P. G., Williams, R. J., Neal, M., Harrow, M., & Wickham, H. (2000). The water quality of the River Kennet: initial observations on a lowland chalk stream impacted by sewage inputs and phosphorus remediation. *Science of the Total Environment, 251*, 477–495.

Redfield, A. C. (1958). The biological control of chemical factors in the environment. *American Scientist, 46*, 205–221.

Reed-Andersen, T., Carpenter, S. R., & Lathrop, R. C. (2000). Phosphorus flow in a watershed-lake ecosystem. *Ecosystems, 3*, 561–573.

Schindler, D. W. (1977). Evolution of phosphorus limitation in lakes. *Science, 195*, 260–262.

Schindler, D. W. (2006). Recent advances in the understanding and management of eutrophication. *Limnology and Oceanography, 51*, 356–363.

Schrage, L. J., & Downing, J. A. (2004). Pathways of increased water clarity after fish removal from Ventura Marsh; a shallow, eutrophic wetland. *Hydrobiologia, 511*, 215–231.

Sinistro, R. (2009). Top-down and bottom-up regulation of planktonic communities in a warm temperate wetland. *Journal of Plankton Research*. doi: 10.1093/plankt/fbp114.

Sommer, U., Sommer, F., Santer, B., Jamieson, C., Boersma, M., Becker, C., & Hansen, T. (2001). Complementary impact of copepods and cladocerans on phytoplankton. *Ecology Letters, 4*, 545–550.

Steedman, H. F. (1976). *Zooplankton fixation and preservation*. Paris: UNESCO Press.

Wetzel, R. G. (2001). *Limnology: lake and river ecosystems* (3rd ed.). San Diego: Academic.

Withers, P., & Jarvie, H. (2008). Delivery and cycling of phosphorus in rivers: a review. *Science of the Total Environment, 400*, 379–395.

7

Effect of consecutive shoot-cutting for 3 years on saplings' sprouting regeneration ability of six deciduous oak species in Korea

Seung-Yeon Lee, Kyu-Tae Cho, Rae-Ha Jang and Young-Han You[*]

Abstract

Background: The sprouts of oak species play an important role in maintaining the oak community in a disturbed environment. In this study, we cut 1-year-old oak in three times during the 3 years and measured the sprout responses to know sprouting ability of six deciduous oaks in Korea.

Results: Oak sprouts have appeared in spring and fall, and some of the sprouts had lifespan as short as a month. As the number of cutting increases, sprout number of *Quercus acutissima* increased whereas the other oak species decreased or died. The average number of sprouts over the 3 years was from 1.4 (*Quercus mongolica*) to 2.2 (*Q. acutissima*) per individual. *Quercus serrata* died after the second cutting, and *Quercus dentata* died after the third cutting. So, the two species have the lowest sprouting ability among six oak species. The sprouts grew actively during fall and slowly in summer. The sprout length during the 3 years was in the following descending order: *Q. acutissima*, *Quercus aliena*, *Q. dentata*, and *Q. mongolica*. Sprout of *Q. acutissima* and *Q. aliena* generated steadily over the 3 years, and sprout of *Quercus variabilis* and *Q. mongolica* was changed by year. After the 3 years, the number of sprouts increased only in *Q. acutissima* but sprout number of the other five oak species decreased. The sprout length of *Q. acutissima*, *Q. aliena*, and *Q. variabilis* increased, but sprout length of the other three oak species decreased. The average survival rate of saplings over the 3 years was in the following descending order: *Q. acutissima*, *Q. aliena*, *Q. variabilis*, and *Q. mongolica*.

Conclusions: As a result, the sprouting ability of *Q. acutissima* was the highest. Such level of sprouting ability may be the evidence of how *Q. acutissima* community exists as a dominant species in a disturbed environment in lowlands of Korea peninsula.

Keyword: Six deciduous oaks, Sprouting ability, Cutting the saplings, Consecutive cutting

Background

The oak forests in Korea provide food to many wild animals and are considered an important source for silviculture and landscaping. Currently, the genus oaks, which are widely dispersed in Korea, consist of six taxa, namely, *Qeurcus mongolica*, *Qeurcus variabilis*, *Qeurcus aliena*, *Qeurcus acutissima*, *Qeurcus serrata*, Qeurcus dentata, and 12 natural hybrids (Lee 2003).

These oak species have different ecological niche; therefore, they live under very diverse conditions (Lee and You 2009; Lee and You 2012). The existing oak forests in Korea are secondary forests that developed after being disturbed by felling, for firewood and heating, or by natural forest fires (Yang 2002). *Q. acutissima*, which has a high tolerance to cold, dryness, and shade, is typically distributed in mountains, roadsides, and residential areas. *Q. variabilis*, which has a high tolerance to dryness, grows well even in dry regions, and it has a high growth rate as well as high sprouting ability (Lee 2003). *Q. aliena* partially remain in well-reserved secondary understory vegetation in lowlands of Korea. *Q. dentata* are grown well in mountain bases, mountainsides, beaches, and even

* Correspondence: youeco21@kongju.ac.kr
Department of Biology, Kongju National University, Gongju, South Korea

on islands. In addition, it is distributed as a dominant species in limestone zone (Lim et al. 2012). *Q. serrata* grow well on sunny valleys or mountains with altitudes ranging 100~1, 800 m except in the northern regions. *Q. mongolica* generally are found in mountain ridges with altitude over 700 m, and many sprouts grow from the stem when it becomes old (Jo 1989). About 29% of forests in Korea is comprised of oak species that have a high growth rate, sprouting ability, and environment adaptability; and their demand is increasing due to their excellent timber quality (Kwon et al. 1998; Lee et al. 2000; Jung et al. 2013). The study on regeneration of oak forest is based more on sprouts rather than saplings (Kwon et al. 1998; Lee et al. 2000).

On the one hand, sprouting is a method of a vegetative reproduction; it is the main maintenance mechanism of oak individuals and community (Imanishi et al. 2010). The sprouts usually grow from buds near the stump or stems. Generally, plants' sprout wakes up from its dormant state and starts to grow like an individual from the shoot if it gets disturbed by environmental factors or gets cut (Barbour et al. 1980). The sprouts growing from cutting tree have fast growth and high resistance to diverse stress factors than individuals that germinate from the seed because sprout use accumulates from the shoot and roots of the tree (Smith 1986). Thus, sprouting is affected by various factors, such as cutting time, tree size, growth stage, light, moisture level, and type (Griffin 1980; Kim et al. 1991; Kwon et al. 1998).

This study was conducted to examine the sprouting ability of six dominant oak species in Korea in response to artificial disturbances. So, we studied the number of sprouts, sprout length, sprout reduction rate, and survival rate of saplings after the three cuttings over the 3 years.

Methods
Study design and measurements
Oak species used in the experiment were six deciduous species which were collected in the mountains around Gongju from September to October in 2009 and kept in cold storage. The species were *Q. acutissima* (Qa), *Q. mongolica* (Qm), *Q. variabilis* (Qv), *Q. dentata* (Qd), *Q. aliena* (Qa), and *Q. serrata* (Qs). Fifty acorns, similar in

size and condition, were selected for each species, and they were sowed in March 2010. The germinated saplings were grown in the experimental field with consistent soil and moisture until February 2011. Among the saplings, only eight individuals that have stable growth condition, without any withered leaf or stem, were selected for each oak to be used for the experiment (Fig. 1).

The sprouting ability was determined by measuring the number of sprouts and the sprout length after cutting the saplings at about 5 cm aboveground. The cuttings were performed three times over 3 years in the same manner, and the measurements were taken at the end of the growth period which was 8–10 months after the cutting. The first cutting was performed in March 2011 whereas other two cuttings were performed at the end of the growth period each following year. The number of sprouts and the sprout length were recorded each season (spring, summer, and fall) until November 2011 to analyze the periodic characteristics of sprouting. The growth period was a year before the first and the third cutting, but it was 2 years before the second cutting. Almost all the saplings of *Q. serrata* died after the first cutting, so we increased the growth period to 2 years before the second measurement for increasing the survival of saplings. The cutting was performed either in spring or in fall because the number of sprouts and the sprout growth rate of oaks is usually higher in spring or fall than summer period (Kim 1995; Lee et al. 2000).

The number of sprouts and the sprout length measured after the first cutting was compared with those measured after the third cutting to analyze the changes in the sprouting ability. The mean value of measurements after the first cutting and that of the third cutting were used to calculate the changes in sprouting ability [(third value - first value)/first value*100] in order to examine which oaks have high sprouting ability. In the calculation, the positive value refers to the increase in the number of sprouts and the sprout length over the measured period and the negative value refers to the reduction of those parameters.

The survival rates were calculated for each year by dividing the remaining number of saplings that survived after each cutting by the initial number of saplings ($n = 8$) to find out which oaks had the highest survival rate of saplings under consecutive cuttings.

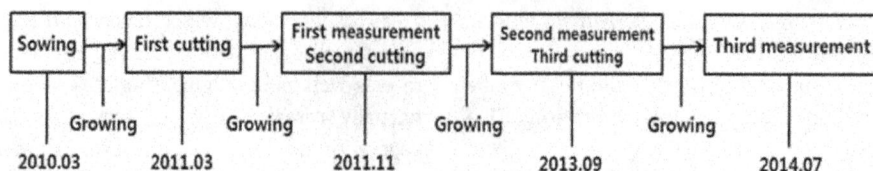

Fig. 1 Experimental procedure of sprout response to sapling cutting of six oak species for 4 years

Statistical analyses

We applied normal distribution test (Kolmogorov-Smirnov test) on the number of sprouts, sprout length, and survival rate because the number of sample is small. The significance between groups was confirmed by performing Kruskal-Wallis post hoc test ($p < 0.05$).

All the statistical analysis was conducted using Statistica Statistics Package (Statsoft CO. 2007) (No and Jeong 2002).

Results and discussion

Number of sprouts

The sprouts of six oaks all started growing from spring, none appeared during summer, and only sprouts of *Q. dentata* showed development in fall (Fig. 2, left). Our result was different from the report of Kwon (2002) who stated that sprouts of mature oak trees continued to appear from spring to autumn because it seems that the saplings of oaks used in this study contain much less assimilates than mature trees used in his experiment. The number of sprouts in *Q. variabilis* that appeared in spring after the first cutting was greater than other five oaks. The number of sprouts that appeared on *Q. dentata* in fall was the same as in spring. In case of *Q. dentata*, the sprouts that appeared in the first year all survived displaying the highest survival rate after the first cutting (Fig. 2, right). On contrary, 20–60% of sprouts in other five oaks died in spring and autumn displaying a low initial survival rate. The differences in sprouting ability among oaks in the same genus seem to be caused by individual variation, such as assimilate storage (Mroz et al. 1985).

The number of sprouts that survived for a year after the first cutting was high in *Q. dentata* and low in *Q. aliena*, but there was no significance among six oaks as they showed similar trends (Fig. 3). Only *Q. acutissima* showed a consistent increase in the number of sprouts after the cuttings whereas other five oaks showed a decrease after each cutting (Qv, Qm) or an increase slightly

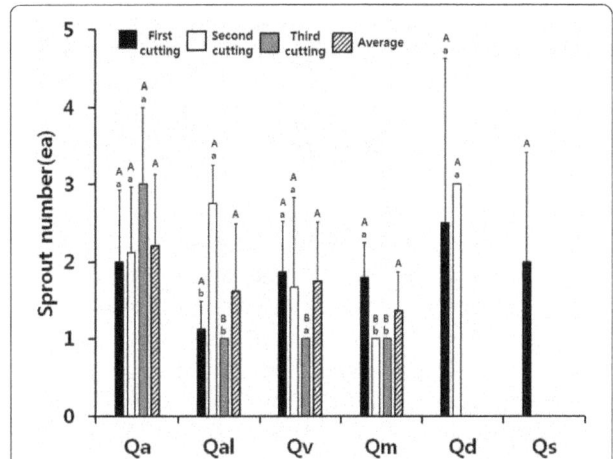

Fig. 3 The number of sprouts according to cutting time of six oaks. The *capital letter above the bar* represents comparison between oaks according to cutting time, and *the small letter* represents comparison within the oaks (*different letters* represent statistical differences, $p < 0.05$) (*Qa Q. acutissima, Qm Q. mongolica, Qal Q. aliena, Qv Quercus varibilis, Qd Q. dentata,* and *Qs Q. serrata*)

after the second cutting (Qal, Qd). After the third cutting, the number of sprouts in *Q. acutissima* was higher than those of *Q. variabilis*, *Q. aliena*, and *Q. mongolica* ($p < 0.05$) but it was the same in the later three oaks (Qv, Qal, Qm). *Q. acutissima* showed relatively high average number of sprouts, which appeared three times over 3 years, among four oaks, but there was no significant differences ($p < 0.05$). However, the sprouts of *Q. serrata* and *Q. dentata* did not appear after the second and the third cutting because all their individuals died. This result proves that *Q. serrata* have the lowest sprouting ability and *Q. dentata* have lower sprouting ability among six oaks in this study.

The fact that the number of shoot in *Q. acutissima* is higher than other oaks, even after consecutive cuttings, does not agree with the result of studies on regions disturbed with frequent forest fires (Jung et al. 2013). This

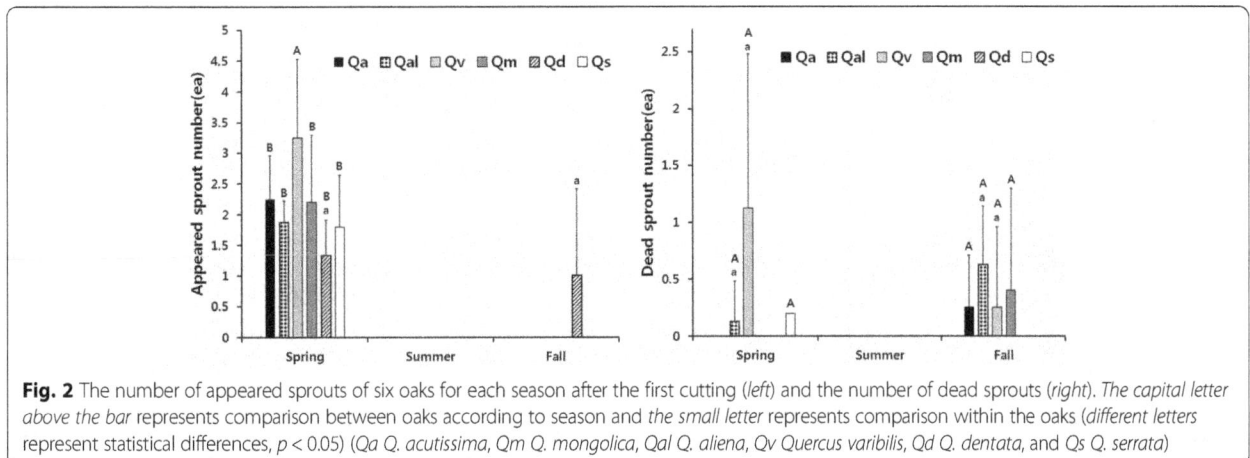

Fig. 2 The number of appeared sprouts of six oaks for each season after the first cutting (*left*) and the number of dead sprouts (*right*). *The capital letter above the bar* represents comparison between oaks according to season and *the small letter* represents comparison within the oaks (*different letters* represent statistical differences, $p < 0.05$) (*Qa Q. acutissima, Qm Q. mongolica, Qal Q. aliena, Qv Quercus varibilis, Qd Q. dentata,* and *Qs Q. serrata*)

difference in result can be attributed to the difference between the development stage of trees used in our study and their trees, which were mature trees growing in the region frequently affected with forest fires (Barbour et al. 1980). In other words, the sprouts can grow from bud located high up in the stem if the fire is weak but even the location of bud distributed on stem and number of bud can be different according to the development stage of the tree and state of growth. Q. acutissima, which has thinner bark than other oaks such as Q. variabilis, is presumed to have relatively low sprout regeneration ability in regions frequently affected with forest fires considering the fact that thick bark of the trunk plays an important role in trees surviving through the forest fires (Griffin 1980; Hengst & Dawson 1994; Pinard & Huffman 1997; Odhiambo et al. 2014).

Sprout length

The sprout length of six oaks after the first cutting grew the most in fall than other two seasons in all six oaks but the growth was low in all the oaks during summer except for Q. serratai (Fig. 4). Q. acutissima grew the most during spring whereas grow length of Q. aliena and Q. variabilis were lower. Q. mongolica, Q. dentata, and Q. serrata have intermediate level. The sprouts grew the most over the year in Q. acutissima whereas Q. dentata grew the least. The other four oaks had similar growth level.

After the first cutting of six oaks, the sprout length of Q. acutissima was the longest, Q. aliena and Q. variabilis were lower. Q. serrata, Q. mongolica, and Q. dentata have intermediate level (Fig. 5). The reason for such

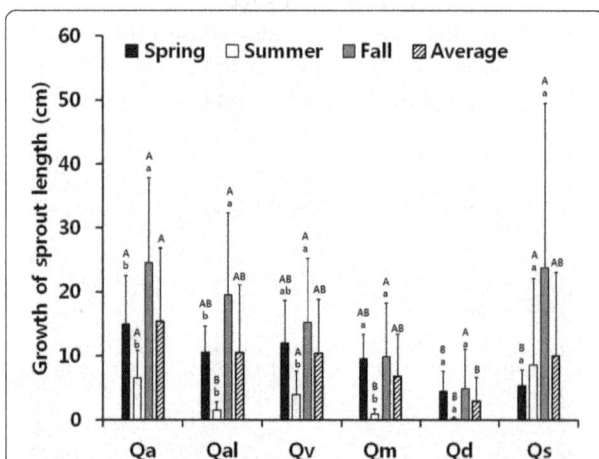

Fig. 5 The sprout length of six oaks according to cutting time. *The capital letter above the bar represents comparison between oaks, and the small letter represents comparison within the oaks (Qa Q. acutissima, Qm Q. mongolica, Qal Q. aliena, Qv Q. varibilis, Qd Q. dentata, and Qs Q. serrata)*

growth in Q. acutissima can be explained by relatively acorn of large size in comparison to other oaks (Shin et al. 2011). After the second cutting, there were no statistical differences in the sprout length among five oaks. Only Q. serrata died. After the third cutting, there were no statistical differences among Q. variabilis, Q. acutissima, and Q. aliena. But Q. serrata and Q. dentata died. The sprout length is the longest after 2 years from its initial cutting in all five surviving oaks. This can be explained by the increase in photosynthetic assimilates that plants accumulated over the 2 years of growth period (Barbour et al. 1980).

In our study, the sprout growth of Q. acutissima was higher than that of Q. mongolica and Q. dentata. Such result is similar to the experiment on sprout regeneration of mature oak trees after the first cutting (Kwon et al. 2002; Lee et al. 2000), but it is different from the experiment conducted in regions frequently affected by forest fires (Jung et al. 2013).

On the other hand, the sprout growth is affected not only among oaks but also by cutting time, stump diameter when cutting in the same oaks. It is expected that most of the regeneration of oaks comprises of sprouting in regions where stem cutting rarely happens while they are still saplings as in our study. However, it seems that maintaining the community of Q. serrata and Q. dentata may not be possible if the disturbance, such as felling that continually has been done over decades in Korea for firewood or heating. Currently, Q. acutissima is distributed widely in lowlands of central regions of Korea whereas Q. dentata community has the least distribution (Yang 2002; Song 2007; Kim et al. 2009). This may have

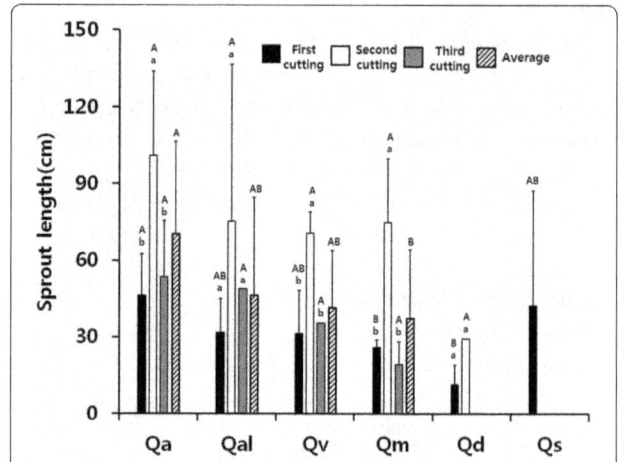

Fig. 4 The sprout length of six oaks after the first cutting. *The capital letter above the bar represents comparison between oaks, and the small letter represents comparison within the oaks (Qa Q. acutissima, Qm Q. mongolica, Qal Q. aliena, Qv Quercus varibilis, Qd Q. dentata, and Qs Q. serrata)*

been caused by the poor sprouting ability of *Q. dentata* in response to artificial disturbances.

The number of sprouts showed positive correlation with the sprout growth (Fig. 6). The study of Lee et al. (2000) on mature oak trees reports that the number of sprouts and the sprout growth has an inverse relationship. Such contrasting result can be attributed to the fact that the mature trees only have chlorophyll in their leaves whereas the saplings used in our study have chlorophyll even in their stems. The more sprout of saplings has more chlorophyll content that can photosynthesize. So, it is able to store more assimilates that can be used for sprout growth. But most of mature trees cannot photosynthesize as their stems have gone through lignification and that have a high rate of energy consuming organism. So, if mature trees have more sprouts, those will be not grown well.

Sprout reduction rate

The number of sprouts only increased in *Q. acutissima* after the cutting whereas other five oaks all decreased (Fig. 7). The sprout length increased in *Q. acutissima, Q. aliena,* and *Q. variabilis* as time passed whereas it decreased in *Q. mongolica, Q. dentata,* and *Q. serrata.* Hence, the only oak that increased in both the number of sprouts and the sprout length was *Q. acutissima* among six oaks. The number of sprouts decreased in *Q. aliena* and *Q. variabilis,* but their sprout length increased. The other three oaks decreased in both the number of sprouts and the sprout length. This result means that *Q. acutissima* has the highest sprouting ability under a disturbed environment where consecutive cuttings occur. Nonetheless, the sprouting ability of oaks will vary according to the change of ecological niche responding to environmental conditions, such as light, moisture, nutrients, and climate change (Mroz et al. 1985).

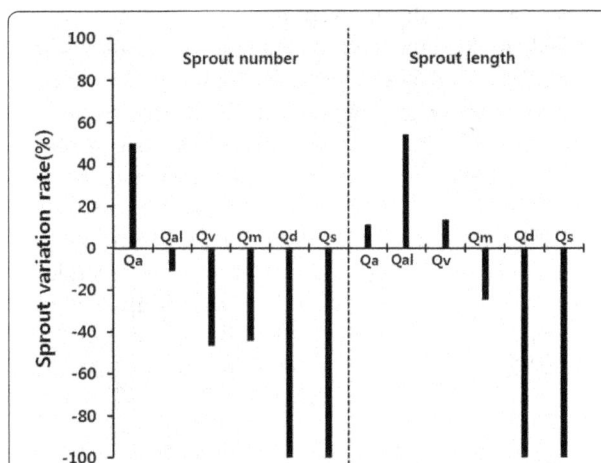

Fig. 7 The relative change in the number of sprouts and the sprout length of six oaks. It shows the ratio of the third measurement to the first measurement. Positive value shows that the number of sprouts and the sprout length increased over time whereas negative value shows a reduction (*Qa Q. acutissima, Qm Q. mongolica, Qal Q. aliena, Qv Q. varibilis, Qd Q. dentata,* and *Qs Q. serrata*)

Survival rate

All saplings of *Q. acutissima, Q. aliena,* and *Q. variabilis* survived after the first cutting but only 62.5% of *Q. mongolica* and 25% of *Q. mongolica* and *Q. serrata* survived (Fig. 8). All saplings of *Q. acutissima* survived after the second cutting, all saplings of *Q. serrata* died, and other four oaks showed 10–50% survival rate. But, the survival rate of *Q. acutissima* and *Q. mongolica* was 37.5% after the third cutting and it was 12.5% for *Q. aliena* and *Q. variabilis.* The saplings of *Q. mongolica* and *Q. serrata* all died by this time. The average survival rate over the 3 years was as follows: *Q. acutissima* (79.2%), *Q. aliena* (54.2%), *Q. variabilis* (50.0%), *Q. mongolica* (45.8%), *Q. dentata* (12.5%), and *Q. serrata* (8.3%). *Q. acutissima,*

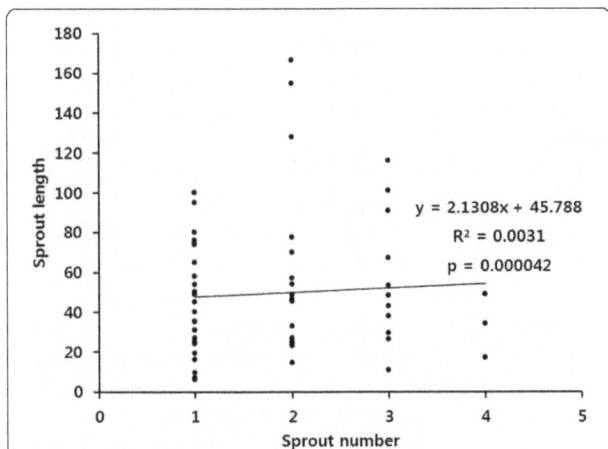

Fig. 6 Regression analysis of the number of sprouts and the sprout length

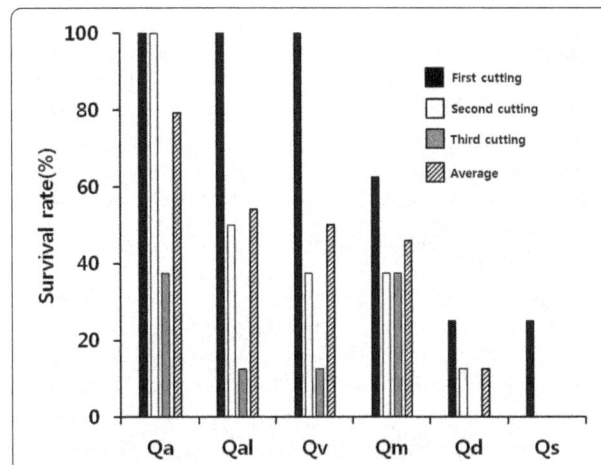

Fig. 8 The survival rate of six oaks according to cutting time (*Qa Q. acutissima, Qm Q. mongolica, Qal Q. aliena, Qv Q. varibilis, Qd Q. dentata,* and *Qs Q. serrata*)

whose saplings all survived until the second cutting, showed the highest survival rate, and *Q. serrata*, whose saplings all died after the second cutting, showed the lowest survival rate. The early death of saplings of the above two oaks (*Q. dentata* and *Q. serrata*) is probably caused by more rapid decrease of assimilates resulted in consecutive cutting than others (Griffin 1980; Kim 1995). Based on the above result, the saplings of *Q. acutissima* had the highest survival rate under a condition where a frequent disturbance, such as cutting, occurred. This could be attributable to the acorn size of *Q. acutissima* which has greater amount of assimilates than other oaks.

Conclusions

The regeneration of sprouts plays a crucial role in the regeneration of oaks community. Our study result revealed that even the saplings of six oaks that are less than a year can develop sprouts, grow, and survive. This means that oak community can still continue to exist/ grow even after being affected by a disturbance which could totally destroy the aboveground part, shoot. Moreover, we could conclude that *Q. acutissima*, among six oaks, had the highest sprouting ability. Such difference in the sprouting ability of oak saplings is an important evidence that could explain why *Q. acutissima* community currently dominates the forests in the lowlands of Korea where artificial disturbances continually occurred in the past.

Abbreviations
Q. acutissima: Quercus acutissima; Q. aliena: Quercus aliena; Q. dentata: Quercus dentata; Q. mongolica: Quercus mongolica; Q. serrata: Quercus serrata; Q. variabilis: Quercus variabilis

Acknowledgements
This study was supported by the Mid-career Researcher Program (NRF-2016R1A2B1010709) through NRF grant funded by the MEST.

Funding
The institute is not involved in any way in the preparation of this manuscript or the decision to submit it.

Authors' contributions
All authors conducted a survey together during the study period. LSY wrote the manuscript. YYH participated in the design of the study and examined the manuscript. All authors read and approved the final manuscript.

Competing interests
The authors declare that they have no competing interests.

References
Barbour, M. G., Burk, J. H., & Pitts, W. D. (1980). Terrestrial plant ecology (p. 634). Menlo Park: The Benjamin Cummings.

Griffin JR. Sprouting in fire damaged valley oaks, Chews Ridge, California. Forest service, US Department of Agriculture. 1980. p. 216-219.

Hengst, G. E., & Dawson, J. O. (1994). Bark properties and fire resistance of selected tree species from the central hardwood region of North America. *Canadian Journal of Forest Research, 24*(4), 688–696.

Imanishi, A., Morimoto, J., Imanishi, J., Shibata, S., Nakanishi, A., Osawa, N., & Sakai, S. (2010). Sprout initiation and growth for three years after cutting in an abandoned secondary forest in Kyoto, Japan. *Landscape and Ecological Engineering, 6*(2), 325–333.

Jo, M. H. (1989). Coloured woody plants of Korea (p. 498). Academic Publisher.

Jung, S. C., Seo, Y. O., & Kim, K. M. (2013). Study on growth and sprouts of oak forest for forest fire site in South Korea. *Life Science Journal, 10*(2), 1256–1260.

Kim, S. K. (1995). Tending method for regenerative afforestation of Q. acutissima. *Forest, 52*, 84–87.

Kim, D. G., Hwang, G. Y., Kim, M. S., & Hong, H. P. (1991). Growth and development of stump sprout of twenty deciduous broadleaf trees. *The Research Reports Forestr Research Institude, 42*, 20–35.

Kim, I. T., Ms, S., & Jung, S. H. (2009). Analysis of distribution and association structure on the sawtooth oak (*Quercus accutissima*) forest in Korea. *Journal of Life Science, 19*(3), 356–361.

Kwon, K. W., Jung, J. C., & Choi, J. H. (1998). Studies on coppice regeneration of oak stands 1—sprouts and their growth of Quercus variabilis and Quercus mongolica. *J Life Sci & Res Wonkwang Univ., 20*, 19–26.

Kwon KW, Choi JH, Song HG. Studies on regeneration strategy establishment of oak species—biomass production, sprouts and their growth of Quercus mongolica, Quercus variabilis and quercus acutissima. J Aca Res. 2002. p. 177-179.

Lee, T. B. (2003). Coloured flora of Korea (p. 910). Hyangmun Publisher.

Lee, H. J., & You, Y. H. (2009). Ecological niche breadth of Q. mongolica and overlap with Q. acutissima and Q. variabilis along with three environment gradients. *Korean journal of environmental biology, 27*, 191–197.

Lee, S. H., & You, Y. H. (2012). Measurement of ecological niche of Quercus aliena and Q. serrata under environmental factors treatments and its meaning to ecological distribution. *Journal of Ecology and Environment, 35*(3), 227–234.

Lee, D. K., Kwon, K. C., Kim, Y. H., & Kim, Y. S. (2000). Sprouting and sprout growth of four Quercus species—at natural stands of Quercus mongolica, Q. variabilis, Q. acutissima and Q. dentata growing at Kwangju-gun, Kyonggi-Do. *J Kor For En, 19*, 61–68.

Lim, H., Kim, H. R., & You, Y. H. (2012). Growth difference between the seedlings of Quercus serrata and Q. aliena under light, moisture and nutrient gradients. *J Wetl Res, 14*, 237–242.

Mroz, G. D., Frederick, D. J., & Jurgensen, M. F. (1985). Site and fertilizer effects on northern hardwood stump sprouting. *Can J For Res., 15*, 535–543.

No, H. J., & Jeong, H. Y. (2002). Well defined statistica analysis according to Statistica (p. 627). Hyeongseol Publisher.

Odhiambo, B., Meincken, M., & Seifert, T. (2014). The protective role of bark against fire damage: a comparative study on selected introduced and indigenous tree species in the Western Cape, South Africa. *Trees, 28*(2), 555–565.

Pinard, M. A., & Huffman, J. (1997). Fire resistance and bark properties of trees in a seasonally dry forest in eastern Bolivia. *Journal of Tropical Ecology, 13*(05), 727–740.

Shin, J. H., & You, Y. H. (2011). Effects of seed size on the rate of germination, early growth and winter survival in four oak species. *Korean journal of environmental biology., 29*, 274–279.

Smith, D. M. (1986). The practice of silviculture (p. 527). John wiley and Sons, Inc.

Song, M. S. (2007). Analysis of distribution and association structure on the sawtooth oak (*Quercus acutissima*) forest in Korea. Doctor's Thesis. Changwon: University of Chanwon.

Yang, K. C. (2002). Classification of major habitats based on the climatic conditions and topographic features in Korea. *Doctor's thesis.* Seoul: University of Chungang.

8

Assessing the phytotoxicity of cetrimonium bromide in plants using eco-physiological parameters

Uhram Song[*] and Han Eol Kim

Abstract

Background: Although cetrimonium bromide is widely used for its bactericidal effects, the safety of cetrimonium bromide remains controversial. Therefore, the phytotoxicity of cetrimonium bromide was tested to evaluate its acute toxicity to plants and possible toxicity to other organisms and the ecosystem.

Results: The germination rates of two test species, Lactuca sativa and Brassica campestris, were significantly decreased after cetrimonium bromide treatment. Furthermore, cetrimonium bromide treatment at over 1 mg/L concentration significantly affected root elongation immediately after germination. In pot experiments with semi-mature plants, significantly decreased shoot elongation and chlorophyll content were detected in both species following cetrimonium bromide treatment. Cetrimonium bromide treatment also significantly increased the antioxidant enzyme activities of plants.

Conclusion: Our results show that cetrimonium bromide is phytotoxic, and since phytotoxicity testing can imply potential toxicity in the environment, further studies of the environmental toxicity of cetrimonium bromide should be performed.

Keywords: Cetrimonium bromide, Toxicity, Germination, Elongation, Chlorophyll, Antioxidant enzyme activity

Background

Cetrimonium bromide, $(C_{16}H_{33})N(CH_3)_3Br$ (CAS no. 57-09-0), also known as hexadecyl-trimethyl-ammonium bromide (International Union of Pure and Applied Chemistry [IUPAC] name), is an amine-based cationic quaternary surfactant that is widely used in components of antiseptic materials (Andersen 1997). Cetrimonium bromide is used in hygienic goods and cleaning agents for bactericidal effects (Oh et al. 2014) and in many cosmetics. However, recently there has been much debate about the safety of cetrimonium bromide (Woo 2015). Since cetrimonium bromide, which was present in many cosmetic products, was believed to be associated with human disease, the safety of this chemical has been questioned (Kim 2014). Furthermore, cetrimonium bromide is used in baby wet tissue wipes, resulting in much debate on its safety (Woo 2015). However, despite these problems, there are a very limited number of studies on the toxicity of this chemical.

Among the few studies, one research group reported an increase in fetal deaths in mice treated with 35.0 mg/kg cetrimonium bromide, although other studies of cetrimonium bromide did not show any acute toxicity (Andersen 1997). However, as there are only a few reports, further studies are required to determine the acute toxicity of cetrimonium bromide. Moreover, although cetrimonium bromide is commonly believed to be dangerous for the environment, especially in aquatic ecosystems, the toxicity of cetrimonium bromide to aquatic organisms is not well documented (Tišler et al. 2004). Therefore, further research on the ecotoxicity of cetrimonium bromide is required.

Phytotoxicity testing is often used to estimate potential toxicity in the environment and for risk assessment of chemicals and formulations of human relevance (Kristen 1997). Phytotoxicity tests are relatively simple but precise toxicological assays with sensitive results (Kristen 1997). Since the toxicity of cetrimonium bromide is still undetermined and its ecotoxicity is almost unknown, testing the toxicity of cetrimonium bromide with plants would supply further information on the toxicity and ecotoxicity of

* Correspondence: uhrami@jejunu.ac.kr
Department of Biology, Jeju National University, Jeju 63243, South Korea

cetrimonium bromide. Therefore, in this study, the phyto-toxicity of cetrimonium bromide was tested by evaluating seed germination rates, root elongation, and the ecophysi-ological responses of mature vegetable crops.

Methods

Cetrimonium bromide was purchased from Daejung Chemicals and Materials (Daejung C&M, Gyonggi Province, Korea). The purity was above 99%, and loss on drying at 100 °C was less than 2%. We used 0.01 mg/L, 0.1 mg/L, 1 mg/L, 10 mg/L, 100 mg/L, and 1000 mg/L cetrimonium bromide solutions for testing by adding measured weight of cetrimonium bromide to distilled water. Although cetrimonium bromide is used at up to 10% concentration in cosmetics (Andersen 1997), we tested up to 1000 mg/L because animal toxicity was observed at a 10 mg/L concentration (Andersen 1997), and concentrations of approximately 29 mg/L can be found in wet tissues (Oh et al. 2014). Therefore, 0.01 mg/L, 0.1 mg/L, and 1 mg/L treatments were selected for testing of environmentally realistic concentrations, and 10 mg/L, 100 mg/L, and 1000 mg/L treatments were selected to test acute toxicity.

Seeds of *Brassica campestris* ssp. *napus* var. *nippo-oleifera* Makina (oilseed rape) and *Lactuca sativa* L. (lettuce) were selected to test the toxicity of cetrimonium bromide. These species were selected because they are common, easy to obtain, and included among the species recommended for the testing of chemicals in the Organisation for Economic Co-operation and Development guidelines (OECD, 2003). Seeds were purchased from a local Syngenta agent (Syngenta AG, Switzerland). The seeds were vernalized for 2 weeks and sterilized for 10 min in 10% sodium hypochlorite solution (USEPA, 1996) before application.

For germination rate tests, seeds were soaked in cetrimonium bromide solutions for 24 h (Zheng et al. 2005) in the dark at room temperature with gentle shaking on an orbital shaker at 60 rpm to improve mixing. Subsequently, the seeds were washed with distilled water. Most seeds were transferred to 100-mm Petri dishes containing a piece of filter paper (90 mm) and 6 mL of distilled water (Lin and Xing 2007). The seeds were tested for germination in a growth chamber under a range of conditions established by the OECD guidelines (OECD, 2003): temperature 25 °C, humidity $70 \pm 25\%$, photoperiod 18 h light, light intensity $300\ \mu E \cdot m^{-2} \cdot s^{-1}$ with protection from drying. Each Petri dish ($n = 5$) contained five seeds, and germination rates were investigated for 1 week.

For root elongation studies, seeds were germinated in Petri dishes. After 2 days, germinated seeds were moved to new Petri dishes. Each Petri dish contained five seedlings and 5 mL of the test medium ($n = 3$). The root lengths of the seedlings were measured every 3 days (six times altogether). Other conditions, including solution concentrations and

chamber conditions, were the same as those of the germination study, described above.

For pot experiment, plants were germinated and grown in a 50-hole pot tray with each hole filled with 10 g of commercial growing soil (Pro-100, Chamgrow, Korea). After a 5-week growing period from the seedling stage, 10 mL of cetrimonium bromide solution was added to each pot (0.01 mg/L, 0.1 mg/L, 1 mg/L, 10 mg/L, 100 mg/L, and 1000 mg/L; seven replicates for each treatment). Solutions were administered several times with a 1-mL pipette to avoid leaching. Therefore, the growing soil contained exactly 0.01 mg/kg, 0.1 mg/kg, 1 mg/kg, 10 mg/kg, 100 mg/kg, and 1000 mg/kg of cetrimonium bromide. The chlorophyll content of leaves was measured using a SPAD 502 system (Minolta Co., Japan), 1 week after treatment. The antioxidant enzyme activities (total antioxidant capacity [TAC] and superoxide dismutase [SOD] activity) of the plants were measured using the protocols of Song and Lee, 2010, 1 week after treatment (five replicates). The average height of the shoots of the plants was 6.4 ± 0.5 cm for *Lactuca sativa*, and 6.7 ± 0.3 cm for *Brassica campestris* (values represent mean $\pm$ SE of 49 replicates). Plant growth after treatment was also measured for 1 week.

A one-way ANOVA was performed to identify significant differences between treatments. Upon detection of a significant difference, Tukey's studentized range (honest significant difference) test was applied post hoc and assessed using SAS 9.3 software (SAS Institute Inc., USA). Differences were considered significant when $p < 0.05$.

Results and discussion

Table 1 shows that cetrimonium bromide treatment significantly decreased the germination rates of both species. Specially for *Lactuca sativa*, the germination rate was significantly affected by cetrimonium bromide, even at 0.1 mg/L, and germination was totally inhibited at 1000 mg/L. Total inhibition of germination has not previously been observed in the authors' prior germination tests

Table 1 Germination rates (%) of plants after cetrimonium bromide treatment

Species	*Lactuca sativa*		*Brassica campestris*	
Treatment/days	3 days	6 days	3 days	6 days
Control	98.0 ± 2.0^a	100.0 ± 0.0^a	96.0 ± 2.4^a	98.0 ± 2.0^a
0.01 mg/L	96.0 ± 2.4^a	100.0 ± 0.0^a	86.0 ± 2.4^{ab}	86.0 ± 2.4^{ab}
0.1 mg/L	88.0 ± 5.8^a	92.0 ± 4.9^a	74.0 ± 5.1^{abc}	78.0 ± 4.9^{ab}
1 mg/L	86.0 ± 2.4^a	92.0 ± 3.7^a	74.0 ± 6.0^{abc}	76.0 ± 5.1^{ab}
10 mg/L	86.0 ± 5.1^a	92.0 ± 3.7^a	72.0 ± 5.8^{bc}	76.0 ± 7.5^{ab}
100 mg/L	46.0 ± 6.8^b	54.0 ± 6.8^b	72.0 ± 4.9^{bc}	76.0 ± 5.1^{ab}
1000 mg/L	0.0 ± 0.0^c	0.0 ± 0.0^c	64.0 ± 6.8^c	66.0 ± 6.8^b

Values represent the mean ± SE of five replicates
Means in a column with the same letter are not significantly different ($p > 0.05$)

of toxins (Song, Jun et al. 2013; Song, Shin et al. 2013; Song et al. 2014); thus, cetrimonium bromide is considerably toxic to plants, in a similar manner to herbicides (Zonno and Vurro 2002). Although cetrimonium bromide significantly decreased the germination of *Brassica campestris*, this species was less affected than *Lactuca sativa* (Table 1). As the seed coat of *Brassica campestris* is harder and thicker than that of *Lactuca sativa*, the short time of exposure would not have been enough to penetrate the seed coat. Nevertheless, *Brassica campestris* showed a significantly reduced germination rate even at the lowest treatment concentration (0.01 mg/L), indicating that cetrimonium bromide is phytotoxic. Therefore, even for environmentally realistic conditions, cetrimonium bromide would likely damage plants when released into the environment. Figure 1 shows that cetrimonium bromide treatment significantly affected root elongation. In both species, treatment at over 1 mg/L concentrations significantly decreased root elongation when compared with that of the control. At 10 mg/L, both species showed definite growth for the first 3 days

(especially for *Brassica campestris*, where there were no significant differences between the 10 mg/L treatment and control) but began to show significantly decreased root length over time, indicating that extended exposure to cetrimonium bromide was more toxic, probably because of accumulation. Furthermore, treatment at over 100 mg/L resulted in no root growth (Fig. 1), indicating that the seedlings were dead. These results show that cetrimonium bromide is phytotoxic at certain concentrations.

Cetrimonium bromide was also phytotoxic to plants beyond the seed and seedling stage. Table 2 shows that plants grown over 5 weeks showed a significant reduction in shoot elongation after cetrimonium bromide treatment. *Lactuca sativa* exhibited a significantly decreased shoot growth with treatments over 1 mg/kg in concentration, and *Brassica campestris* exhibited a significantly decreased shoot growth with treatments over 0.1 mg/kg in concentration. Moreover, the highest concentration treatment resulted in almost no growth for a week (Table 2). By contrast, 0.1 mg/kg treatment of *Lactuca sativa* and 0.01 mg/kg treatment of *Brassica campestris* resulted in approximately 10–20% greater growth than controls. This result is hard to explain because physiological parameters, such as chlorophyll content (Table 3) and antioxidant enzyme activity (Table 4), show that the plants were stressed even at these concentrations. One possible explanation is that since the plants were grown for 5 weeks, the soil became contaminated by microorganisms, which stressed the plants. At low concentrations, cetrimonium bromide would likely decrease the activity of these microorganisms and provide an advantage to the plants. However, this is only a hypothesis and the exact reason remains unknown. Overall, cetrimonium bromide significantly affected the growth of semi-mature plants.

Chlorophyll content was more sensitive to treatment, as both plant species showed significantly decreased chlorophyll content even at the lowest (0.01 mg/kg) concentration (Table 3). Especially with treatments over 100 mg/kg in concentration, both species lost chlorophyll and began to fade. As the environmentally realistic treatments all resulted

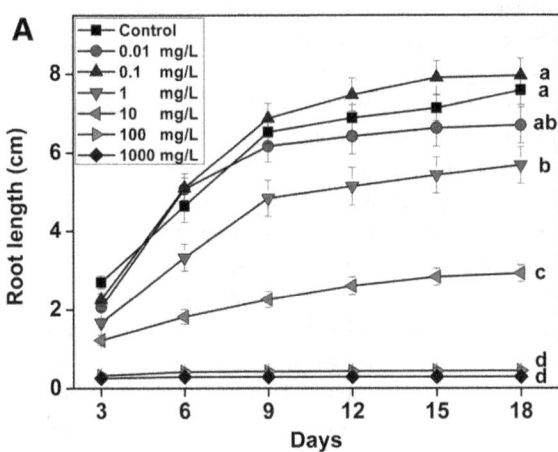

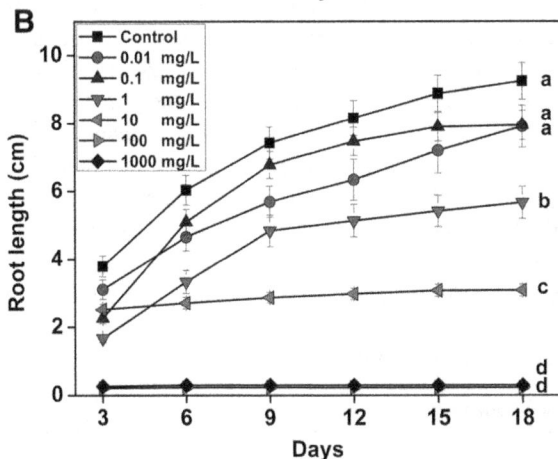

Fig. 1 Root elongation of **a** *Lactuca sativa* and **b** *Brassica campestris* after cetrimonium bromide treatment. *Symbols* and *error bars* represent the mean ± SE of 20 replicates. *Symbols* with the same letters are not significantly different (*p* > 0.05)

Table 2 Shoot elongation (mm) of plants 1 week after cetrimonium bromide treatment

Treatment/species	Lactuca sativa	Brassica campestris
Control	3.4 ± 0.6[ab]	8.1 ± 1.7[ab]
0.01 mg/kg	3.3 ± 0.7[abc]	10.1 ± 1.7[a]
0.1 mg/kg	3.7 ± 1.0[a]	5.0 ± 1.5[ab]
1 mg/kg	1.0 ± 0.2[bcd]	5.0 ± 1.7[ab]
10 mg/kg	1.3 ± 0.2[abcd]	6.8 ± 2.1[ab]
100 mg/kg	0.8 ± 0.3[cd]	1.8 ± 0.8[b]
1000 mg/kg	0.0 ± 0.0[d]	0.2 ± 0.1[c]

Values represent the mean ± SE of seven replicates
Means in a column with the same letter are not significantly different (*p* > 0.05)

Table 3 Chlorophyll content (SPAD-502 units) of plant leaves 1 week after cetrimonium bromide treatment

Treatment/species	Lactuca sativa	Brassica campestris
Control	14.2 ± 0.2^a	23.2 ± 1.1^a
0.01 mg/kg	11.2 ± 0.3^b	21.5 ± 0.3^{ab}
0.1 mg/kg	10.2 ± 0.3^{bc}	18.8 ± 0.3^{bc}
1 mg/kg	9.3 ± 0.4^c	17.2 ± 0.8^c
10 mg/kg	9.4 ± 0.2^c	16.7 ± 0.6^c
100 mg/kg	7.0 ± 0.3^d	11.5 ± 0.7^d
1000 mg/kg	4.4 ± 0.5^e	6.8 ± 0.9^e

Values represent the mean ± SE of seven replicates
Means in a column with the same letter are not significantly different ($p > 0.05$)

in significantly decreased chlorophyll content, cetrimonium bromide will likely affect plants in the field when it is released into the surrounding environment (ecosystem).

Notably, these results indicate that when the cetrimonium bromide used by humans enters aquatic ecosystems via sewage, the impact on aquatic plants could be considerable, and therefore the results should be monitored and investigated. The original experimental design intended to monitor plant growth and chlorophyll content for a few weeks, but since the plants treated at high concentrations began to fade and lose chlorophyll within 1 week (Table 3), we harvested the plants for antioxidant enzyme activity assays before the plants died. Table 4 shows the antioxidant enzyme activities of Lactuca sativa after cetrimonium bromide treatment. We intended to measure the antioxidant enzyme activities of both species; however, there was not enough Brassica campestris leaf biomass (over 1 g fresh weight) for protein extraction via a phosphate buffer and Bradford assay (Song and Lee 2010). The above ground biomass allocation of Brassica campestris was over 75% on stems, even in the control, and there was not enough leaf biomass. Cetrimonium bromide treatment significantly increased both the TAC and SOD values (Table 4), indicating that the plants were under stress (Song, Jun, et al. 2013). The TAC results

Table 4 Zinc accumulation in plants after exposure to zinc oxide nanoparticles for 5 weeks

Zinc content (mg/kg)	Hydrilla verticillata	Phragmites australis
Control	ND	0.02 ± 0.01^c
0.01	0.08 ± 0.01^c	0.22 ± 0.01^c
0.1	0.10 ± 0.01^c	0.57 ± 0.31^{bc}
1	0.11 ± 0.01^c	1.17 ± 0.03^{ab}
10	0.20 ± 0.03^b	1.43 ± 0.03^a
100	0.35 ± 0.01^a	1.47 ± 0.03^a
1000	0.34 ± 0.01^a	1.67 ± 0.22^a

Values represent mean ± SE of three replicates
Values with different letters are significantly different at the $p < 0.05$ level, whereas those with the same letters are not

in particular indicate that the plants were under overall stress, which may be reflected in the growth and chlorophyll content of the plants (Tables 2 and 3). SOD values are frequently used as an indicator of pollutant stress (Koricheva et al. 1997). Since plants treated at over 0.1 mg/kg showed significantly increased SOD values, cetrimonium bromide should be treated as a potential pollutant. As bromine residues in the soil can cause phytotoxicity (Lear 1975), cetrimonium bromide clearly shows phytotoxicity. Overall, all of the above results, including the germination rate, root and shoot elongation, chlorophyll content, and antioxidant enzyme activity, consistently show that cetrimonium bromide is phytotoxic.

Conclusion

The germination rates of both Lactuca sativa and Brassica campestris species were significantly decreased after cetrimonium bromide treatment. Notably, both species showed total inhibition of germination with the 1000 mg/L treatment. Furthermore, cetrimonium bromide treatment significantly affected root elongation immediately after germination. In semi-mature plants, significant reductions in shoot elongation and chlorophyll content were detected in both species after cetrimonium bromide treatment. Antioxidant enzyme activities of the plants were also significantly increased by cetrimonium bromide. These results indicate that cetrimonium bromide is markedly phytotoxic. However, surprisingly there are no related articles that report the phytotoxicity of cetrimonium bromide. Since phytotoxicity testing can be used to estimate potential toxicity in the environment, our results show that cetrimonium bromide could also be toxic to other organisms and ecosystems when released into the surrounding environment. Also, as cetrimonium bromide is likely to be mostly released by hydrologic system as the chemical is mainly used for water and cosmetic treatment, the growth of vegetables also could be affected by irrigation. As only a few articles report the toxicity of cetrimonium bromide, and these are limited to human (Momblano et al. 1984) and mammal (Andersen 1997) toxicity, further studies to define the acute toxicity to other organisms and potential environmental toxicity should be performed. Furthermore, cetrimonium bromide should be carefully monitored for its effects after release into the surrounding environment and into edible crops. Therefore, further studies of the toxicity of cetrimonium bromide at both the species level and environment-ecosystem level are required.

Acknowledgements
This research was supported by the 2015 scientific promotion program funded by Jeju National University.

Authors' contributions
US designed the experiment, participated the chamber experiment and drafted the manuscript. HE participated the chamber experiment. All authors read and approved the final manuscript.

Competing interests
The authors declare that they have no competing interests.

References
Andersen, F. (1997). Final report on the safety assessment of cetrimonium chloride, cetrimonium bromide, and steartrimonium chloride. *Int J Toxicol, 16*, 195–220.

Kim, SJ. (2014). *Stricter safety standards on wet tissues, shampoo The Korea Times*. https://www.koreatimes.co.kr/www/common/printpreview. asp?categoryCode=116&newsIdx=169060.

Koricheva, J, Roy, S, Vranjic, JA, Haukioja, E, Hughes, PR, & Hänninen, O. (1997). Antioxidant responses to simulated acid rain and heavy metal deposition in birch seedlings. *Environ Pollut, 95*, 249–258.

Kristen, U (1997). Use of higher plants as screens for toxicity assessment. *Toxicol in Vitro, 11*, 181–191.

Lear, B (1975). Phytotoxicity associated with bromide uptake in plants grown in soil fumigated with brominated hydrocarbon. *Nematologica, 5*, 24.

Lin, D, & Xing, B (2007). Phytotoxicity of nanoparticles: inhibition of seed germination and root growth. *Environ Pollut, 150*, 243–250.

Momblano, P, Pradere, B, Jarrige, N, Concina, D, Bloom, E (1984). Metabolic acidosis induced by cetrimonium bromide. *Lancet, 324*, 1045.

Oh, J, Kim, K, Pyo, H, Chung, BC, Lee, J (2014). *External standard addition method development of benzalkonium chloride, cetrimonium bromide and cetylpyridinium chloride in wet-tissues by liquid chromatography-electrospray ionization/mass spectrometry* (pp. 232–232). Proceedings of 53th symposium of the Korean society of analytical sciences.

Organization for Economic Cooperation and Development (OECD). (2003). *OECD Guidelines for the testing of chemicals: Proposals for updating guideline 208 - Terrestrial Plant Test: Seedling Emergence and Seedling Growth Test*. http://www.oecd.org/dataoecd/11/31/33653757.pdf.

Song, U, Lee, E (2010). Ecophysiological responses of plants after sewage sludge compost applications. *J Plant Biol, 53*, 259–267.

Song, U, Jun, H, Waldman, B, Roh, J, Kim, Y, Yi, J, Lee, EJ. (2013). Functional analyses of nanoparticle toxicity: a comparative study of the effects of TiO2 and Ag on tomatoes (Lycopersicon esculentum). *Ecotoxicol Environ Safety, 93*, 60–67.

Song, U, Shin, M, Lee, G, Roh, J, Kim, Y, Lee, E (2013). Functional analysis of TiO2 nanoparticle toxicity in three plant species. *Biol Trace Elem Res, 155*, 93–103.

Song, U, Mun, S, Waldman, B, Lee, E (2014). Effects of three fire-suppressant foams on the germination and physiological responses of plants. *Environ Manage, 54*, 865–874.

Tišler, T, Zagorc-Končan, J, Cotman, M, Drolc, A (2004). Toxicity potential of disinfection agent in tannery wastewater. *Water Res, 38*, 3503–3510.

U.S. Environmental Protection Agency (USEPA). 1996. Ecological effects test guidelines (OPPTS 850.4200): Seed Germination/Root Elongation Toxicity Test. http://www.epa.gov/opptsfrs/publications/OPPTS_Harmonized/850_Ecological_Effects_Test_Guidelines/Drafts/850-4200.pdf

Woo, HC (2015). *Baby wipes raise health concerns The Korea Times*. http://www.koreatimes.co.kr/www/news/biz/2014/09/123_164982.html. Accessed 21 Oct 2015.

Zheng, L, Hong, F, Lu, S, Liu, C (2005). Effect of nano-TiO(2) on strength of naturally aged seeds and growth of spinach. *Biol Trace Elem Res, 104*, 83–91.

Zonno, MC, Vurro, M (2002). Inhibition of germination of Orobanche ramosa seeds by Fusarium toxins. *Phytoparasitica, 30*, 519–524.

9

Coexistence between *Zostera marina* and *Zostera japonica* in seagrass beds of the Seto Inland Sea, Japan

Kenji Sugimoto[1][*], Yoichi Nakano[1], Tetsuji Okuda[2], Satoshi Nakai[3], Wataru Nishijima[4] and Mitsumasa Okada[5]

Abstract

Background: There have been many studies on the growth conditions of *Zostera marina* and *Zostera japonica*, but few studies have examined how spatial and temporal factors affect growth in established seagrass beds or the distribution range and shoot density. This study aims to clarify the factors that determine the temporal and spatial distribution of *Zostera marina* and *Zostera japonica* in the Seto Inland Sea east of Yamaguchi Prefecture.

Methods: The study site is in Hiroshima Bay of the Seto Inland Sea, along the east coast of Yamaguchi Prefecture, Japan. We monitored by diving observation to confirm shoot density, presence or absence of both species and observed water temperature, salinity by sensor in study sites.

Results: The frequency of occurrence of *Zostera marina* was high in all seasons, even in water depths of D.L. + 1 to −5 m (80 ± 34% to 89 ± 19%; mean ± standard deviation), but lower (as low as 43 ± 34%) near the breakwall, where datum level was 1 to 2 m, and it was further reduced in datum level −5 m and deeper. The frequency of occurrence of *Zostera japonica* was highest in water with a datum level of +1 to 0 m. However, in datum level of 0 m or deeper, it became lower as the water depth became deeper. Datum level +1 m to 0 m was an optimal water depth for both species. The frequency of occurrence and the shoot density of both species showed no negative correlation. In 2011, the daily mean water temperature was 10 °C or less on more days than in other years and the feeding damage by *S. fuscescens* in the study sites caused damage at the tips.

Conclusions: We considered that the relationship between these species at the optimal water depth was not competitive, but due to differences in spatial distribution, *Zostera marina* and *Zostera japonica* do not influence each other due to temperature conditions and feeding damage and other environmental conditions. *Zostera japonica* required light intensity than *Zostera marina*, and the water depth played an important role in the distribution of both species.

Keywords: *Zostera marina*, *Zostera japonica*, Coexistence of distribution, Low water temperature, Herbivorous fishes

Background

Zostera beds are major primary producers in estuarine and coastal areas, and they support large and diverse faunal assemblages. These beds are excellent habitats for many commercial fishes, providing hatcheries and nurseries for juvenile fish (Kikuchi 1980; Pollard 1984; Orth et al. 1984). Zostera beds in the Seto Inland Sea, Japan are made up of *Zostera marina* and *Zostera japonica* (Biodiversity Center of Japan 2008). In Japan, *Z. marina* is found from Hokkaido to Kyushu, and *Z. japonica* is distributed from Hokkaido to the Ryukyu Islands (Aioi 1998; Omori 2000; Aioi and Nakaoka 2003). The distribution of species and shoot density varies with the life history of the seagrass, environmental conditions, and by year (Sugimoto et al. 2008; Harrison 1982; Abe et al. 2004). *Z. marina* is taller and has longer and wider leaves than *Z. japonica*, while the shoot density of *Z. japonica* is higher than that of *Z. marina* (Arasaki 1950a, b). Therefore, the dominant species and distribution of seagrass is an important factor in determining the character of the seagrass bed. In particular, there are reports in recent years of *Z. japonica* increasing its presence in seagrass beds (Mach et al. 2014; Shafer et al. 2014).

* Correspondence: k-sugimoto@ube-k.ac.jp
[1]National Institute of Technology, Ube College, Ube, Japan
Full list of author information is available at the end of the article

Physical factors also play a role in the distribution and growth of Zostera beds. The upper part of seagrass is affected by water turbulence, and the lower part is affected by photosynthesis inhibition (Dennison and Alberte 1985; Duarte 1991; Mach et al. 2010; Kendrick et al. 2002). The habitats of *Z. marina* and *Z. japonica* are affected by salinity and depth (Greve and Krausen-Jensen 2005; Morita et al. 2010; Abe et al. 2009; Shafer et al. 2011). For example, in the summer, *Z. marina* growth becomes difficult when the daily mean water temperature is 28 °C or more. Seagrass reproduction occurs by branching of rhizomes or seed germination. In reproduction by seed, the germination rate of seagrass is higher when the water temperature is lower, and it is highest when the water temperature has been below about 10 °C.

There have been many studies on the growth conditions of *Z. marina* and *Z. japonica*, but few studies have examined how spatial and temporal factors affect growth in established seagrass beds or the distribution range and shoot density.

This study aims to clarify the factors that determine the temporal and spatial distribution of *Z. marina* and *Z. japonica* in the Seto Inland Sea east of Yamaguchi Prefecture.

Methods
Study sites
The study site is in Hiroshima Bay of the Seto Inland Sea, along the east coast of Yamaguchi Prefecture, Japan (34°00′ N, 132°12′ E) (Fig. 1). The Imazu and Monzen rivers flow into the bay near the study site, and at times of heavy rain about 15 km to the north, the salinity is temporarily low. Salinity in the vicinity of the study site ranged from 28 to 32‰. The seawater in the study area had low transparency, and water mixing is low at the north side of the bay (Kawanishi 1999). At the bottom

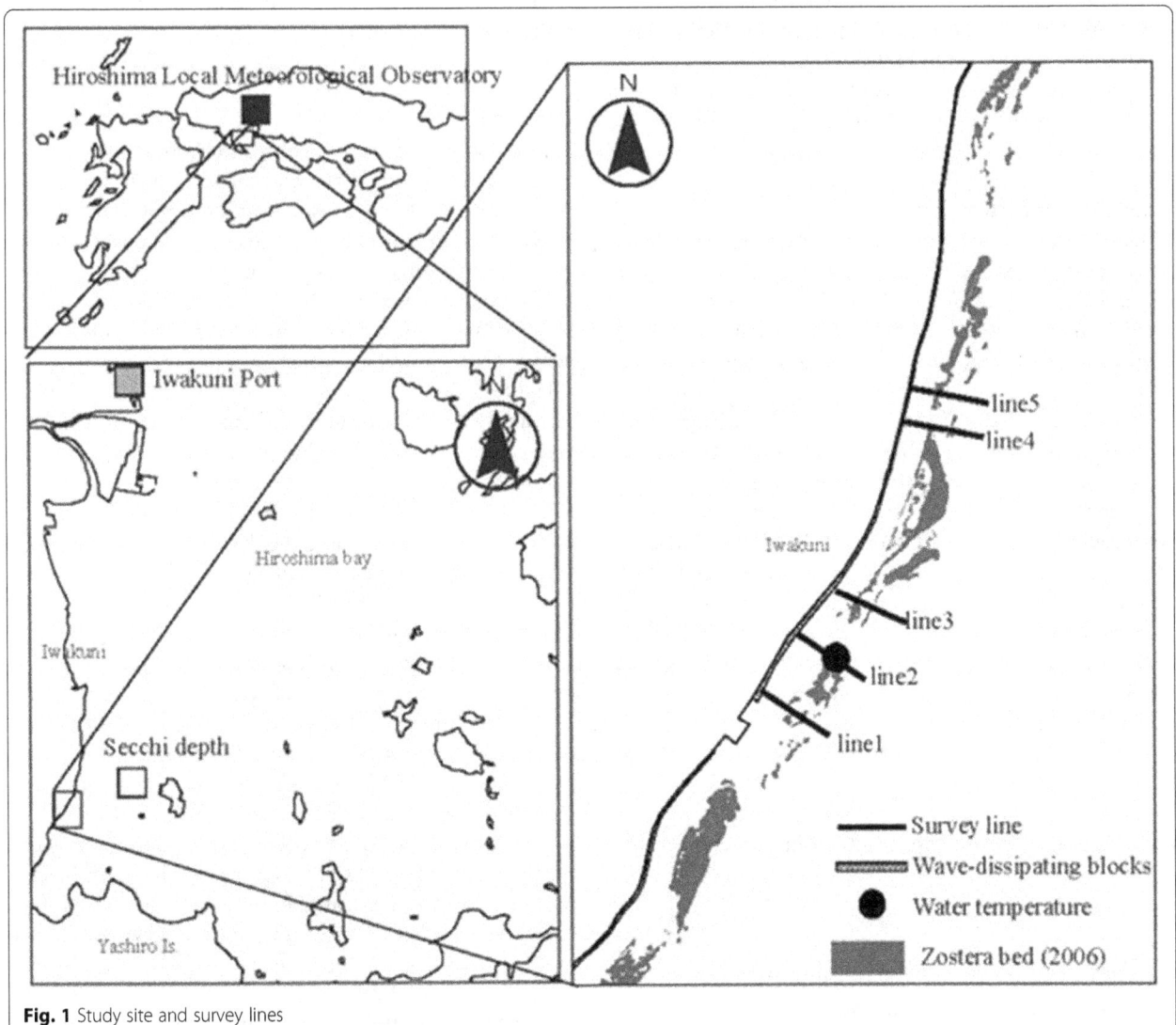

Fig. 1 Study site and survey lines

layer, the flow rate at the study site was a mean of 3 cm·s^{-1}, and a maximum of about 35 cm·s^{-1}, with a back and forth flow along the coastline. The flow at Obatakeseto, which is about 5 km southwest of the study site, is very fast and becomes about 250 cm·s^{-1} during the faster spring tide. In the vicinity of the study site, seagrass grows to a range of about 200 m from the shore, but at about 15 km north, near Iwakuni airport, south Ohbatake seagrass beds have formed almost continuously up to 4 km from the shore. Sea bottom materials in Zostera bed had silt and clay percentage of 5–8% and sand percentage of 90–91%.

Survey lines were designated as starting at the base point of the wave-dissipating blocks (breakwater) and extending offshore perpendicular to the shoreline. The length of each survey line and the number of sampling points along each survey line are given in Table 1. Survey lines at the south side of the study area had more sampling points. In December 2010, water depth along the survey lines was measured relative to a reference point in Iwakuni Port and reported as depth limit (datum level) recorded in meters (Fig. 2).

Monitoring of seagrass beds was conducted from December 2010 to October 2013 at a frequency of 2 to 5 months.

Monitoring of Zostera bed

Seagrass beds in the survey area were monitored at the five survey lines near Iwakuni, Yamaguchi Prefecture. Monitoring was carried out by diving observation to confirm the presence or absence of *Z. marina* and *Z. japonica*. Water depth along the monitoring lines was measured relative to a reference point in Iwakuni Port and was reported as D.L. recorded in meters from December 2010 to October 2013. Observations were made every 10 m. Survey findings for D. L. in the range of +1 to 0 m and in the range of 0 to −7 m were examined separately. The number of shoots in quadrats (*Z. marina* 50 × 50 cm, *Z. japonica* 10 × 10 cm) was measured. Shoot height was measured for up to ten shoots of each *Z. marina* and *Z. japonica* in the range of D.L. +1 to 0 m. The shoot height was measured as the longest leaf on the vegetative shoots, and if there were no vegetative shoots, seedling shoots were measured.

Environmental conditions

Temperature data loggers (Onset Computer Corporation; TidbiT v2) were used to take measurements at a

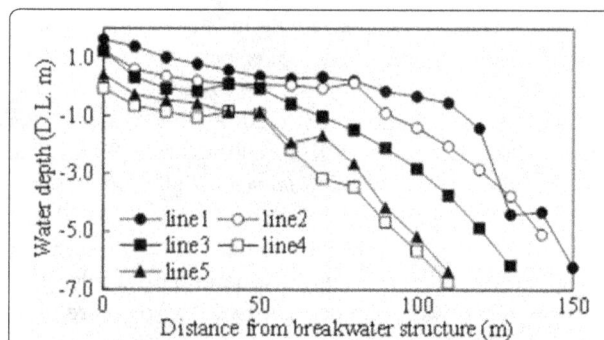

Fig. 2 Water depth based on D.L. along lines 1 to 5 (the reference point in Iwakuni Port)

height of 20 cm from the bottom of the sea at line 2 to 120 m from the breakwater every 10 min from January 2010 to December 2013. Solar irradiation data collected at Hiroshima Local Meteorological Observatory 50 km north–northeast of the study site, and Secchi depth was collected 3 km to the west of the study site in a Yamaguchi Prefecture public waters survey.

Results
Distribution of seagrass, height, and shoot density

Figure 2 shows the horizontal distance from the breakwater and the depth at each sampling point along each survey line in the seagrass beds. The flat water depth was ranged from +1.0 to −1.0 m to a distance of 50 to 80 m along each survey line and then became gradually lower offshore. The survey lines at the south end of the survey area showed a steeper drop in water depth.

Figure 3 shows the frequency of occurrence of *Z. marina* and *Z. japonica* along the survey lines in different water depths by season. The occurrence of both species was in the range of D.L. +1 to −7 m.

The frequency of occurrence of *Z. marina* was high in all seasons, even in water depths of D.L. + 1 to −5 m (80 ± 34% to 89 ± 19%; mean ± standard deviation), but lower (as low as 43 ± 34%) near the breakwall, where D.L. was from 1 to 2 m, and it was further reduced in D.L. −5 m and deeper. In June 2011, there was no difference in the frequency of occurrence of *Z. marina* for D.L. in the range of +1 to −5 m, in August 2011, the frequency of occurrence in June 2013 was 95 ± 9%, which was higher than 80 ± 23% of other times, and this was a statistically significant difference ($P < 0.05$).

The frequency of occurrence of *Z. japonica* was highest in water with a D.L. of +1 to 0 m. However, in D.L. of 0 m or deeper, it became lower as the water depth became deeper. For *Z. japonica*, the seasonal difference was less ($P > 0.85$). When the frequency of occurrence for both species throughout the year is more than 50% at a water depth, the optimal water depth for *Z. marina*

Table 1 Length of survey lines and number of sampling points

	Line 1	Line 2	Line 3	Line 4	Line 5
The length of the survey line (m)	150	140	130	110	110
The number of research point	16	15	14	12	12

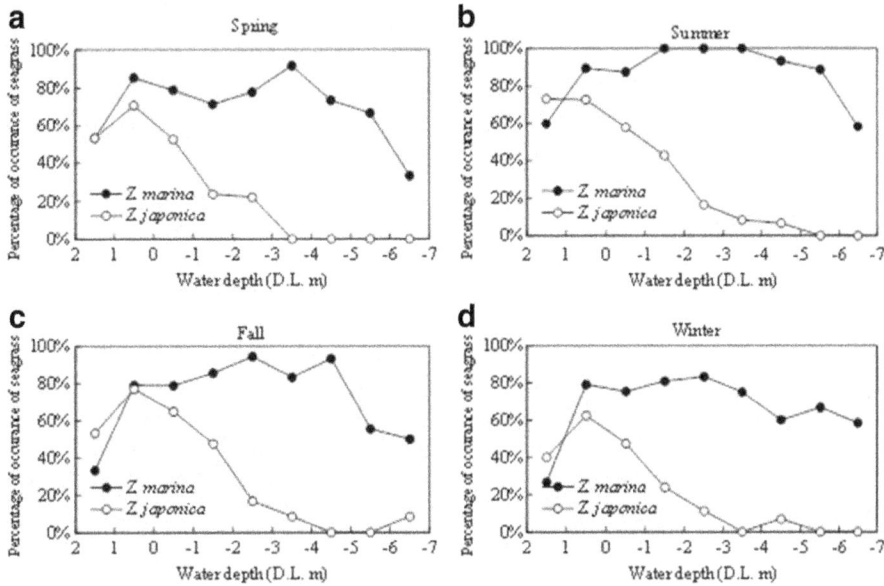

Fig. 3 Seagrass distribution at water depth based D.L. in each season

and *Z. japonica* is taken as D.L. from +1 to –5 m and +1 to 0 m, respectively.

Figure 4 shows the change in the frequency of occurrence at D.L. +1~0 m, which was an optimal water depth common to *Z. marina* and *Z. japonica* from December 2010 to October 2013, but in deeper water (D.L. 0~ –7 m), *Z. marina* was more common.

The frequency of occurrence was 83 ± 11% for *Z. marina* at the common optimal water depth, and the frequency of occurrence for *Z. japonica* was 70 ± 10%; there were no significant changes throughout the period of the survey. In the D.L. from 0 to –7 m, the frequency of occurrence of *Z. marina* was 77 ± 31%, and the frequency of occurrence of *Z. japonica* was 17 ± 22%, and in December 2012 and later, it was on the increase.

Figure 5 shows the changes in shoot density of *Z. marina* and *Z. japonica* at DL +1 to 0 m for both species from December 2010 to October 2013. The shoot density of *Z. marina* and *Z. japonica* decreases rapidly from

April or June 2011 to February 2012, and then shoot density was restored.

Figure 6 shows the correlation between the frequency of occurrence and shoot density for *Z. marina* and *Z. japonica* from DL +1 to 0 m using data from December 2010 to October 2013. The frequency of occurrence and the shoot density of both species showed no negative correlation, as would be expected in a competitive relationship. Therefore, we considered that the relationship between these species at the optimal water depth was not competitive.

Water temperature and solar irradiation

There were no significant variations in the frequency of occurrence of both species at the optimal water depth, and the shoot density showed great large variation of rapid decline and subsequent recovery from the summer of 2011 to the winter of 2012. We examined shoot density and photosynthesis data for factors that affect growth

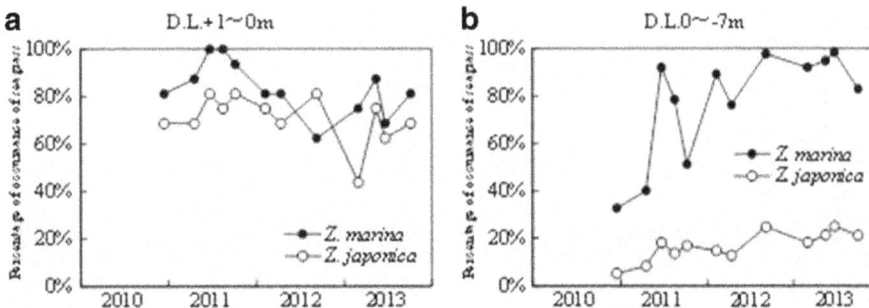

Fig. 4 Change in occurrence distribution of *Z. marina* and *Z. japonica* at two optimal water depths

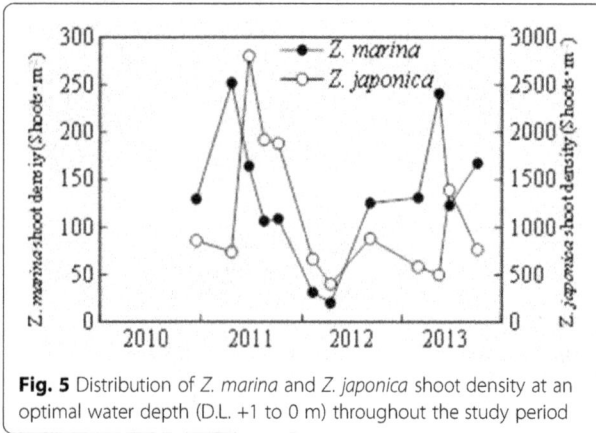

Fig. 5 Distribution of *Z. marina* and *Z. japonica* shoot density at an optimal water depth (D.L. +1 to 0 m) throughout the study period

and germination. Figure 7 shows the changes in mean daily water temperature from January 2010 to December 2013. In October 2011 or later, the shoot density of both species decreased rapidly. However in 2011, there were no observations of 28 °C (Tiina et al. 2014) or higher water temperatures which affect the growth of seagrass, and annual mean water temperature ranged from 17.1 to 17.9 °C. Since there was no statistically significant difference by year, annual water temperature was not considered to be a factor for the decrease in shoot density, and annual water temperature was not a direct cause of the decrease in shoot density in both species after October 2011.

Figure 8 shows the number of days in each year from 2010 to 2013 for which the daily mean water temperature was lower than 10 °C; a temperature that may affect the germination of seagrass (Abe et al. 2009; Kawasaki et al. 1986). In 2011, the shoot density of *Z. marina* and *Z. japonica* was higher until June before the summer, and the number of days on which the daily mean water temperature was lower than 10 °C was higher than in any other year at 43 days. In 2011, the daily mean water temperature was lower than 10 °C from January 16 to March 26, and 2011 was the only year in which

the water temperature was less than 10 °C at the end of March.

Figure 9 shows the daily mean solar irradiation at the Hiroshima Local Meteorological Observatory from January 2010 to December 2013. Annual mean of daily mean solar irradiation ranged from 13.6 to 14.6 $MJ \cdot m^{-2}$, and there was no significant difference by year. Similar to the mean water temperature, it was considered that there was no direct effect of solar irradiation changes on the rapid decline of seagrass since October 2011.

The estimated lower depth limit of *Z. marina* and the Secchi depth (S_d, m) from January 2010 to December 2013 at a point 3 km east of the study site is shown in Fig. 10. The estimated lower depth limit distribution of *Z. marina* (Z_c, m) was calculated using the light attenuation coefficient (K, m^{-1}) as follows (Duarte 1991).

$$Z_c = 1.86K \quad \text{and} \quad K = 1.7/S_d.$$

The annual mean S_d and Z_c are reported in Table 2. The S_d and Z_c in 2010 were 6.3 ± 1.6 m and 6.8 ± 1.8 m, respectively. These values gradually increased after 2010, the S_d and Z_c in 2013 were 8.9 ± 2.9 m and 9.8 ± 3.1 m, respectively. Therefore, since 2010, the daily mean solar irradiation is due to having a deeper S_d, and every year, the light conditions for the growth of *Z. marina* and *Z. japonica* were considered to improve.

Figure 11 shows the changes in salinity from December 2010 to October 2013. The salinity in the study period ranged from 29.3 to 32.4 PSU and never decreased below 10 PSU (Kawasaki et al. 1986); a level that affects the growth of *Z. marina*.

Feeding damage

In October 2011, the upper portions of many of the *Z. marina* appeared to have had feeding damage, as shown in Fig. 12. Therefore, shoots height of *Z. marina* and *Z. japonica* was examined from December 2010 to October 2013, as an indicator of feeding damage (Fig. 13). The

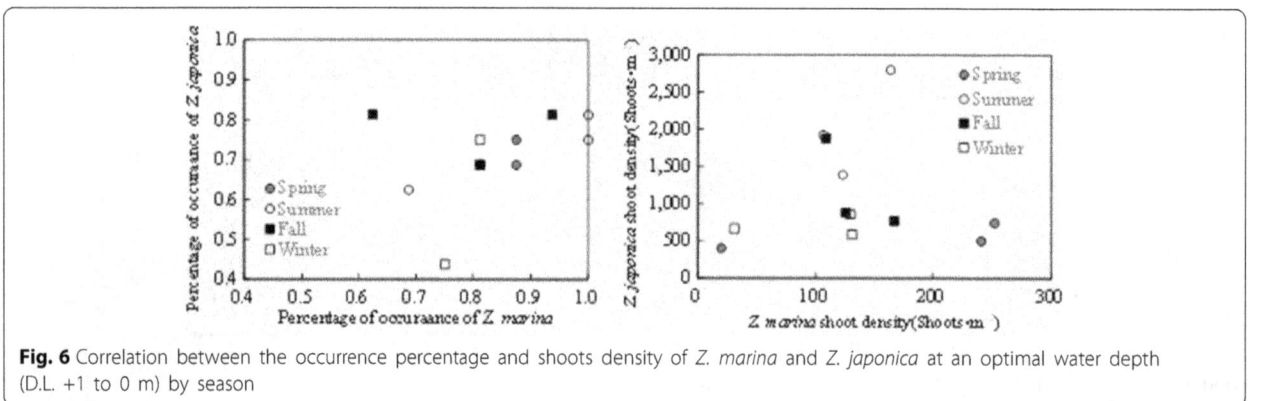

Fig. 6 Correlation between the occurrence percentage and shoots density of *Z. marina* and *Z. japonica* at an optimal water depth (D.L. +1 to 0 m) by season

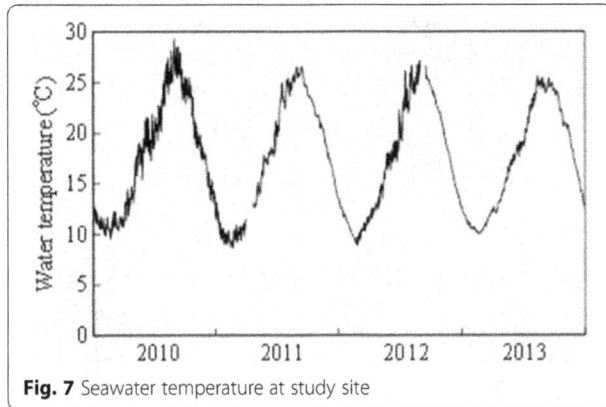

Fig. 7 Seawater temperature at study site

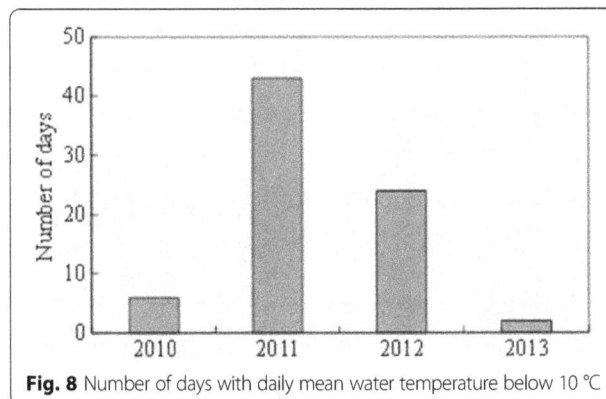

Fig. 9 Solar irradiation recorded at Hiroshima Local Meteorological Observatory

mean shoot height (stomach contents in fish were not confirmed to be *Z. marina*) in August 2011 was 50 cm, after which feeding damage reduced it to 23 cm. The mean shoot height of *Z. japonica* in August 2011 and in October 2011 was 23 and 15 cm, respectively. The shoot height of *Z. japonica* also decreased in the same way as the shoot height of *Z. marina*. In the Seto Inland Sea, herbivorous fishes, Siganus fuscescens, were observed in 2011 to cause feeding damage to the seagrass bed, and feeding damage to *Z. marina* was also observed in Hiroshima Bay (Sugimoto et al.). Although not directly observed, the feeding damage by *S. fuscescens* in the study sites caused damage at the tips, and damage was assumed to cause a decrease in shoot density for both species.

Discussion

The distribution of *Z. japonica* is mostly in the intertidal zone, while that of *Z. marina* is mostly in the subtidal zone (Mach et al. 2010), but the factors affecting these differences in distribution were not clarified. *Z. marina* and *Z. japonica* in D.L. −4 m and shallower had overlapping distribution, and for D.L. from +1 to 0 m showed high occurrence rate for both species. In other words, *Z. japonica* grew at water depths that were also included in the range of *Z. marina*. *Zostera marina* had greater

shoot height than *Z. japonica*, and these species are assumed to have a competitive relationship based on *Z. marina* growing higher and shading *Z. japonica*. However, in the optimal water depths, occurrence frequency and shoot densities are similar for both species, and the lack of a negative correlation suggests that the growth of both species is independent.

The subtidal is deeper than the optimal water depth of both species, and the seagrass distribution, and thick seagrass (*Z. marina*) which grows tall and has wide leaves shades short seagrass (*Z. japonica* and *Ruppia maritima*), inhibiting the photosynthesis and possibly the distribution of short seagrass. A negative correlation was observed in the biomass of tall *Z. marina* and short *Ruppia maritima* in the Chesapeake Bay subtidal zone (Orth 1977). *Z. marina* has limited light due to competition from algae (Sugimoto et al. 2007). When *Z. japonica* of the subtidal is subjected a light limitation due to shading by leaves of *Z. marina* at the study site, there is little difference in the light compensation point of *Z. marina* and *Z. japonica*, the difference between the plant heights is low, and the distribution of *Z. japonica* by water depth in October 2011, April and June 2012 will surpass that of *Z. marina*. However, even in this

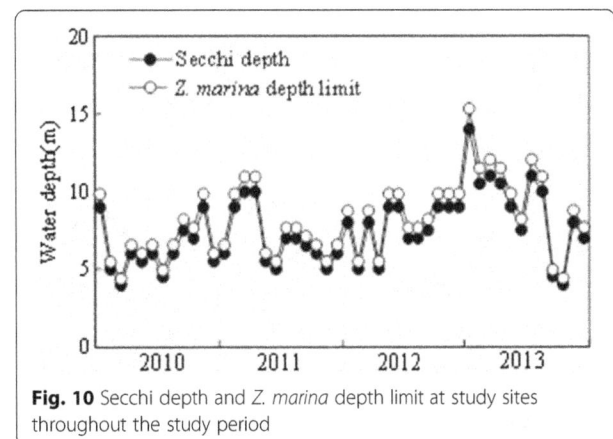

Fig. 8 Number of days with daily mean water temperature below 10 °C

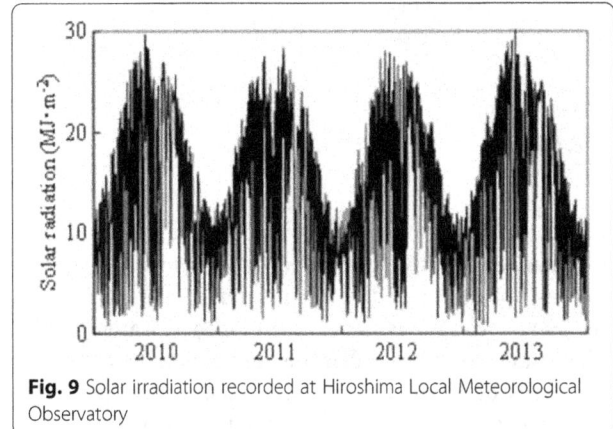

Fig. 10 Secchi depth and *Z. marina* depth limit at study sites throughout the study period

Table 2 The annual mean value of Secchi depth and estimated lower depth of *Z. marina*

	2010	2011	2012	2013
Secchi depth (m)	6.3 ± 1.6	6.9 ± 1.8	7.7 ± 1.5	8.9 ± 2.9
Z. marina depth limit (m)	6.8 ± 1.8	7.6 ± 2.0	8.4 ± 1.6	9.8 ± 3.1

Fig. 12 Losses at the upper end of the seagrass leaves (October 2011:line 2)

period, the water depth of *Z. japonica* distribution had not changed greatly. Compared to *Z. marina*, *Z. japonica* was not distributed in deep water. In the study sites, the shear stress was reduced in deeper water. Under calm water conditions, diatoms and sediment adhere to the leaves of *Z. marina* and *Z. japonica*, and moderate disturbance is necessary to avoid growth inhibition (Shimaya et al. 2004; Sugimoto et al. 2003). *Z. japonica* habitats in the deep water of Tokyo Bay had higher bottom shear stress values than in non-vegetated areas. Epiphytic *Ectocarpus siliculosus* (Fig. 14), a mobile brown algae and *Diffalaba picta picta* (Fig. 15), a conch, are epiphytic and both were observed in the study site. No quantitative data on the occurrence frequency or levels of *E. siliculosus* and *D. picta picta* were collected; however, there is a possibility that they affected the distribution of *Z. japonica* in the subtidal zone. We could not confirm the influence on the growth distribution due to the coexistence between *Z. marina* and *Z. japonica*. However, it was clear that the water depth was an important factor determining the distribution of these species. The frequency of occurrence of *Z. japonica* was decreasing from D.L. 0 m. In addition, the frequency of occurrence of *Z. marina* was decreasing from D. L. −4 m. Therefore, it was considered that the frequency of occurrence of these species were lowered in deep points due to the influence of the light condition. *Z. japonica* required light intensity than *Z. marina*, and the water depth played an important role in the distribution of these species.

In this study, no large annual variations in the frequency of occurrence of *Z. marina* and *Z. japonica* were found. On the other hand, great variation was observed

in the shoot density. *Z. marina* and *Z. japonica* occurrence frequency were determined by the required growth light conditions (depth); however, they were less affected by other factors, and shoot density was sensitive to environmental factors. During the study period, in particular, feeding damage and low water temperature in winter were factors affecting shoot density.

When the seeds of *Z. marina* were kept at 14 °C and buried in sea sand in laboratory experiments, germination was 23%, but germination at 9 °C or less was 66% or greater (Tiina et al. 2014). When the same test for the *Z. japonica* was conducted, germination rate at 23 °C and at 4 °C was 40 and 80%, respectively (Kishima et al. 2011). Thus, a water temperature of 10 °C or less in the winter increases the germination rate and distribution of both species in spring. In the study site in 2011, the daily mean water temperature was 10 °C or less on more days than in other years. In particular, water temperature was lower in shallow water at low tide. Therefore, there is a possibility that a high germination rate of seagrass occurs in depths from D.L. +1 to 0 m.

In Gozensaki-cho, Shizuoka Prefecture, leaves of *Ecklonia cava* were lost due to feeding damage by *S. fuscescens*, and

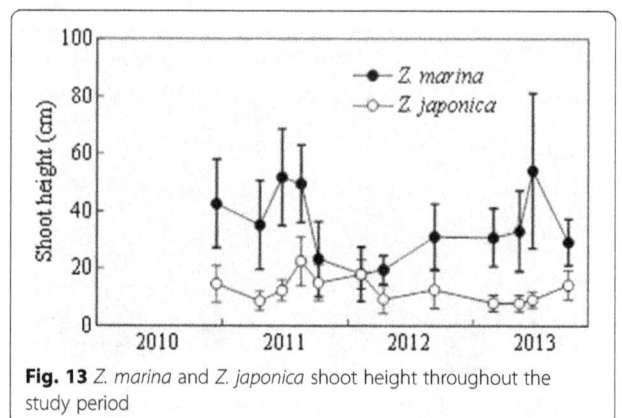

Fig. 11 Salinity at study site throughout study period

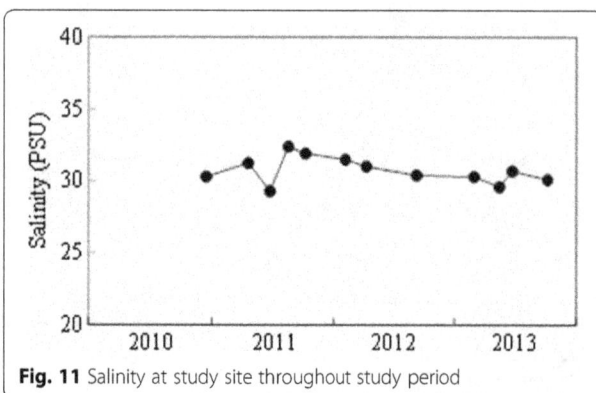

Fig. 13 *Z. marina* and *Z. japonica* shoot height throughout the study period

Fig. 14 *E. siliculosus* adhering to the leaves of *Z. marina* and *Z. japonica* (February 2013:line 2)

seagrass beds were in decline in a short period from October to November in 1998 (Masuda et al. 2000). The grazing of algae by *S. fuscescens* rose along with the rise in water temperature, and algae grazing became higher after the spawning from July to August. Examination of the amount of predation of *E. cava* by *S. fuscescens* using a water tank showed significantly increased predator weight at 25 °C before and after the heating has been observed (Kawamata and Masegawa 2006). The water temperature exceeded 25 °C in 2011 when *S. fuscescens* were thought to be grazing on eelgrass from August 2 to September 16, which is before and after spawning. In the period of August 17 to October 6, shoot height of *Z. marina* was shorter. The number of days for which the water temperature was higher than or equal to 25 °C over the period of 2010 to 2013 was 38 ± 16 days, and in 2011 it was 42 days. Therefore, it is considered that water temperature in 2011 was not a factor in the increased

Fig. 15 *Diffalaba picta picta* adhering to the leaves of *Z. marina* (February 2013:line 4)

predation pressure by *S. fuscescens* compared with other years.

The seagrass beds in the Seto Inland Sea, Japan east of Yamaguchi Prefecture are comprised of *Z. marina* and *Z. japonica*, which are competing species, but due to differences in spatial distribution, *Z. marina* and *Z. japonica* do not influence each other due to temperature conditions and feeding damage and other environmental conditions.

Conclusions

This study aims to clarify the factors that determine the temporal and spatial distribution of *Z. marina* and *Z. japonica* in the Seto Inland Sea east of Yamaguchi Prefecture. The specific conclusions derived from this study are as follows:

1) When the frequency of occurrence for both species throughout the year is more than 50% at a water depth, the optimal water depth for *Z. marina* and *Z. japonica* is taken as D.L. from +1 to −5 m and from +1 to 0 m, respectively. Thus, D.L. from +1 to 0 m was an optimal water depth for both species.

2) The frequency of occurrence and the shoot density of both species showed no negative correlation, as would be expected in a competitive relationship. Therefore, we considered that the relationship between these species at the optimal water depth was not competitive.

3) *Z. japonica* required light intensity than *Z. marina*, and the water depth played an important role in the distribution of both species.

4) In 2011, the mean water temperature, salinity, and solar irradiation were no difference between 2010 and 2013, however, the daily mean water temperature was 10 °C or less on more days than in other years. Therefore, there is a possibility that a high germination rate of seagrass occurs.

5) Since 2010, the daily mean solar irradiation is due to having a deeper S_d, and every year, the light conditions for the growth of *Z. marina* and *Z. japonica* were considered to improve.

6) In October 2011, the feeding damage by *S. fuscescens* in the study sites caused damage at the tips, and damage was assumed to cause a decrease in shoot density for both species.

Abbreviations
D.L.: Datum level; *Z. japonica: Zostera japonica; Z. marina: Zostera marina*

Acknowledgements
Not applicable.

Funding
The design of the study and collection; National Institute of Technology, Ube College.

Authors' contributions

KS carried out sea survey and performed the statistical analysis and drafted the manuscript. YN carried out sea survey. TO, SN, WN, and MO participated in the design of the study and helped to draft the manuscript. All authors read and approved the final manuscript.

Competing interests

The authors declare that they have no competing interests.

Author details

[1]National Institute of Technology, Ube College, Ube, Japan. [2]Faculty of Science and Technology, Ryukoku University, Kyoto, Japan. [3]Department of Chemical Engineering, Hiroshima University, Hiroshima, Japan. [4]Environmental Research and Management Center, Hiroshima University, Hiroshima, Japan. [5]The Open University of Japan, Chiba, Japan.

References

Abe, M., Hashimoto, N., Kurashima, A., & Maegawa, M. (2004). Structure and seasonal change of the *Zostera marina* population on the coast of Matsunase, Mie Prefecture, central Japan. *Nippon Suisan Gakkai, 70*(4), 523–529.

Abe, M., Yokota, K., Kurashima, A., & Maegawa, M. (2009). Temperature characteristics in seed germination and growth of *Zostera japonica* Ascherson & Graebner from Ago Bay, Mie Prefecture, central Japan. *Fisheries Science, 75*, 921–927.

Aioi, K. (1998). On the red list species of Japanese seagrasses. *Aquabiology, 20*(1), 7–12.

Aioi, K., & Nakaoka, M. (2003). The seagrass of Japan. In E. P. Green & F. T. Short (Eds.), *World Atlas of Seagrasses* (pp. 185–192). Berkley: University of California Press.

Arasaki, S. (1950a). Studies on the ecology *Zostera marina* and *Z. nana* (1). *Nihon Suisan Gakkai, 15*(10), 567–572.

Arasaki, S. (1950b). Studies on the ecology *Zostera marina* and *Z. nana*. *Nihon Suisan Gakkai, 16*(2), 70–76.

Biodiversity Center of Japan (2008) *7th Natural Environment Basic Research Shallow water ecosystem research (Seagrass bed research) Report* (pp. 354–357)

Dennison, W. C., & Alberte, R. C. (1985). Role of daily light period in the depth distribution of *Zostera marina* (eelgrass). *Marine Ecology Progress Series, 25*, 51–61.

Duarte, C. M. (1991). Seagrass depth limits. *Aquatic Botany, 40*, 363–377.

Greve, T. M., & Krausen-Jensen, D. (2005). Predictive modeling of eelgrass (*Zostera marina*) depth limit. *Marine Biology, 146*, 849–858.

Harrison, P. G. (1982). Seasonal and year-to-year variations in mixed intertidal populations of *Zostera japonica* Aschers. & Graebn. and *Ruppia maritima* L. S. L. *Aquatic Botany, 14*, 357–371.

Kawamata, S., & Masegawa, M. (2006). Effects of waves and temperature on feeding by the Rabbitfish *Siganus Fsuscescens* on Kelps *Eisenia Bicyclis* and *Ecklonia cava*. *Fisheries Engineering, 46*(1), 69–79.

Kawanishi, K. (1999). Characteristics of Water Exchange and Current Structures in North Area of Hiroshima Bay. *Proceedings of Coastal Engineering, 46*, 1041–1045.

Kawasaki, Y., Iituka, T., Goto, H., Terawaki, T., Shimo, S. (1986) Effects of temperature on *Zostera marina* L. 1. Seed Germination and Seedling Development, Central Research Institute of Electric Power Industry, Rep. No. 485028

Kendrick, G. A., Aylward, M. T., Hegge, B. J., Cambridge, M. L., Hillman, K., Wyllie, A., & Lord, D. A. (2002). Changes in seagrass coverage in Cockburn Sound, Western Australia between 1967 and 1999. *Aquatic Botany, 73*, 75–87.

Kikuchi, T. (1980). Faunal relationships in temperate seagrass beds. In R. C. Phillips & C. P. McRoy (Eds.), *Holland book of Seagrass Biology* (An Ecosystem Perspective, pp. 153–172). New York and London: Garland STPM Press.

Kishima, J., Harada, S., & Sakurai, R. (2011). Suitable water temperature for seed storage of *Zostera japonica* for subtropical seagrass bed restoration. *Ecological Engineering, 37*, 1416–1419.

Mach, M.E., Wyllie-Echeverria S., Ward J. R. (2010) Distribution and potential effects of a non-native seagrass in Washington State *Zostera japonica* Workshop, Friday Harbor Laboratories, San Juan Island, WA, (pp. 14–16)

Mach, M. E., Wyllie-Echeverria, S., & Chan, K. M. A. (2014). Ecological effect of a nonnative seagrass spreading in the Northeast Pacific: a review of *Zostera japonica*. *Ocean & Coastal Management, 102*, 375–382.

Masuda, M., Tsunoda, T., Hayashi, Y., Nishio, S., Mizui, H., Horiuchi, S., & Nakayama, Y. (2000). Decline of afforested *Ecklonia Cava* community by grazing of Herbivorous fish *Siganus fuscescens*. *Fisheries Engineering, 37*(2), 135–142.

Morita, T., Kokubu, H., Miyamatsu, A., Fujii, M., Kurashima, A., & Maegawa, M. (2010). The effect of water temperature in the sediment and the desiccation period on the growth and survival of transplant seagrass *Zostera japonica* shoots in Mesocosm tank culture. *Aquaculture Science, 58*(2), 261–267.

Omori, Y. (2000). Japanese seagrasses. Distirbution and morphology. *Aquabiology, 22*(6), 524–532.

Orth, R. C. (1977). Effect of nutrient enrichment on growth of the eelgrass *Zostera marina* in the Chesapeake Bay Virginia USA. *Marine Biology, 44*, 187–194.

Orth, R. J., Heck, K. L. H., Jr., & Montfrans, J. V. (1984). Faunal communities in seagrass beds: a review of the influence of plant structure and prey characteristics on predator–prey relationships. *Estuaries, 7*(4A), 339–350.

Pollard, D. A. (1984). A review of ecological studies on seagrass–fish communities, with particular reference to recent studies in Australia. *Aquatic Botany, 18*, 3–42.

Shafer, D. J., Kaldy, J. E., Sherman, T. D., & Marko, K. M. (2011). Effects of salinity on photosynthesis and respiration of the seagrass *Zostera japonica* : a comparison of two established populations in North America. *Aquatic Botany, 95*, 214–220.

Shafer, D. J., Kaldy, J. E., & Gaeckle, J. L. (2014). Science and management of the introduced seagrass *Zostera japonica* in North America. *Environmental Management, 53*, 147–162.

Shimaya, M., Sato, K., Nakase, K., Kuwae, T., & Nakamura, Y. (2004). A study on physical environment suitable for *Zostera Japonica* growth. *Annual Journal of Coastal Engineering, 51*, 1031–1035.

Sugimoto, K., Asaoka, Y., Matsunami, D., Terawaki, T., Okada, M. (2012). Extinction of eelgrass off the coast of Iwakuni, Seto Inland Sea in decline period, *Annual Journal of Japan Society on Water Environment, 46*, 236.

Sugimoto, K., Hiraoka, K., Tamaki, H., Terawaki, T., Okada., M., (2003) Effects of Breakwater Construction on the Distribution of Transplanted Eelgrass Beds (*Zostera marina* L.), Proceedings of the 6th International EMECS Conference, (pp. 295–302)

Sugimoto, K., Hiraoka, K., Ohta, S., Niimura, Y., Terawaki, T., & Okada, M. (2007). Effects of ulvoid (*Ulva* spp.) accumulation on the structure and function of eelgrass (*Zostera marina* L.) bed. *Marine Pollution Bulletin, 54*, 1582–1585.

Sugimoto, K., Hiraoka, K., Tanida, K., Terawaki, T., & Okada, M. (2008). Restration of eelgrass (*Zostera marina* L.) Beds by Filling up Borrow Pits with Natural Sediment. *Journal of Japan Society on Water Environment, 31*(4), 217–221.

Tiina, S., Morten, F. P., & Christoffer, B. (2014). Population specific salinity tolerance in eelgrass (*Zostera marina*). *Journal of Experimental Marine Biology and Ecology, 461*, 425–429.

10

Effects of habitat differences on the genetic diversity of *Persicaria thunbergii*

Bo Eun Nam[1], Jong Min Nam[1] and Jae Geun Kim[1,2]* (iD)

Abstract

To understand the effects of habitat characteristics on the genetic diversity of *Persicaria thunbergii*, three sites of different environmental conditions in a water system were surveyed. Site A was the closest to the source of the water system, and there was a dam between sites A and B. Site C is located on the lowest downstream in the water system. Vegetation survey of four quadrats at each site was performed, and soil samples were collected for physicochemical analysis. Random amplification of polymorphic DNA (RAPD) analysis of ten *P. thunbergii* individuals at each site was conducted to calculate population genetic diversity and genetic distance among populations. Soil was sterile sand at site A, whereas loamy soil at sites B and C. A pure stand of *P. thunbergii* appeared at site A, while other species occurred together (such as *Humulus japonicus* and *Phragmites australis*) at sites B (Shannon-Wiener index; $H_B = 0.309$) and C ($H_C = 0.299$). Similar to the species diversity, genetic diversity (Nei's gene diversity; h) within population of site A ($h_A = 0.2381$) was relatively lower than sites B ($h_B = 0.2761$) and C ($h_C = 0.2618$). However, site C was separated from sites A and B in genetic distance rather than the geographical distance (Nei's genetic distance; A~B, 0.0338; B~C, 0.0685; A~C, 0.0833).

Keywords: Cleistogamic flower, Genetic distance, Genetic diversity, RAPD-PCR

Introduction

Persicaria thunbergii (Siebold & Zucc.) H. Gross is an annual Polygonaceae herb that commonly appears at diverse wetland ecosystems such as streamlets, abandoned paddy fields, montane fens, and floating mats in Korea (Kim 2009; Kim et al. 2012; Park et al. 2013). *P. thunbergii* inhabits widely and sometimes dominate in wetland ecosystems where the water level is not much deep (Kawano 2008; Kim et al. 2013). In Korea, *P. thunbergii* is occasionally considered as a management object due to its rapid recovery through asexual reproduction after harsh flooding disturbances (Kim et al. 2012).

P. thunbergii has the amphicarpic trait that produces different types of seeds only shown in a few species (Cheplick 1987). *P. thunbergii* produces not only aerial chasmogamous flower but subterranean cleistogamic flower (Kawano et al. 1990; Choo et al. 2014). Two types of seeds from different types of flower (aerial and subterranean seed) have significant differences in seed and seedling stage

properties (Choo et al. 2015). In addition, the ratio of aerial and subterranean flower and seed number varies under different hydrological conditions and soil fertility in *P. thunbergii* (Choo et al. 2014; Kim et al. 2016).

There might be differences in genetic diversity between two seed types in *P. thunbergii* as well as physiological differences in seed and seedling stage as a result of differences in flower type (aerial and subterranean). However, studies about the genetic diversity of the wetland plants mainly have focused on clonal, perennial species (Piquot et al. 1996; Koppitz 1999; Honnay et al. 2010; Min et al. 2012; Raabová et al. 2015). Previous studies on the genetic diversity of *P. thunbergii* only focused on their spatiotemporal genetic distribution (Konuma and Terauchi 2001; Kim et al. 2008). Genetic diversity of *P. thunbergii* could be affected by phenotypic plasticity in the ratio of aerial and subterranean flower production, as a result of the plasticity caused by their environment.

The aim of this study is to assess the genetic diversity of *P. thunbergii* populations with different environmental characteristics. Some ecological characteristics of *P. thunbergii* also would be considered in population genetic diversity.

* Correspondence: jaegkim@snu.ac.kr
[1]Department of Biology Education, Seoul National University, Seoul 08826, South Korea
[2]Center for Education Research, Seoul National University, Seoul 08826, South Korea

Materials and methods

Study site

The study was conducted at the Han River and Gyeongan-cheon stream which is a tributary to the Han River, Seoul and Gyeonggi Province, Republic of Korea. Field study was conducted in July 2014. Three sites where the coverage of *P. thunbergii* was more than 50% with relatively few artificial disturbances were chosen for the field study and sampling (Fig. 1).

Site A (Yongin, Gyeonggi Province) is the upper region and adjacent to the source of Gyeongan-cheon (<2 km), where the width of the stream is 2~3 m. Before Gyeongan-cheon joins the Han River at Paldangho Lake, in site B (Gwangju, Gyeonggi Province), *P. thunbergii* has been inhabited at the floodplain of stream which width is about 80-m wide. Also, site C (Gangdong-gu, Seoul) is also a floodplain of the Han River with approximately 700-m width.

Field survey and sampling of *P. thunbergii*

Four 1 m × 1 m quadrats were installed at each site. Vegetation survey was conducted at each quadrat. Relative coverage of emerged plant species and species diversity were calculated (*H'*; Shannon-Weiner index).

Relative coverage = cover of species/total cover

$$H' = -\sum\nolimits_{i=1}^{N} p_i \log p_i \; (N = \text{no. of emerged species;}$$
$$p_i = \text{relative coverage of species } i)$$

Soil sampling was conducted closely by each quadrat. In case of site A, only one soil sample was collected because *P. thunbergii* population rooted on the rock with a little soil medium. For assessment on population genetic diversity of *P. thunbergii*, leaves of ten individuals in 2-m-interval nearby quadrats were sampled at each site. Leaf samples were stored at −20 °C until DNA extraction.

Soil analysis

Soil samples were passed through a 2-mm sieve (standard sieve #10) before analysis. Water content was determined after drying samples at 105 °C in an oven, and analysis of organic matter content followed by the loss-on-ignition method at 550 °C furnace (Boyle 2004). Soil pH (AP63, Fisher, Hampton, USA) and conductivity (EC; Corning Checkmate II, Corning, Lowell, MA) were measured from the soil solution which is a mixture of soil samples with distilled water at a mass ratio of 1 to 5. NO_3-N and NH_4-N were extracted with 2 M KCl solutions and measured by a hydrazine method and indophenol method, respectively (Kamphake et al. 1967; Kim et al.

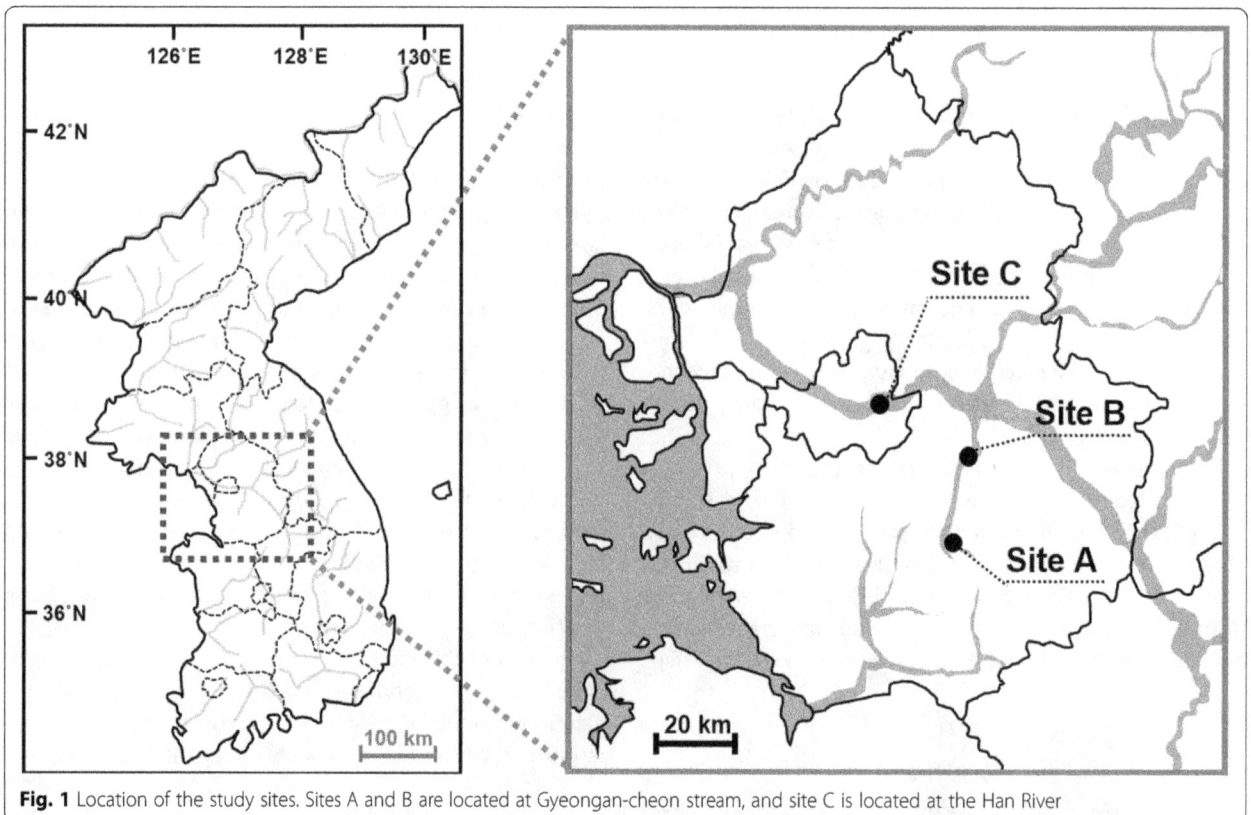

Fig. 1 Location of the study sites. Sites A and B are located at Gyeongan-cheon stream, and site C is located at the Han River

2004). PO_4-P was extracted and analyzed by a Bray No. 1 method (Bray and Kurtz 1945). Available K^+, Ca^{2+}, Na^+, and Mg^{2+} were extracted with 1 N ammonium acetate solution (Allen et al. 1974) and measured with an atomic absorption spectrophotometer (AA240FS, Varian, Palo Alto, USA). Soil texture was determined using the hydrometer method and the soil texture triangle of USDA (Carter 1993).

RAPD-PCR

Extraction of individual genomic DNA was conducted by DNeasy Plant Mini Kit (Qiagen, Hilden, Germany). PCR was followed by using AccuPower PCR premix (Bioneer, Daejeon, Korea). The total reaction volume of PCR was 20 μL, composed of 1 μL of extracted DNA, 1 μL of primer (10 pmol) and 18 μL of deionized water. Four random primers which showed the polymorphism in electrophoresis were selected for random amplification of polymorphic DNA (RAPD) analysis (Table 1). Reaction cycle of PCR began with 2 min of denaturation at 94 °C. Forty cycles were followed with 45 s of denaturation at 94 °C, 45 s of annealing at the T_m of each primer, and 1 min and 30 s of elongation at 72 °C. Lastly, 2 min of the final extension step at 72 °C was followed.

Electrophoresis of amplified fragments was conducted in GelRed™ (BIotium, Heyward, CA) stained 1% agarose gel. Under the UV light, bands were observed and recorded by binary character matrix with 1 (presence) and 0 (absence).

Data analysis

Genetic diversity within the population (number of polymorphic loci; Nei's gene diversity, h; Shannon index, i) and genetic distance (Nei's genetic distance) among populations were calculated by Popgen32 (Nei 1973; Yeh and Boyle 1997). Principal coordinate analysis (PCoA) with the presence and absence of each band was conducted by R version 3.1.1.

Results

Habitat characteristics of P. thunbergii populations

Relative coverage of *P. thunbergii* at sites A, B, and C were 100, 52.5, and 77.9%, respectively. Five species emerged at site C, whereas three species emerged at site B (Table 2). At site A, only *P. thunbergii* emerged in the surveyed

Table 1 Random primer sequences and number of amplified RAPD fragments

Primer	Sequence	T_m (°C)	No. of observed bands
N-8002	CAATCGCCGT	32	13
N-8004	TCGGCGATAG	32	7
N-8007	GTGACGTAGG	32	9
N-8010	CTGAGACGGA	32	11

Table 2 Plant species, relative coverage (%), and species diversity

	Site A	Site B	Site C
Perennials			
Phragmites australis	–	–	4.1
Equisetum arvense	–	–	0.9
Artemisia princeps var. *orientalis*	–	–	0.4
Annuals			
Persicaria thunbergii	100	52.5	77.9
Humulus japonicas	–	47.2	16.7
Persicaria perfoliata	–	0.3	–
Shannon-Weiner index	–	0.309	0.299

quadrats although the other emergent macrophytes occurred nearby. Species diversity index (Shannon-Wiener index) at sites B and C were 0.309 and 0.299, respectively.

At field survey, *P. thunbergii* population at sites B and C had a distance from the surface of the stream or river. However, the average water depth was 2 cm at site A. Soil characteristics at site B and site C were similar except NH_4-N and NO_3-N content (Table 3). Soil textures were sandy loam~silt loam at site B, silt loam~loam at site C, whereas there was sand at site A. EC (A, 37.2 μS/cm; B, 52.5~102.3 μS/cm; C, 61.8~92.6 μS/cm) and organic matter content (A. 2.25%; B, 7.21~9.26%; C, 7.54~9.23%) at site A were considerably lower than sites B and C. Similarly, at site A, NO_3-N (A, 0.53 mg/kg; B, 2.01~3.51 mg/kg; C, 4.57~5.35 mg/kg) and PO_4-P content (A, 6.75 mg/kg; B, 8.08~15.74 mg/kg; C, 11.91~17.29 mg/kg) were lower than sites B and C. However, NH_4-N content at site A was relatively higher than sites B and C except one plot at site B (A, 12.5 mg/kg; B, 0.60~16.32 mg/kg; C, 0.49~1.63 mg/kg). K^+, Ca^{2+}, and Mg^{2+} content were relatively higher at sites B and C

Table 3 Physicochemical soil characteristics (Mean ± 1 SD)

	Site A ($n = 1$)	Site B ($n = 4$)	Site C ($n = 4$)
pH	6.86	6.62 (±0.06)	6.66 (±0.09)
EC (μS/cm)	37.2	75 (±20.9)	79.2 (±12.8)
Water content (%)	28.2	27.9 (±3.8)	30.1 (±1.4)
LOI (%)	2.25	8.24 (±1.12)	8.06 (±0.79)
NH_4-N (mg/kg)	12.5	6.56 (±6.95)	0.94 (±0.52)
NO_3-N (mg/kg)	0.53	2.76 (±0.69)	4.95 (±0.33)
PO_4-P (mg/kg)	6.75	11.9 (±3.9)	14.6 (±2.2)
K^+ (mg/kg)	64.1	543.8 (±109.5)	599.8 (±94.4)
Na^+ (mg/kg)	19.9	18.1 (±1.4)	16.8 (±1.7)
Ca^{2+} (mg/kg)	928.4	2467 (±240)	2327 (±129)
Mg^{2+} (mg/kg)	94.4	272.8 (±37.8)	322.1 (±40.6)
Soil texture	Sand	Sandy loam or silt loam	Loam or silt loam

than site A, but Na$^+$ content was similar at the three sites (Table 3). Soil water content and pH showed just a little difference between sites A, B, and C.

Genetic diversity

A total of 40 band loci were observed in four primers, and 35 of them were polymorphic. Each number of polymorphic loci at sites A, B, and C were 29, 28, and 29, respectively. The total Nei's gene diversity (h) was 0.2969 and Shannon's information index (i) was 0.4472. The highest genetic diversity was observed at site B ($h = 0.2761$, $i = 0.4100$; Table 4).

In a 2-D plot of PCoA based on 35 RAPD bands, individuals from three sites were mixed (Fig. 2). However, individuals from site C were relatively separated from sites A and B on PCoA plot. Nei's genetic distance between sites A and B was 0.0338, between sites B and C was 0.0685, and between sites A and C was 0.0833.

Discussion

The total potential genetic diversity of *P. thunbergii* population seemed to decrease when harsh flooding occurs in a natural habitat. In addition, high soil fertility increases the number of aerial seed production despite the decrease of mass allocation in aerial seeds (Kim et al. 2016). Considering the mass allocation, genetic diversity of *P. thunbergii* population is expected to be relatively higher in relatively fertile habitat than sterile habitat.

Both two sites with fertile soil are affected by annual harsh flooding in July and August (Han River Flood Control Office 2014). Although site A is located on the same water system as sites B and C, the flooding pattern is different being closer to the source of the stream. The frequency of flooding disturbance seemed to be relatively higher, but the intensity seemed to be less harsh in site A than sites B and C. Weathering and sediment deposition by flowing water frequently occurred in sites B and C than site A, where soil is relatively sterile. Number of aerial seed production might be higher in sites B and C, where the annual harsh flooding occurs with fertile soil (Choo et al. 2015; Kim et al. 2016). It also could contribute to the relatively higher genetic diversity than the habitat with oligotrophic soil.

While genetic diversity of sites B and C were similar, site C was genetically separated from sites A and B. It

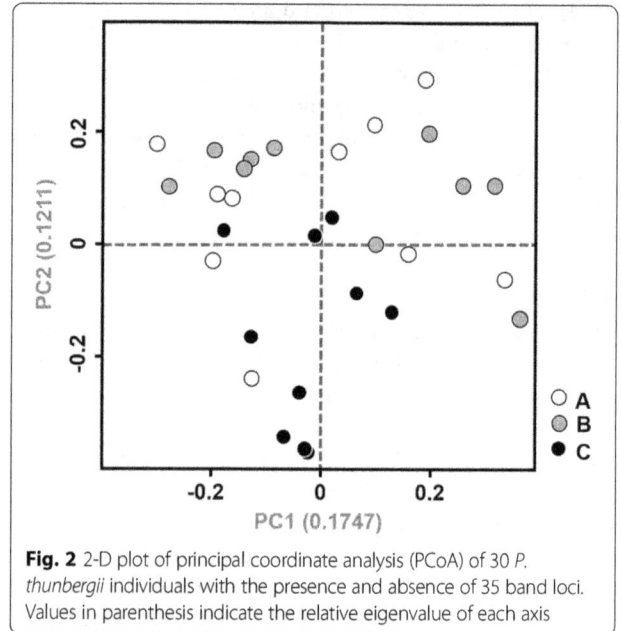

Fig. 2 2-D plot of principal coordinate analysis (PCoA) of 30 *P. thunbergii* individuals with the presence and absence of 35 band loci. Values in parenthesis indicate the relative eigenvalue of each axis

could be inferred from the geographical barrier, a dam between two sites. In a previous study, *P. thunbergii* seemed to have a spatial genetic structure (isolation-by-distance) by reason of the limited seed dispersal (Konuma and Terauchi 2001). However, in this study, geographical barrier appeared to accelerate the genetic isolation between sites B and C.

Abbreviations
PCR: Polymerase chain reaction; RAPD: Random amplification of polymorphic DNA; USDA: United States Department of Agriculture

Acknowledgements
Not applicable.

Funding
This study was supported by the Basic Science Research Program through the National Research Foundation of Korea (NRF-2012R1A1A2001007 and NRF-2015R1D1A1A01057373).

Authors' contributions
NBE participated in the design of the study, carried out the molecular studies, and wrote the manuscript draft. JGK conceived of the study, participated in the design of the study, edited the manuscript draft, and secured the funding. JMN participated in the field work. All authors read and approved the final manuscript.

Competing interests
The authors declare that they have no competing interests.

Table 4 Number of polymorphic loci and percentage, Nei's gene diversity (h), and Shannon index (i) in the three sites

Site	No. of polymorphic loci (%)	h	i
Site A	29 (72.5)	0.2381	0.3592
Site B	28 (70.0)	0.2761	0.4100
Site C	29 (72.5)	0.2618	0.3925
Total	35 (87.5)	0.2969	0.4472

Total does not mean the sum of polymorphic loci numbers of three populations but the total loci number which considers overlapped loci among populations

References

Allen, S. E., Grimshaw, H. M., Parkinson, J. A., & Quarmby, C. (1974). *Chemical analysis of ecological materials*. Oxford: Blackwell Scientific Publication.

Boyle, J. (2004). A comparison of two methods for estimating the organic matter content of sediments. *Journal of Paleolimnology, 31*, 125–127.

Bray, R. H., & Kurtz, L. T. (1945). Determination of total, organic and extracted forms of phosphorus in soil. *Soil Science, 59*, 39–45.

Carter, M. R. (1993). *Soil sampling and methods of analysis*. Boca Raton: Lewis Publishers.

Cheplick, G. P. (1987). The ecology of amphicarpic plants. *Trends in Ecology & Evolution, 2*, 97–101.

Choo, Y. H., Kim, H. T., Nam, J. M., & Kim, J. G. (2014). Flooding effects on seed production of the amphicarpic plant *Persicaria thunbergii*. *Aquatic Botany, 119*, 15–19.

Choo, Y. H., Nam, J. M., Kim, J. H., & Kim, J. G. (2015). Advantages of amphicarpy of *Persicaria thunbergii* in the early life history. *Aquatic Botany, 121*, 33–38.

Han River Flood Control Office. (2014). Annual report on hydrology in Republic of Korea—water level. http://hrfco.go.kr/web/sumun/floodgate.do Accessed 16 Nov 2016.

Honnay, O., Jacquemyn, H., Nackaerts, K., Breyne, P., & Looy, K. V. (2010). Patterns of population genetic diversity in riparian and aquatic plant species along rivers. *Journal of Biogeography, 37*, 1730–1739.

Kamphake, L. J., Hannah, S. A., & Cohen, J. M. (1967). Automated analysis for nitrate by hydrazine reduction. *Water Research, 1*, 205–216.

Kawano, S. (2008). 11: *Polygonum thunbergii* Sieb. Et Zucc. (Polygonaceae). *Plant Species Biology, 23*, 222–227.

Kawano, S., Hara, T., Hiratsuka, A., Matsuo, K., & Hirota, I. (1990). Reproductive biology of an amphicarpic annual *Polygonum thunbergii* (Polygonaceae): spatio-temporal changes in growth, structure and reproductive components of a population over an environmental gradient. *Plant Species Biology, 5*, 97–120.

Kim, J. G. (2009). Ecological characters of *Sphagnum* fens in Mt. Odae: I. Sowhangbyungsan-neup. *Journal of Korean Wetland Society, 11*, 15–27.

Kim, J. G., Park, J. H., Choi, B. J., Shim, J. H., Kwon, G. J., Lee, B. A., Lee, Y. W., & Ju, E. J. (2004). *Methods in ecology*. Seoul: Bomoondang.

Kim, Y. H., Tae, K. H., & Kim, J. H. (2008). Genetic variations and relationships of *Persicaria thunbergii* (Sieb. & Zucc.) H. Gross ex Nakai (Polygonaceae) by the RAPD analysis. *Korean Journal of Plant Resources, 21*, 66–72.

Kim, D. H., Choi, H., & Kim, J. G. (2012). Occupational strategy of *Persicaria thunbergii* in riparian area: rapid recovery after harsh flooding disturbance. *Journal of Plant Biology, 55*, 226–232.

Kim, D. H., Kim, H. T., & Kim, J. G. (2013). Effects of water level and soil type on the survival and growth of *Persicaria thunbergii* during early growth stages. *Ecological Engineering, 61*, 90–93.

Kim, J. H., Nam, J. M., & Kim, J. G. (2016). Effects of nutrient availability on the amphicarpic traits of *Persicaria thunbergii*. *Aquatic Botany, 131*, 45–50.

Konuma, A., & Terauchi, R. (2001). Population genetic structure of the self-compatible annual herb; *Polygonum thunbergii* (Polygonaceae) detected by multilocus DNA fingerprinting. *American Midland Naturalist, 146*, 122–127.

Koppitz, H. (1999). Analysis of genetic diversity among selected populations of *Phragmites australis* world-wide. *Aquatic Botany, 64*, 209–221.

Min, S. J., Kim, H. T., & Kim, J. G. (2012). Assessment of genetic diversity of *Typha angustifolia* in the development of cattail stands. *Journal of Ecology and Field Biology, 35*, 27–34.

Nei, M. (1973). Analysis of gene diversity in subdivided populations. *Proceedings of the National Academy of Sciences of the United States of America, 70*, 3321–3323.

Park, J., Hong, M. G., & Kim, J. G. (2013). Relationship between early development of plant community and environmental conditions in abandoned paddy terraces at mountainous valleys in Korea. *Journal of Ecology and Environment, 36*, 131–140.

Piquot, Y., Saumitou-Laprade, P., Vernet, P. P., & Epplen, J. T. (1996). Genotypic diversity revealed by allozymes and oligonucleotide DNA fingerprinting in French populations of the aquatic macrophyte *Sparganium erectum*. *Molecular Ecology, 5*, 251–258.

Raabová, J., Rossum, F. V., Jacquemart, A., & Raspé, O. (2015). Population size affects genetic diversity and fine-scale spatial genetic structure in the clonal distylous herb *Menyanthes trifoliata*. *Perspectives in Plant Ecology, Evolution and Systematics, 17*, 193–200.

Yeh, F. C., & Boyle, T. J. B. (1997). Population genetic analysis of codominant and dominant markers and quantitative traits. *Belgian Journal of Botany, 129*, 157.

11

Effect of pH on soil bacterial diversity

Sun-Ja Cho[1], Mi-Hee Kim[2] and Young-Ok Lee[3*]

Abstract

Background: In order to evaluate the effect of pH, known as a critical factor for shaping the biogeographical microbial patterns in the studies by others, on the bacterial diversity, we selected two sites in a similar geographical location (site 1; north latitude 35.3, longitude 127.8, site 2; north latitude 35.2, longitude 129.2) and compared their soil bacterial diversity between them. The mountain soil at site 1 (Jiri National Park) represented naturally acidic but almost pollution free (pH 5.2) and that at site 2 was neutral but exposed to the pollutants due to the suburban location of a big city (pH 7.7).

Methods: Metagenomic DNAs from soil bacteria were extracted and amplified by PCR with 27F/518R primers and pyrosequenced using Roche 454 GS FLX Titanium.

Results: Bacterial phyla retrieved from the soil at site 1 were more diverse than those at site 2, and their bacterial compositions were quite different: Almost half of the phyla at site 1 were Proteobacteria (49 %), and the remaining phyla were attributed to 10 other phyla. By contrast, in the soil at site 2, four main phyla (Actinobacteria, Bacteroidetes, Proteobacteria, and Cyanobacteria) composed 94 %; the remainder was attributed to two other phyla. Furthermore, when bacterial composition was examined on the order level, only two Burkholderiales and Rhizobiales were found at both sites. So depending on pH, the bacterial community in soil at site 1 differed from that at site 2, and although the acidic soil of site 1 represented a non-optimal pH for bacterial growth, the bacterial diversity, evenness, and richness at this site were higher than those found in the neutral pH soil at site 2.

Conclusions: These results and the indices regarding diversity, richness, and evenness examined in this study indicate that pH alone might not play a main role for bacterial diversity in soil.

Keywords: Pyrosequencing, pH, Bacterial diversity, Biodiversity indices

Background

Soil environment, such as the water/sediment environment, are the most complex of all microbial habitats that include water and air (Horner-Devine et al. 2004). The first attempt at quantifying microbial diversity was undertaken in 1990 (Amann et al. 2001). However, questions regarding microbial diversity and geographic distribution have yet to be completely answered (Green et al. 2008; Maier and Pepper 2009; Madigan et al. 2010). One of the main difficulties associated with obtaining overviews of microbial diversity is the method of analysis used. Traditionally, for over a century, microbes have been cultured using media in microbial diversity studies, but it was realized comparatively recently that microbial diversities determined using a cultivation-based method due to the restriction of culture condition and media accounted for only 0.1 to 1 % of total bacterial communities (Torsvik et al. 1990). For example, the diversity of DNA extracted from bacteria in 1 g of soil is about 200 times that in same samples analyzed using a cultivation-based method (Torsvik et al. 1990), which substantially reduces reported numbers and diversities of microorganisms (Smit et al. 2001; Torsvik and Øvreås 2002). Accordingly, it is believed that results obtained using cultivation-based methods do not reflect the true diversities and compositions of bacterial communities (Smit et al. 2001; Sandaa et al. 2001). To solve this problem, ribosomal RNA (rRNA) (ribosomal DNA (rDNA)) molecules with highly conserved sequence region which is in all of organisms have been introduced (Amann et al. 2001; DeLong et al. 1989; Ludwig and Klenk 2001), and their use has increased rapidly in parallel with the development of molecular techniques, such as fluorescent in situ hybridization (FISH), polymerase chain reaction-denaturing gradient gel electrophoresis (PCR-DGGE), and bar-coded pyrosequencing for the study of microbial diversity in

* Correspondence: ecolomi@daegu.ac.kr
[3]Department of Biological Sciences, Daegu University, Daegu 38453, South Korea
Full list of author information is available at the end of the article

various environments (Amann et al. 2001; Torsvik and Øvreås 2002; DeLong et al. 1989; Ludwig and Klenk 2001; Schramm et al. 1998; Liang et al. 2011; Li et al. 2012).

In addition, recent studies have shown relations between microbes and environmental factors, such as geographical location (Fierer and Jackson 2006; Lauber et al. 2009), soil texture (Sessitsch et al. 2001), land use (Smit et al. 2001; Cookson et al. 2007; Brons and van Elsas 2008; Hartman et al. 2008; Will et al. 2001), pH (Fierer and Jackson 2006; Lauber et al. 2009; Hartman et al. 2008), nutrients (Cookson et al. 2007; Will et al. 2001; Han et al. 2008), contaminants like oil (Liang et al. 2011), and heavy metals (Sandaa et al. 2001; Kelly et al. 1999; Roane and Pepper 2000; Lee et al. 2008). Of the abovementioned factors, pH is regarded one of the most important in terms of shaping biogeographical patterns (Fierer and Jackson 2006; Lauber et al. 2009). Lauber et al. (2009) analyzed soils collected from 88 sites across North and South America that represented a wide range of ecosystems on a continental scale by pyrosequencing and postulated that five dominant bacterial groups, that is, Acidobacteria, Actinobacteria, Bacteroidetes, α-Proteobacteria, and β-γ-Proteobacteria dominated, and their abundances were independent of geographical location but dependent on pH, which was found to be significantly correlated with bacterial diversity at 66 sites. These authors concluded pH could be considered a predictor of biogeographical microbial patterns.

If pH is so critical to microbial community and its growth, the following question could be raised. How would microbial community structure differ between a site with naturally acidic conifer forest soil (low pH) that is ecologically stable and a (sub-)urban forest with neutral pH but probably loaded by several contaminants?

Furthermore, between ecological stability achieved in pollution-free environment and pH, which factor has the greater effect on microbial community dynamics? To answer these questions, we selected two sites for study.

Therefore, the purpose of this study was to compare bacterial community structure in natural acidic forest soil collected from the lodging area of Mt. Jiri, a Korean national park, with neutral mountain soil obtained in a suburban area using bar-coded pyrosequencing. In addition, to better understand the soil properties in the study areas, geochemical parameters, such as pH, concentration of ions, cation exchange capacity (CEC), and heavy metal concentration, were measured concurrently.

Methods
Site description and sampling
Soil samples were collected to determine the bacterial community structures of soils obtained from two mountains with different pHs and ecological stability. The first

sample was collected from a lodging area of Mt. Jiri (site 1), a Korean national park, representative of acidic pH due to its dominant conifer vegetation (Korea Forest Service: http://www.forest.go.kr. Accessed 24 June 2015). The second sample was obtained from Mt. Jang (site 2), representing a less stable ecosystem and located close to the city of Busan (Fig. 1). At each site, about 200 g of minimally disturbed soil with high litter contents was obtained from a depth of 10 cm at 5 to 10 randomly selected locations within an area of 100 m^2. Soil samples were composited and stored in an ice box at 4 °C for transportation. On receipt, samples were sieved and mixed by passing them through a 2-mm mesh (10 mesh). Homogenized soils were frozen at –80 °C until subjected to microbial DNA extraction.

Geochemical property of soils
The pH values of air dried soil were measured using a pH Meter (Horiba, Japan) after mixing them with distilled water in a ratio of 1:5 (w:v) for 30 min. To measure water-soluble anion (NO_3^-, Cl^-, SO_4^{2-}, F^-) concentrations, 50 g soil samples was mixed with 500 mL of deionized water under continuous agitation for 12 h at 200 rpm on a reciprocal shaker (Ahn et al. 2007), and filtered through an 8-μm-pore-size filter (Whatman 40). Concentrations of anions in filtrate were determined by IC (Dionex ICS-3000, USA). Concentrations of exchangeable cations were measured by ICP (Varian 720-ES, USA) after extracting the filtrate with 1 M NH_4OAc and expressed in milliequivalents of negative charge per 100 g of soil (meq/100 g).

Heavy metals
The concentrations of heavy metals, such as Pb, Zn, Cd, Cu, As, Ni, Cr^{6+}, and Hg, were measured using the Korean standard method for soil pollution (Environment standard methods 2009). Briefly, 3 g of air dried soil was suspended in a Teflon vessel containing a mixture of 7 mL of 0.7 M nitric acid and 21 mL of 0.36 M hydrochloric acid. The solution was then adjusted to a final volume of 100 mL with 0.5 M nitric acid. The vessel was heated to 200 °C for 2 h on a heating mantle (Echofree, Korea) to ensure a complete digestion of the sediment. Concentrations of heavy metals were measured by ICP (Varian 720-ES, USA) and results are expressed as milligrams per kilogram.

Microbial analysis of soils
Enumeration of heterotrophic bacteria and fungi using culture media
To get a glimpse of the biodegradability of organic matter present at each study site, numbers of cultivable fungi and bacteria, that is, the so-called heterotrophic plate count (HPC), were enumerated by three replicates

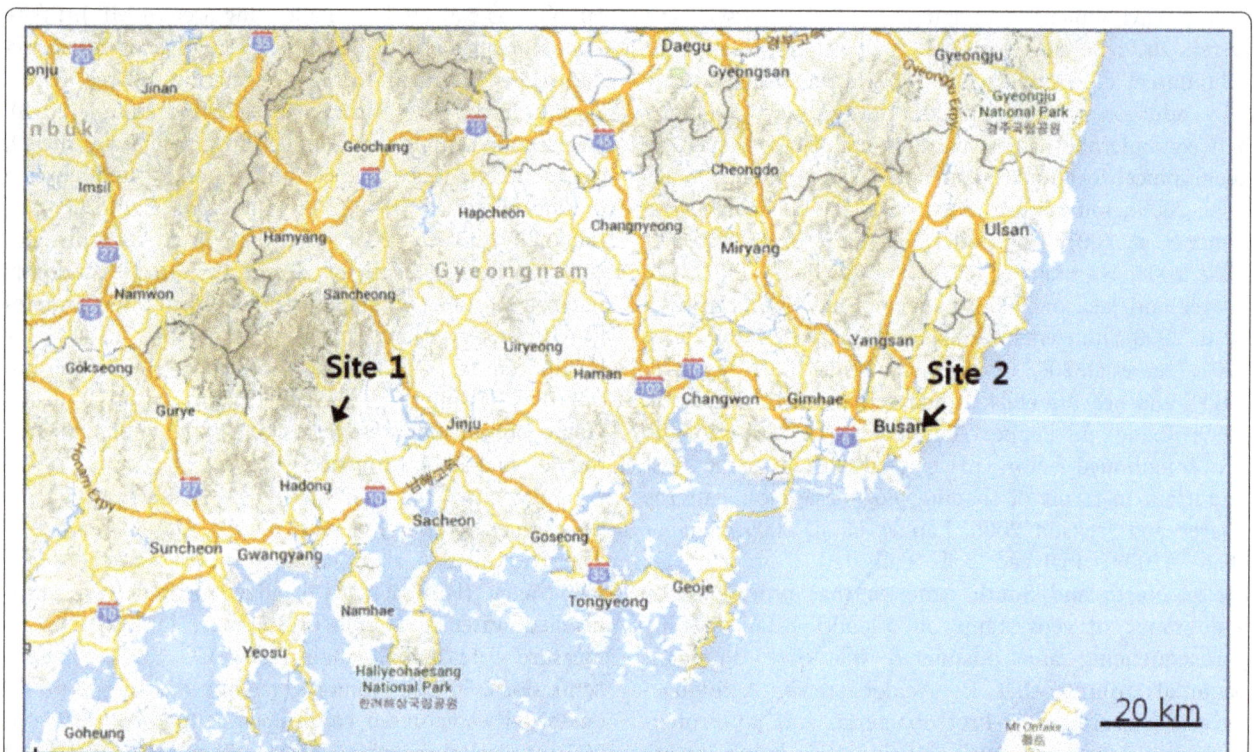

Fig. 1 This map of South Korea shows the two sampling sites. Site 1 which is the lodging area at Mt. Jiri is located southwest (north latitude 35.3; longitude 127.8) and site 2 which is Mt. Jang is located near the metropolitan city of Busan (north latitude 35.2; longitude 129.2)

using conventional plating techniques from soil slurries decimally diluted with sterile distilled water. Bacteria and fungi were cultured by Nutrient Agar (Difco, Baltimore, MD) or Potato Dextrose Agar (Difco, Baltimore, MD), respectively, and incubated at 30 °C for 48 h. Counting revealed 30 to 300 colonies per plate, and mean values were expressed in colony-forming unit (CFU) per gram of soil (dry weight).

Analysis of 16S rDNA for bacterial diversity by pyrosequencing
DNA extraction and bar-coded pyrosequencing
Metagenomic DNAs were extracted from 0.5 g soil samples using a FastDNA Spin Kit (MP, USA), with an additional 2 min of bead beating step to limit DNA shearing. Eluted DNAs were stored at −20 °C before a pyrosequencing procedure. A 20 ng aliquot of the 16S rDNA gene from the extracted DNA samples was amplified using universal primer sets, 27F (5′-GAGTTTGAT CMTGGCTGG-3′) primer with a Roche 454 A pyrosequencing adapter and 518 R (5′-GTATTACCGCG GCTGCTGG-3′) with a Roche 454 B sequencing adapter. The targeted gene region is believed as the most reliable for the accurate taxonomic classification of bacterial sequences (Liu et al. 2007). Pyrosequencing library was prepared using PCR products according to the GS FLX titanium library prep guide.

Fast Start High Fidelity PCR System (Roche) was used for PCR under the following conditions: 94 °C for 3 min followed by 35 cycles of 94 °C for 15 s; 55 °C for 45 s; 72 °C for 60 s; and a final elongation step at 72 °C for 8 min. After PCR, the products were purified using AMPure beads (Beckman-Coulter, USA). Sequencing was performed on a Roche 454 GS FLX Titanium from Macrogen Ltd. (Seoul, Korea).

Processing of pyrosequencing data
After sequencing, reads were sorted by tag sequence (bar-coded sequence). Then the sequences were trimmed to remove any non-16S bacterial reads such as barcodes, primers, chimeras, and sequences less than 350 bp by the in-house software of Macrogen (Korea). Extracted sequences were clustered by OTU (operational taxonomic unit) sequences which were obtained by calculating at a 97 % similarity cut-off (Hwang et al. 2014). OTU clustering was performed using software of mothur (v.1.27.0) and CD-HIT-OUT (OTU clustering, mothur (version 1.27.0). (http://www.mothur.org), CD-HIT-OTU (http://weizhong-lab.ucsd.edu/cd-hit-otu/). Accessed 29 Aug 2013). Representative sequences from each phylotype were blasted, and the taxonomic identity of each phylotype was determined using the Silva rRNA database (Silva rRNA database (http://www.arb-silva.de/). Accessed 29 Aug 2013). Results were calculated by

expressing same reads as percentages of total classifiable reads.

Diversity indices

To assess bacterial diversities, we calculated Shannon-Weaver indices (H'), Simpson indices, evenness indices (E) developed by Pielou, and richness by Margalef were calculated with the extracted OTU sequences as described by Kennedy and Smith (1995). To unify the measurement criterion regarding diversity indices of Shannon-Weaver and Simpson, the reciprocal of the Simpson index was used because Shannon-Weaver index is a high number for high diversity, but Simpson index is a low number for high diversity.

Results and discussion

Geochemical properties of soils in the studied area

Anions and exchangeable cation concentrations, pH values, and CEC at both sites are summarized in Table 1. The soil from site 1 (lodging area of Jiri National Park) had an acidic pH of 5.2, while that from site 2, which was located close to the metropolitan city of Busan, was almost neutral (pH 7.7), which is in the optimal range (pH 6–8) required for microbial growth (Maier and Pepper 2009). The pH value of soil affects the solubilities of chemicals by influencing ionization degrees (Maier and Pepper 2009). It should be added that the pH values at the two sited mentioned above are integrated results due to numerous interactions between cations and anions in the soil solution (Fierer and Jackson 2006). Just the large difference in pH values at the two sites implies that the geochemical environment of both sites differed. As has been reported by others (Fierer and Jackson 2006; Lauber et al. 2009), we presumed that pH played a definite role on the diversities and compositions of bacterial community.

The total concentration of anions at site 1 was greater than at site 2. In particular, the concentration of NO_3^-, which can be utilized immediately by microbes and plants, was much higher at site 1 (12.7 cmol/kg) than at site 2 (0.04 cmol/kg). For total exchangeable cations (Ca^{2+}, Mg^{2+}, K^+, Na^+), their summed concentration at site 1 was almost the same as that at site 2 (Table 1). CEC, a measure of the capacity of soils and organic colloids to remove cations from solution, varies depending on the type of soil, and its value increases in line with

the decomposition rate of organic matter by microorganisms (Alexander 1977). At the time of the sampling in November of 2012, the sites were already densely covered by litter to be degraded by microbes; determined values of CEC to an extent reflect the decomposing of organic matter (leaves).

The CEC of 30.7 meq/100 g determined for soil from site 1 indicates relatively better conditions for microbial growth than at site 2 (13.0 meq/100 g). According to Maier and Pepper (2009), the average of CEC of soils range from 15 to 20 meq/100 g, and that CEC values of <15 meq/100 g leads to low nutrient levels in soil because of a reduced capacity to retain cations and essential nutrients, such as NO_3^- and PO_4^{3-}. Other factors, such as soil particle size, water, and nutrient availability not investigated in this study, might also have influenced the soil environment in various, but not fully explored, ways, as reported by others (Cookson et al. 2007; Brons and van Elsas 2008; Hartman et al. 2008; Will et al. 2001; Han et al. 2008). Particle size might change chemical properties by changing adsorption affinities (Maier and Pepper 2009). For example, small particle soil (silt and clay) allow more diverse microbial inhabitants than large particle. In terms of dominant bacterial groups related to particle size, bacteria belonging to the phylum Acidobacteria and the genus *Prosthecobacter* sp. (phylum Verrucomicrobia) were found to be more diverse in soils with small particles, whereas α-Proteobacteria dominated in large particle soils (Sessitsch et al. 2001). In addition, the availabilities of nutrients and organic matter also strongly influence bacterial abundances and diversities (Smit et al. 2001; Hartman et al. 2008; Will et al. 2001; Han et al. 2008).

Heavy metals

Of the eight heavy metals detected in this study, Cr^{6+} and Hg were almost undetectable (less than 0.0001 mg/kg) at both sites. Addressing the remaining six metals in decreasing order, Zn was detected at the highest concentration at both sites, although its value at site 1 (47.26 mg/kg) was lower than at site 2 (58.43 mg/kg). Pb and Ni were present at site 1 at slightly higher concentrations (19.43 and 13.27 mg/kg vs. 13.04 and 11.44 mg/kg), which was unexpected. Cu had concentrations of 11.67 and 14.25 mg/kg at

Table 1 Chemical properties of the soil analyzed

Sample	pH	CEC (meq/100 g)	Anion (cmol/kg DW)					Cation (cmol/kg DW)				
			F^-	Cl^-	NO_3^-	SO_4^{2-}	Total	Ca^{2+}	K^+	Mg^{2+}	Na^+	Total
Site 1	5.2	30.7	0.03	0.51	1.27	1.18	2.99	10.0	0.75	0.99	0.19	13.90
Site 2	7.7	13.0	0.12	0.46	0.04	0.99	1.61	12.1	0.18	0.63	0.16	13.00

DW (dry weight)

sites 1 and 2, respectively. As had concentrations of 4.64 and 5.11 mg/kg, respectively, and Cd had concentrations of 0.40 and 1.35 mg/kg, respectively. So the concentrations of heavy metals, with the exception of Pb and Ni, were higher at site 2 than in soil from the Korean National Park (site 1), as depicted in Fig. 2. It has been reported that heavy metals can not only inhibit microbial growth and activity but also shift bacterial populations from heavy metal non-resistant to resistant populations over time (Kelly et al. 1999; Roane and Pepper 2000). Kelly and coworkers (1999), in a laboratory investigation on the effects of a Zn smelter on microbes, added 6000 mg/kg of Zn to soil, and 15 days later found Zn level in soil had reduced to 4660 mg/kg, which was ascribed to adsorption on the surfaces both of soil and microbes (Lee et al. 2008), and that cultured bacteria (isolates) had reduced by 87 %. However, over the course of the experiment, it was found that the bacterial composition had changed from a non-resistant to a resistant population. In addition, contamination by Cd or Pb at concentrations of 5–55 mg/kg and 75–1660 mg/kg, respectively, reduced of bacterial numbers by up to 1 %. Accordingly, it would appear the relatively low concentration of heavy metals found at both sites was insufficient to have affected microbial growth. However, the slightly higher heavy metal levels at site 2 might have had a negative effect on the dynamics of the bacterial community.

Moreover, in accord with the abovementioned results concerning the geochemical property of soils, the number of heterotrophic bacteria/fungi was higher at site 1 than at site 2 (Table 2). In the site 1 soil sample, numbers of heterotrophic bacteria (HPC) which is considered as an indicator of easily degradable organic compounds (Maier

Table 2 Microbiological characteristics evaluated by culturing bacteria and fungi

Sample	HPC (bacteria, CFU/g DW)	Fungi (CFU/g DW)
Site 1	9.2×10^7	3.6×10^3
Site 2	8.0×10^6	6.0×10^2

HPC heterotrophic plate count, *CFU* colony-forming unit, *DW* dry weight

and Pepper 2009) were more than ten times greater than that at site 2.

Taken all together, the geochemical environment at site 1 seems to be more favorable for microbial growth than that at site 2. Because of the greater heterogeneity of soil per se that was revealed from millimeters of a micro scale, the so-called micro-environment (Madigan et al. 2010; Schramm et al. 1998; Hartman et al. 2008), maybe up to a continental scale, soil investigations are very time-demanding and costly, as compared with similar investigation on water or air. Nevertheless, overall but precise knowledge regarding the physico-geochemical properties of soil gathered through sophisticated methods is an important prerequisite to the understanding of its microbiological characteristics.

Comparison of bacterial compositions at the two sites

From the 7614 retrieved sequences, 344 chimeric sequences were removed using ChimeraSlayer (Haas et al. 2011) to avoid misreading, and the remaining 7270 classifiable sequences were analyzed by pyrosequencing (Table 3). For checking the validity of data used in this study, rarefaction curves, which were generated using the relationship between the OTU numbers and the sequence reads, were created using mothur output (data not shown) (OTU clustering, mothur (version 1.27.0). (http://www.mothur.org), CD-HIT-OTU (http: //weizhong-lab.ucsd.edu/cd-hit-otu/). Accessed 29 Aug 2013). Based on the rarefaction curve, the numbers of reads obtained were sufficient to assess the bacterial diversity at both sites.

Bacteria at both sites were affiliated with 12 phyla across the entire data set. Regarding the bacterial diversity observed on the phylum level in Table 4, with the exception of unclassified (1.4 %), the 1428 classifiable sequences retrieved from site 1 were distributed widely to 11 different phyla, while those for site 2 belonged to only six phyla. Three phyla, Proteobacteria (site 1; 49.2 %, site 2; 21.8 %), Actinobacteria (site 1; 21.8 %, site 2; 29.8 %), and Cyanobacteria (site 1; 9.8 %, site 2; 17.5 %) dominated both sites. The next most abundant phylum, Planctomycetes (7.2 %), at site 1 was not observed at site 2, whereas Bacteroidetes, which is regarded as a typical inhabitant of soil (Madigan et al. 2010; Lauber et al. 2009; Will et al. 2001), was much more abundant at site 2 (site 1; 0.5 %, site 2; 24.3 %,). Acidobacteria, known as bacteria occurring frequently in

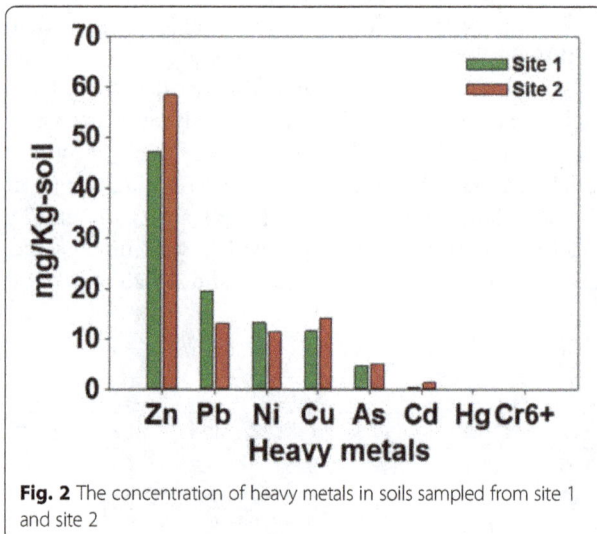

Fig. 2 The concentration of heavy metals in soils sampled from site 1 and site 2

Table 3 Data sets used for pyrosequencing and bacterial diversity indices

Samples	No. of reads		No. of OTUs	Richness (Margalef)	Evenness (Pielou)	Diversity index	
	Raw	Chimera removed				Simpson $(D = 1/\lambda)$[a]	Shannon-Weaver (H')
Site 1	1465	1428	165	22.58	0.93	0.99	4.76
Site 2	6149	5842	66	7.49	0.64	0.85	2.65

[a]Reciprocal of Simpson index and Shannon-Weaver index, higher numbers represent greater diversity

not only acidic soil but in all kinds of soil (Smit et al. 2001; Lauber et al. 2009; Will et al. 2001; Han et al. 2008), accounted only for a small proportion at both sites (site 1; 2.7 %, site 2; 1.1 %). On the other hand, somewhat unexpectedly, Cyanobacteria made up high proportion (site 1; 9.8 %, site 2; 17.5 %) at both sites. In a study on the bacterial diversity based on 16S rDNA clone from Korean acidic pine (pH 4.1) and oak wood (pH 5.3) soil, Proteobacteria was found to be the most dominant, followed by Firmicutes, Acidobacteria, Actinobacteria, Bacteroidetes, Verrucomicrobia, and Plantomycetes, but not Cyanobacteria (Han et al. 2008). In a loamy sand soil with a little acidic pH (5.5–6.5), Brons and van Elsas (2008) observed the bacterial community by clone analysis and found Cyanobacteria in a minor proportion.

In agreement with studies by others (Madigan et al. 2010; Brons and van Elsas 2008; Han et al. 2008), Proteobacteria, the most abundant phylum in soil, composed almost half of total bacteria at site 1, and of the classes of this phylum, Alpha(α)·Beta(β)·Gamma(γ)·-Delta(δ)·Zeta(ζ)·Epsilon(ε)-Proteobacteria, the first three classes α·β·γ-Proteobacteria dominated with proportions of 28.7, 5.3, and 13.6 % at site 1, while those at site 2 accounted for 18.3, 5.2, and 2.0 %, respectively (Fig. 3). In a comprehensive study, bacterial composition based on the analysis of 287,933 sequences obtained from soil across the large spatial scale revealed that the phyla

Proteobacteria, Acidobacteria, Bacteroidetes, Verrucomicrobia, and Planctomycetes dominated, but in different proportions depending on soil characteristics and geographical location (Madigan et al. 2010).

Agricultural use of land also causes changes in bacterial diversity and composition (Smit et al. 2001; Will et al. 2001). Smit et al. (2001) analyzed the bacterial community in a wheat field using a cultivation-based method and by analyzing 16S rDNA clone sequences and found that high GC Gram-positive bacteria were mainly detected by a cultivation-based method and not by clone analysis. Instead, based on clone analysis, Acidobacteria Proteobacteria, Nitrospira, Cyanobacteria, and green sulfur bacteria dominated, and they were found to be more evenly distributed. This result demonstrates the limitation of the cultivation-based method. Moreover, the abundance of γ-Proteobacteria, which is regarded as a fast growing bacteria in nutrient-rich environment like *Pseudomonas* sp. (Smit et al. 2001), was unexpectedly greater at site 1 than at site 2 (Fig. 4), which was probably due to litter decomposition. It was reported that plant root showed a selective effect towards γ-Proteobacteria (Marilley and Aragno 1999), which comprise the majority of fast growing decomposer for easily degradable substrates (Madigan et al. 2010). In addition, the phylum Elusimicrobia, previously known as "Termite Group 1" and occurring in various environments

Table 4 Comparison of soil bacterial composition between two sites on the phylum level

Phylum	Site 1 (%)	Site 2 (%)
Proteobacteria	49.2,	21.8
Actinobacteria	21.8	29.8
Cyanobacteria	9.8	17.5
Bacteriodetes	0.5	24.3
Planctomycetes	7.2	0
Acidobacteria	2.7	1.1
Firmicutes	1.3	5.5
Chloroflexi	2.5	0
Verrucomicrobia	2.1	0
Elusimicrobia	1.2	0
Gemmatimonadates	0.3	0
Unidentified	1.4	0.1

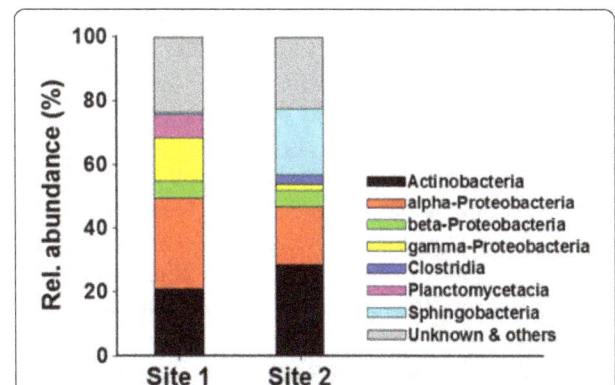

Fig. 3 Difference in bacterial composition on class level obtained from soils at site 1 and site 2. Their relative abundance was assessed by grouping the OTUs derived from 16S rDNA retrieved from each soil. Based on classifiable sequences, the OTUs were determined from the mothur clustering. Others are the sum of minor classes which individually show a relative abundance of less than 3 %

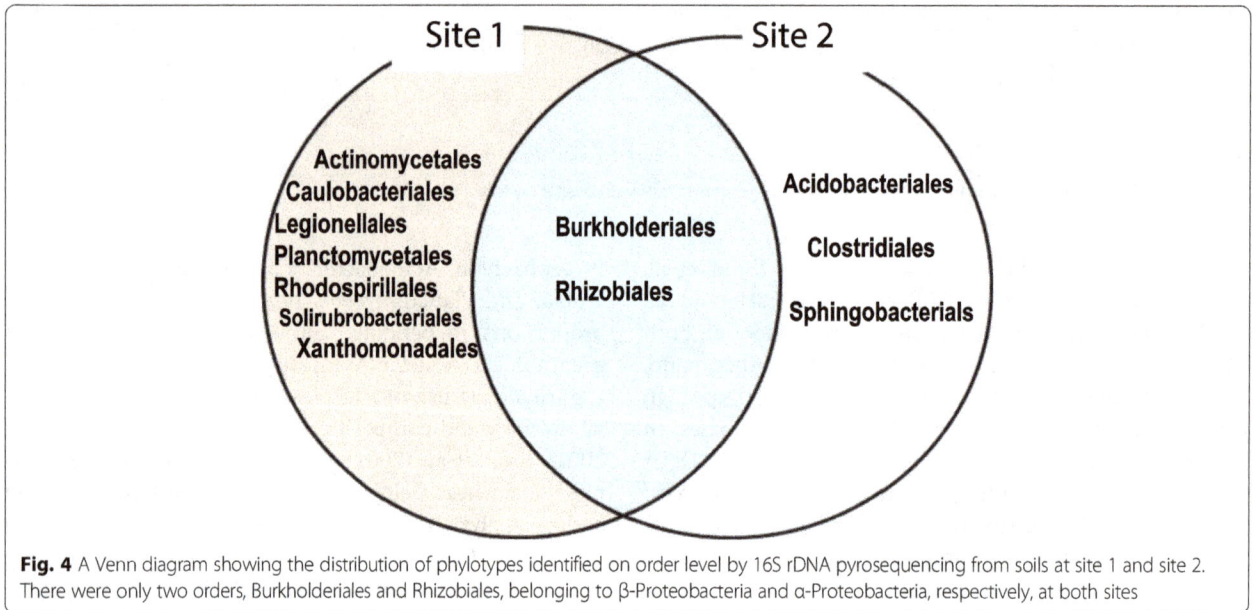

Fig. 4 A Venn diagram showing the distribution of phylotypes identified on order level by 16S rDNA pyrosequencing from soils at site 1 and site 2. There were only two orders, Burkholderiales and Rhizobiales, belonging to β-Proteobacteria and α-Proteobacteria, respectively, at both sites

(Herlemann et al. 2007), was present in a lower proportion at site 1 (1.2 %), but not at site 2 (Table 4).

Differences between the bacterial compositions at the two sites became clearer when the distribution of phylotypes were examined on the order level using a Venn diagram, as shown in Fig. 4. To make the Venn diagram, orders which represented less than 3 % of relative abundance were discarded. Only two orders, Burkholderiales and Rhizobiales, were found at both sites. The common order Burkholderiales belongs to β-Proteobacteria and was found in similar proportions at both sites (Fig. 4). The other order Rhizobiales belongs to α-Proteobacteria and includes genus *Rhizobium* that is able to fix nitrogen and is associated with the roots of legumes (Madigan et al. 2010). The genus *Rhizobium* is a typical soil inhabitant in the rhizosphere, which is considered a nutrient-rich niche (Marilley and Aragno 1999). Rhizobiales constituted a high proportion at both sites [site 1; 18.2 %, site 2; 10.3 % (Fig. 4)]. At site 2, bacteria belonged to three different orders, Acidobacteriales, Clostridiales, and Sphingomonadales, whereas at site 1, seven different orders were identified (Actinomycetales, Caulobacteriales, Legionellales, Planctomycetales, Rhodospirillales, Solirubrobacterales, Xanthomonadales). Based on our assessment of the bacterial community, it could be concluded that the soil ecosystem at site 1 may have a higher degree of bacterial diversity than at site 2. Of course, the degree of variability resulted from phylogenetic assemblages does not reflect degree of functional or ecological diversity. Therefore, more detailed studies, including its functional capability, are needed to determine why these bacteria are present in given soil.

Status of bacterial communities as evaluated using different indices

Originally, the diversity index was developed to assess diversity and stability of plant and animal communities (Kennedy and Smith 1995), but nowadays it is also used in bacterial community (Smit et al. 2001; Liang et al. 2011; Han et al. 2008; Hwang et al. 2014). In the present study, the Shannon-Weaver index and the reciprocal of Simpson index were used to assess diversity, and the Margalef and Pielou indices were used to assess richness and evenness (Kennedy and Smith 1995). As shown in Table 3, the index of diversity refers to the number of different phylotypes, the index of richness refers to the abundance of the same phylotype, and the index of evenness refers to how close in numbers phylotypes are to each other in the bacterial community. The numbers were much higher for site 1, even showing an acidic pH (5.2).

This implies that at site 1 bacterial communities, which act primarily as decomposers, were more diverse than at site 2. Accordingly, it would appear that nutrient cycling, which is achieved by the interactions between many types of microbes, would be smoother at site 1 than at site 2 (Cookson et al. 2007; Hartman et al. 2008; Will et al. 2001). As compared with the results obtained for bacterial diversities in Korean mountain pine and oak woods, for which the Shannon-Weaver index was estimated to fall in a range of from 3.42 to 3.63 (Han et al. 2008), our results (site 1; 4.76, site 2; 2.65) show that in terms of bacterial biodiversity, the soil ecosystem at site 1 was more stable than that at site 2.

Conclusions

In order to determine the effect of pH on bacterial diversity, we compared the bacterial communities of acidic soil (pH 5.2) from Jiri National Park's lodging area (site 1) with neutral mountain soil (pH 7.7, site 2) by bar-coded pyrosequencing of 16S rDNA.

Regarding the question, which factor has the greater effect on microbial community dynamics either pH or ecological stability of studying site, pH may have an effect on bacterial diversity. So depending on pH, the bacterial community in soil at site 1 was found to differ from that at site 2, and although the acidic soil of site 1 represented a non-optimal pH for bacterial growth, the bacterial diversity, evenness, and richness at this site were higher than those found in the neutral pH soil at site 2. Accordingly, these results imply that pH might not be a critical factor for shaping bacterial diversity and its stability. However, this study was performed over a short period without regard of the effects of other environmental factors, such as precipitation, oxygen, and some other primary nutrients such as N and P. So nothing could be considered as conclusive proof. Therefore, we suggest more advanced detailed studies in a long term should be conducted to identify the environmental factors responsible for the establishment of particular bacterial community structure.

Abbreviations

CEC: Cation exchange capacity; CFU: Colony-forming unit; DW: Dry weight; FISH: Fluorescent in situ hybridization; OTU: Operational taxonomic unit; PCR-DGGE: Polymerase chain reaction-denaturing gradient gel electrophoresis

Acknowledgements

Not applicable.

Funding

This study was funded by Daegu University Research Grant 2012 (grant number 20110599).

Authors' contributions

SJC and YOL carried out in the design of the study and the molecular genetic studies, participated in the sequence alignment, and drafted the manuscript. MHK analyzed the chemical properties of the sampled soil. All authors read and approved the final manuscript.

Competing interests

The authors declare that they have no competing interests.

Author details

[1]Department of Microbiology, Pusan National University, Busan 46269, South Korea. [2]Busan Metropolitan City Institute of Health and Environment, Busan 46616, South Korea. [3]Department of Biological Sciences, Daegu University, Daegu 38453, South Korea.

References

Ahn, BK, Kim, DH, & Lee, JH (2007). Post harvest cropping impacts on soil properties in continuous Watermelon (Citrullus lanatus Thunb.) cultivation plots (in Korean). Korean Journal of Soil Science and Fertilizer, 40, 98–107.

Alexander, M (1977). Introduction to soil microbiology (2nd ed., Vol. 142, pp. 12–13). New York: John Wiley & Sons.

Amann, R, & Schleifer, KH (2001). Nucleic acid probes and their application in environmental microbiology. In DR Boone & R Castenholz (Eds.), Bergey's Manual of Systematic Bacteriology (Vol. I, pp. 67–82).: Springer.

Brons, JK, & van Elsas, JD (2008). Analysis of bacterial communities in soil by use of denaturing gradient gel electrophoresis and clone libraries, as influenced by different reverse primers. Applied and Environmental Microbiology, 74, 2717–2727.

Cookson, WR, Osman, M, Marschner, P, Abaye, DA, Clark, I, Murphy, DV, Stockdale, EA, & Watson, CA (2007). Controls on soil nitrogen cycling and microbial community composition across land use and incubation temperature. Soil Biology and Biochemistry, 39, 744–756.

DeLong, EF, Wickham, GS, & Pace, NR (1989). Phylogenetic strains; ribosomal RNA-based probes for the identification of single microbial cells. Science, 243, 1360–1363.

Environment standard methods. (2009). 2009-255, ES07400.2. Metals by inductively coupled plasma-atomic emission spectrometry. Korea: Ministry of Environment standard.

Fierer, N, & Jackson, B (2006). The diversity and biogeography of soil bacterial communities. PNAS, 103, 626–631.

Green, JL, Bohannan, BJM, & Whitaker, RJ (2008). Microbial biogeography: from taxonomy to traits. Science, 320, 1039–1043.

Haas, BJ, Gevers, D, Earl, AM, Feldgarden, M, Ward, DV, Giannoukos, G, Ciulla, D, Tabbaa, D, Highlander, SK, & Sodergren, E (2011). Chimeric 16S rRNA sequence formation and detection in Sanger and 454-pyro- sequenced PCR amplicons. Genome Research, 21, 494–504.

Han, SI, Cho, MH, & Whang, KS (2008). Comparison of phylogenetic characteristics of bacterial populations in a Quercus and pine humus forest soil. Korean Journal of Microbiology, 44, 237–243.

Hartman, WH, Richardson, CJ, Vilgalys, R, & Bruland, GL (2008). Environmental and anthropogenic controls over bacterial communities in wetland soils. PNAS, 105, 17842–17847.

Herlemann, DPR, Geissinger, O, & Brune, A (2007). The termite group I Phylum is highly diverse and wide-spread in the environment. Applied and Environmental Microbiology, 73, 6682–6685.

Horner-Devine, MC, Carney, KM, & Bohannan, JM (2004). An ecological perspective on bacterial diversity. Proceedings of the Biological Sciences, 271(1535), 113–122.

Hwang, OK, Raveendar, S, Kim, YJ, Kim, JH, Kim, TH, Choi, DY, Jeon, CO, et al. (2014). Deodorization of pig slurry and characterization of bacterial diversity using 16S rDNA sequence analysis. Journal of Microbiology, 52, 918–929.

Kelly, JJ, Häggblom, M, & Tate, RL, III (1999). Changes in soil microbial communities over time resulting from one time application of zinc: a laboratory microcosm study. Soil Biology and Biochemistry, 31, 1455–1465.

Kennedy, AC, & Smith, KL (1995). Soil microbial diversity and the sustainability of agricultural soils. Plant and Soil, 170, 75–86.

Korea Forest Service: http://english.forest.go.kr/newkfsweb/eng/idx/Index.do?mn=ENG_01. Accessed 24 June 2015.

Lauber, CL, Hamady, M, Knight, R, & Fierer, N (2009). Pyrosequencing-based assessment of soil pH as a predictor of soil bacterial community structure at the continental scale. Applied and Environmental Microbiology, 75, 5111–5120.

Lee, EY, Lim, JS, Oh, KH, Lee, JY, Kim, SK, Lee, YK, & Kim, K (2008). Removal of heavy metals by an enriched consortium. Journal of Microbiology, 46, 23–8. doi:10.1007/s12275-007-0131-6.

Li, Y, Chen, L, Wen, H, Zhou, T, Zhang, T, & Gao, X (2012). 454 Pyrosequencing analysis of bacterial diversity revealed by a comparative study of soils from mining subsidence and reclamation areas. Journal of Microbiology and Biotechnology, 24, 313–323.

Liang, Y, Van Nostrand, JD, Deng, Y, He, Z, Wu, L, Zhang, X, Li, G, & Zhou, J (2011). Functional gene diversity of soil microbial communities from five oil-contaminated fields in China. ISME Journal, 5, 403–413.

Liu, Z, Lozupone, C, Hamady, M, Bushman, FD, & Knight, R (2007). Short pyrosequencing reads suffice for accurate microbial community analysis. Nucleic Acids Research, 35, e120.

Ludwig, W, & Klenk, HP (2001). Overview: a phylogenetic backbone and taxonomic framework for prokaryotic systematics. In DR Boone & RW Castenholz (Eds.), *Bergey's Manual of Systematic Bacteriology* (Vol. I, pp. 49–65).: Springer.

Madigan, MT, Martino, JM, Stahl, DA, & Clark, DP (2010). *Brock biology of microorganisms. Global 13th ed* (pp. 670–711).: Benjamin Cummings.

Maier, RM, & Pepper, IL (2009). Earth Environments. In R. M. Maier, I. L. Pepper, & C. P. Gerba (Eds.), *Environmental microbiology* (2nd ed., pp. 57–82).: Academic.

Marilley, L, & Aragno, M (1999). Phylogenetic diversity of bacterial communities differing in degree of proximity of Lolium perenne and Trifolium repens roots. *Applied Soil Ecology, 13*, 127–136.

OTU clustering; mothur (version 1.27.0). (http://www.mothur.org), (http://www.mothur.org), CD-HIT-OTU (http: //weizhong-lab.ucsd.edu/cd-hit-otu/) (http://weizhong-lab.ucsd.edu/cd-hit-otu/). Accessed 29 Aug 2013.

Roane, TM, & Pepper, IL (2000). Microbial responses to environmentally toxic cadmium. *Microbial Ecology, 38*, 358–364.

Sandaa, RA, Torsvik, V, & Enger, O (2001). Influence of long term heavy metal contamination on microbial communities in soil. *Soil Biology and Biochemistry, 33*, 287–295.

Schramm, A, DE Beer, D, Wagner, M, & Amann, R (1998). 1998. Identification and activities in situ of Nitrosospira and Nitrospira spp. as dominant populations in a nitrifying fluidized bed reactor. *Applied and Environmental Microbiology, 64*, 3480–3485.

Sessitsch, A, Weilharter, A, Gerzabek, MH, Kirchmann, H, & Kandeler, E (2001). Microbial population structures in soil particle size fractions of long-term fertilizer field experiment. *Applied and Environmental Microbiology, 67*, 4215–4224.

Silva rRNA database (http://www.arb-silva.de/). (http://www.arb-silva.de/). Accessed 29 Aug 2013.

Smit, E, Leeflang, P, Gommans, S, van den Broek, J, van Mil, S, & Wernars, K (2001). Diversity and seasonal fluctuations of the dominant members of the bacterial soil community in a wheat field as determined by cultivation and molecular methods. *Applied and Environmental Microbiology, 67*, 2284–2291.

Torsvik, V, & Øvreås, L (2002). Microbial diversity and function in soil: from genes to ecosystems. *Current Opinion in Microbiology, 5*, 240–245.

Torsvik, V, Salte, K, Sørheim, R, & Goksøyr, J (1990). Comparison of phenotypic diversity and DNA heterogeneity in a population of soil bacteria. *Applied and Environmental Microbiology, 56*, 776–781.

Will, C, Thürmer, A, Wollherr, A, Nacke, H, Herold, N, Schrumpf, M, Gutknecht, J, Wubet, T, Buscot, F, & Daniel, R (2001). Horizon-specific bacterial community composition of German grassland soils, as revealed by pyrosequencing-based analysis of 16S rRNA genes. *Applied and Environmental Microbiology, 76*, 6751–6759.

12

Importance of biomass management acts and policies after phytoremediation

Uhram Song[1] and Hun Park[2]* [iD]

Abstract

Background: Although phytoremediation is a promising method for pollution control, biomass produced by the remediation process must be managed; otherwise, it will eventually return to the environment and cause secondary pollution. Therefore, research and policy development for the post-remediation management of biomass are both required.

Results: While there are many published studies of phytoremediation, research into post-remediation management is very limited. Therefore, a new study using biomass as a co-composting material was conducted and showed positive effects on soil characteristics and plant performance. However, despite its potential, research and policies to promote this form of management are still lacking.

Conclusions: We suggest public engagement in support of "Post-phytoremediation management" legislation that stipulates management of biomass after phytoremediation, promotes recycling of biomass with known environmental risks, and includes specific policies developed for managers. Further research to support and inform such policies and laws is also required.

Keywords: Phytoremediation, Biomass, Post-remediation, Compost, Pollutants, Legal provisions

Background

Modern civilization inevitably produces large quantities of pollutants through industrialization, intensive agriculture, and urbanization (Eapen and D'Souza 2005). Among these pollutants, chemical wastes including agrichemicals, heavy metals (Lu et al. 2015), nitrogen (N) and phosphorus (P) from fertilizers (Smith and Siciliano 2015), nanomaterials (Song et al. 2013a), and even microplastics (Fendall and Sewell 2009) can all degrade water quality. Such pollutants can directly affect living organisms by their own toxicity but can also indirectly affect whole ecosystems, for example, by eutrophication and effects on trophic cascades (Paerl and Whitall 1999). Pollutant management is therefore very important. However, standard pollutant-removal techniques such as the use of sewage and wastewater treatment plants are constrained by factors such as high cost and unsustainability (Abma et al. 2010). Therefore, the development of other treatment systems to compensate for these disadvantages has been actively pursued (Mohan and

Gandhimathi 2009). Among many new methods, phytoremediation and wetland construction for biofiltration have begun to attract interest because of their environmentally friendly and cost-effective traits (Justin and Zupancic 2009). In Korea, phytoremediation studies applied to many streams, rivers, and lakes have received national funding and policy support (Song and Kang 2006, Kim and Yun 2012). However, this form of phytoremediation has associated problems of long-term post-remediation management, for which research and policy development are very limited (Song et al. 2016). Pollutants taken up by plants during phytoremediation can easily be returned to the target environment by postmortem decomposition, thereby degrading the efficiency of the remediation process (Helfield and Diamond 1997). Effective management of plant biomass after remediation is therefore necessary. The use of plant biomass after phytoremediation is suggested for biofuel (Banuelos 2006), and one comprehensive study suggested incineration after comparing multiple methods of biomass management such as compaction, incineration, ashing, pyrolysis, direct disposal, and liquid extraction (Sas-Nowosielska et al. 2004). However, research into

* Correspondence: parkhun@gmail.com
[2]Institute for Legal Studies, Yonsei University, Seoul 03722, South Korea
Full list of author information is available at the end of the article

this aspect of management is still sparse (Song et al. 2016), and in particular, there is no information about domestic or international policies concerning these problems. Since phytoremediation is being applied widely in Korea and in many other countries, there is a need for a review of methods for biomass management after phytoremediation and of the policies used to promote these systems. At least to the authors' knowledge, no papers have yet been published that suggest relevant policies based on direct evidence utilizing scientific experiments. In this study, therefore, we carried out a literature search on current phytoremediation methods and policies and tested our own method for managing biomass after remediation. From a synthesis of these results, we propose the requirements for an effective biomass management method and highlight the need for policies to address the problem of waste produced by the phytoremediation process.

Methods
Research, case studies, and polices relating to phytoremediation
The Research Information Sharing Service (RISS, http://www.riss.kr) and Google Scholar (http://scholar.google.com) were used to locate published studies of phytoremediation from Korea and other countries. Policies were retrieved from online repositories of regulations. For South Korea, the "National Law Information Center" (http://www.law.go.kr/eng/engMain.do) was the principal source. For the United States, regulations were obtained from Cornell University's "Legal Information Institute" (https://www.law.cornell.edu/). Regulations for the European Union were found at EUR-Lex (http://eur-lex.europa.eu/homepage.html?locale=en).

Field experiment into feasibility of post-remediation management
To test the feasibility of using plant biomass after phytoremediation, an experiment was carried out to convert biomass of plants to compost, a form of fertilizer. The Sudokwon landfill in Incheon, Korea, is one of the biggest sanitary landfills in the world and has approximately 5 km of leachate channel that acts as a remediation and buffer zone before leachate is emitted to the sea. However, the plants growing in the channel, reed (*Phragmites australis*) and cattail (*Typha angustifolia*), are not subject to any management and their decomposition may potentially reduce the remediation capacity of the channel and return pollutants to the environment (Song 2010). *P. australis* and *T. angustifolia* were therefore harvested from the channel in October, at the end of the growing season. Harvested macrophytes (20 kg) were chopped to a length of approximately 5 cm and mixed with 10 L water, 1 L EM (effective microorganism) solution, and 100 g brown sugar for windrow composting

(EM-center 2003). The composting materials were covered with several blankets to prevent heat loss. After the composting process, the compost was sealed and left in the shade until the following spring before experimental application.

Compost was applied to the landfill slope in April where, recently, 2-year-old chestnut trees (*Castanea crenata*) were transplanted with either 1400 g macrophyte compost or 150 g commercial fertilizer/m^2 (Osmocote Plus, 13 N + 13 P + 13 K + 2 MgO; Scotts International B.V., Geldermalsen, The Netherlands). The application rates were designed to supply 20 g nitrogen/m^2, which is the optimal fertilizing rate for the landfill slope (Kim 2001). Including the control, four treatments were established and monitored. In summer (August), photosynthetic rates of chestnut trees were analyzed using a portable photosynthesis measurement system (Li-6400, Li-cor Biosciences, USA) to check the condition of the plants. The leaves of the second highest branches were selected for measurement. Every leaves of trees were harvested in the fall and analyzed for nutrition. We collected core samples of soil (5 cm length, 100 cm^3; Eijkelkamp, Giesbeek, The Netherlands). A composite of three cores was made for each sample, and we assessed three samples (nine cores). Nitrogen content of soil and plant leaves was measured using an elemental analyzer (Flash EA 1112, Thermo Electron Co., USA). Heavy metal content of compost and plants was analyzed by ICP emission spectrometry (ICPS-1000IV; Shimadzu, Japan) after acid digestion of samples (Song et al. 2013b).

Statistical analysis
Differences between paired groups were evaluated using either Student's t test for normally distributed variables or the Mann–Whitney U test when normality assumptions were violated (SAS v. 9.1, SAS Institute Inc., USA). For multiple comparisons, data were analyzed using one-way PROC ANOVA. When a significant treatment effect was detected, post hoc comparisons of the means were performed using Tukey's honest significant difference (HSD) test (SAS v. 9.1, SAS Institute Inc., USA). Statistical significance was inferred when $p < 0.05$.

Results and discussion
Research and case studies relating to phytoremediation
There is an extensive literature on phytoremediation. The RISS and Google Scholar retrieved many published case studies, including the use of artificial wetlands and floating islands for remediation. For South Korea alone, the literature search uncovered >100 relevant examples. However, for artificial wetlands, most case studies address their use for educational purposes or for park construction, and research relevant to remediation is more limited. Examples in

South Korea are the Nanji leachate retention wetland in Seoul (Song and Kang 2006), the artificial wetland in the Sihwa development area (No et al. 2002), the Masan wetland in Asan city (Lee et al. 1999), the Jangja stream in Guri city (Lee 2005), the Juam lake in Boseong county (Choi et al. 2012), the artificial wetland in Nonsan city (Kim et al. 2012), and the floating island in Paldang lake, Gangwon province (Choi et al. 2007). Globally, there are too many cases to summarize here but phytoremediation is in wide use because of its cost-effectiveness and efficiency (Salt et al. 1995) for water, soil, and air. In addition, constructed wetlands are employed not only for the pollutant-accumulation capacity of their plants but in holistic systems (Cheng et al. 2002) that maximize provision of ecosystem services. Many studies also report increased remediation capacity through the use of plant-microorganism interactions (Barac et al. 2004) or transgenic plants (Eapen and D'Souza 2005). These examples show the continuing development of techniques in phytoremediation research. For South Korea, Table 1 summarizes published studies reporting pollutant contents of plants (by tissue accumulation) after remediation. Data show that plants after remediation contain high levels of N and/or P and have high biomass. For macrophyte species in particular, tissue decomposition can affect water quality (Asaeda et al. 2000) and reverse the remediation process. Macrophyte tissue decomposes by a factor of >80% within 200 days, returning nutrients to the water (Chimney and Pietro 2006). Furthermore, plant species accumulating heavy metals or toxic wastes would also cause secondary pollution of the remediation site after tissue decomposition. With respect to these problems, we did not locate any South Korean studies involving post-remediation management of plant biomass except research papers written by authors of this manuscript.

Field experiment into feasibility of post-remediation management

The remediation macrophytes in the Sudokwon landfill leachate channel had high biomass and high N content (Table 2). As this is a sanitary landfill, macrophyte metal contents do not exceed the national standards set for agricultural waste (Cd 5.0 mg/kg, Cr 70 mg/kg, Pb 100 mg/kg, Zn 300 mg/kg) (Korea Ministry of Environment KME 2010). However, pollutants released by the decomposition of plant tissues will eventually re-pollute the channel. As the metal contents are within the permitted national standards, the material can be reclaimed as waste. However, the process thereby creates another form of waste and also results in unused biomass. Reuse of biomass will reduce waste production and also save the cost of reclamation. As all metal values in Table 2 are less than 10% of the standards set for sludge recycled compost (Korea Environmental Institute KEI 2003), it will be more efficient to recycle biomass as compost. Otherwise, decomposition of plant tissue will re-contaminate the channel, reducing sustainability of the remediation system. Therefore, we tested a composting method to manage plants after remediation. This method will require harvesting of macrophytes in the channel, thus preventing secondary pollution.

During composting, the temperature of the compost increased and showed 30 °C higher temperature (36 °C) than the surrounding environment (6 °C) after 1 month, before falling to ambient temperature after 2 months. The final product, even after composting with sewage sludge, the macrophyte compost still had less than 10% of the national standards set for sludge recycled compost (Korea Environmental Institute KEI 2003), except Zn contents which does not have standard values (Table 3).

Table 4 shows that soil characteristics were improved more by compost treatment than by commercial fertilizer. As the landfill has very poor soil with low organic matter, low nutrient contents, and semi-arid

Table 1 Location, species, accumulation rate, and biomass of plants after the remediation process

Location	Species	Accumulation rate (mg/g)	Per area biomass (g/m^2)	References
Lake Sihwa	Rees	N 24.9, P 2.5	2515 (DW)	(No et al. 2002)
Masan wetland	Reed, cattail	Reed (cattail)—N 12 (11), P 0.9 (1.7)	Reed 532 cattail 1000	(Lee et al. 1999)
Artificial wetland	Water rice	N 26.9, P 6.8	4032 (DW)	(Kim and Ihm 1998)
Lake Juam	Water-fringe, water-lily	N about* 20 P about 4.0	NA	(Choi et al. 2012)
Artificial wetland	Reed	N 32.1, P 2.4	8100 (FW)	(Kim et al. 2012)
Lake Paldang	Reed, cattail	N 30, P 2.5	NA	(Choi et al. 2007)
Indoor	Duckweed	N 12, P 40.9 Cd 0.2 Cu 0.9	NA	(You 2016)
Indoor	Yellow flag-iris	Cd 0.2	NA	(Lee et al. 2009)

Scientific names of plant species: water rice, *Zizania latifolia*; water-fringe, *Nymphoides peltata*; water-lily, *Nymphaea tetragona*; duckweed, *Spirodela polyrhiza*; flag-iris, *Iris pseudacorus*
*Estimated value from graphs

Table 2 Biomass, nutrient, and metal contents of macrophytes in the leachate channel

Species	Biomass per shoot (g)	N (%)	C (%)	Cd	Cr	Fe	Pb	Zn
T. angustifolia	258	3.53	42.1	0.12	8.1	1021	4.4	184
P. australis	177	2.77	43.1	0.11	6.8	954	3.9	192

The data presented are means of ten replicates for biomass and four replicates for other variables
Metal content is expressed as milligrams per kilogram

condition (Kim 2001, Song 2010), the application of compost with high organic matter and nitrogen contents had huge effect. Also, commercial fertilizer significantly improved soil nitrogen contents but not soil moisture. The N organic matter and moisture contents of soil were all higher than in the Osmocote treatment, as organic matters of compost significantly increased moisture-holding capacity of reclaimed soil (Song and Lee 2010b). As the landfill slope is a semi-arid environment (Song 2010), increased organic matter and moisture content is very important. Also, as compost usually contains more available form of N (NH_4^+ and NO_3^-) by decomposition activities of microorganisms, even supplied with same rate of TN, N sources of compost could be more useful to plants.

The observed improvement was closely related to plant performance. Chestnut trees treated with macrophyte compost showed higher photosynthetic performance (Fig. 1) indicative of improved health (Song and Lee 2010a). Table 5 also shows that N content of leaves from plants treated with compost were higher than control values, indicating that compost-treated plants had more access to nutrients. Nutrient availability is a major obstacle to re-vegetating landfill slopes (Song and Lee 2010b). Therefore, results show that plants after remediation had excellent potential for use as composting materials. The good results for both P. australis and T. angustifolia imply that the effects are not species specific. The very hard stem of P. australis is a partial disadvantage for composting (Song 2010). However, the good results for this species suggest that remediation plants with softer stems and leaves may also be effective. Overall, results show that it is possible to re-use remediation plant biomass, thereby reducing secondary pollution and enhancing production of cost-effective and environmentally friendly fertilizer. Especially when phytoremediation is targeting N and P, the resultant biomass will produce nutrient-rich fertilizer. The post-remediation management and use of

biomass therefore justifies further research and policy development.

Research and policy relating to post-remediation management

Despite the reported limitations and problems of phytoremediation, effective techniques exist for managing biomass after remediation. Simple harvesting will reduce re-entry of pollutants by up to 75% (Asaeda et al. 2000). Generation of energy by incineration (Brooks et al. 1998) or by production of bio-ethanol has also been suggested (Banuelos 2006). Incineration has been considered the most realistic method (Sas-Nowosielska et al. 2004), although its waste products may pose a problem (Witters et al. 2012). Research into post-remediation techniques is still very limited (Song et al. 2016), and more studies such as ours are needed. Although we focused on phytoremediation, any phytoremediation method will produce biomass requiring post-remediation management. Furthermore, as harvesting remediation may be costly (Song et al. 2016), policies promoting effective management are also required.

Measures for the effective utilization of post-remediation biomass could be implemented in two stages. The first stage is to pre-emptively prohibit the use in compost of biomass containing toxic residues. In South Korea, one regulation relevant to the management of post-remediation biomass can be found in the "Wastes Control Act" (Act Number 13038). Article 13 of the act ordains, "Where the Minister of Environment deems that any products or materials, the hazard criteria of […] which are manufactured by recycling wastes require a certain control, he/she may enter into an agreement with the head of the local government concerned, the manufacturer of the said products or materials, etc. that requires them to disclose the use and quantities of

Table 3 Nitrogen and metal contents of macrophyte compost before application

Compost type	N (%)	Cd	Cr	Fe	Pb	Zn
T. angustifolia compost	3.53	2.0	4.8	1779	3.9	1740
P. australis compost	2.77	1.8	5.2	1521	3.7	1720

The data presented are means of three replicates
Metal content is expressed as milligrams per kilogram

Table 4 Soil characteristics 3 months after treatment

Items	Moisture (%)	OM (%)	N (%)	C (%)
Control	13.03 ± 0.23^b	0.52 ± 0.01^b	0.02 ± 0.01^c	0.19 ± 0.03^b
Osmocote	12.83 ± 0.33^b	0.54 ± 0.02^b	0.04 ± 0.00^b	0.47 ± 0.04^a
P compost	17.60 ± 0.49^a	0.78 ± 0.03^a	0.05 ± 0.00^{ab}	0.49 ± 0.00^a
T compost	17.37 ± 0.52^a	0.87 ± 0.03^a	0.06 ± 0.00^a	0.62 ± 0.07^a

The data presented are means of three replicates (mean ± SE). Means within a column followed by the same superscript letter are not significantly different at $p = 0.05$
OM organic matter, P Phragmites australis, T Typha angustifolia

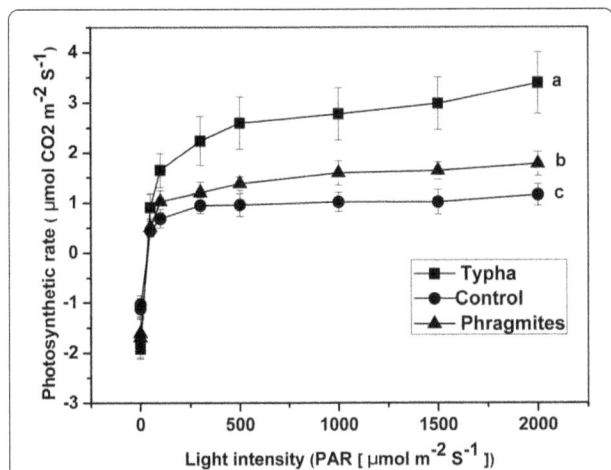

Fig. 1 Photosynthetic rates of chestnut tree (30 °C, 400 ppm CO_2) by treatments. *Symbols* and *bars* represent mean ± SE of four replicates. *Symbols having the same letter* are not significantly different at the 0.05 level. *Typha T. angustifolia, Phragmites P. australis*

each type of wastes, the heavy metal contents of such wastes, and other information". At least until a new law is enacted to regulate the safe and efficient use of post-remediation biomass, local governments can utilize the Wastes Control Act by their own local ordinances. The ordinances may have to classify the biomass as a form of recycled material. In other countries also, the use of post-remediation biomass can be governed by currently existing legal instruments for municipal solid or bio-degradable waste. For example, the USA's "Solid Waste Disposal Act" (Public law 89-272) and its amendments can direct how the post-remediation biomass is used depending on the concentration levels of toxic elements. Similarly, the European Union's "Urban Waste Water Treatment Directive" (91/271/EEC) and "Sewage Sludge Directive" (86/278/EEC) can be quoted to control toxic elements in post-remediation biomass. However, we did not find any acts or laws specifically concerned with post-remediation biomass. Therefore, to improve the management of biomass (or, at least, to not leave post-remediation biomass as it is), additional legislation should be considered.

The second stage of policy implementation is to en-force or promote the use of post-remediation biomass

that is proven safe by the first-stage policy measures. In South Korea, the enforcement decree (Ministry of Environment, No. 2011-64) for the "Wastes Control Act" has a special section titled "Regulations on recycling organic sludge etc. as soil modifier or landfill cover soil." The government may modify the decree and promote the use of composted post-remediation biomass. In the USA, the "Federal Water Pollution Control Act" (Public law 92-500) and its amendments can be consulted for biomass composting. The European Union has a specific law titled "Waste Framework Directive" (2008/98/EC) that requires composting of bio-waste, and which can also be applied to post-remediation biomass. However, these laws are only related to the re-use of post-remediation biomass and are not framed specifically to promote post-remediation management. Therefore, to protect the environment and achieve sustainability, there is a need for the enactment of a "Post-phytoremediation management" or even "Post-bioremediation management" law that sets out the government's obligations to (a) manage biomass after phytoremediation to prevent secondary pollution, (b) ensure that toxic pollutants are not used for recycling, and (c) promote recycling of biomass with environmentally friendly methods.

Conclusions

Post-remediation biomass must be managed; otherwise, it will eventually return to the environment and cause secondary pollution. Research into the post-remediation management of biomass is therefore required. Using biomass as a co-composting material could be one of the best solutions because it is cost-effective and environmentally friendly. In this research, we proved that post-remediation management of biomass is effective by actual experiment, and with this scientific basis, we became more confident of requirements for such policies.

However, despite of its potential, research and policy development to promote this form of management is lacking. There is a need for enactment of a post-phytoremediation management law that stipulates management of biomass after phytoremediation and promotes recycling (using environmentally friendly methods) of biomass known to pose an environmental risk. Specific policies for managers must be developed, and additional research should be carried out to support and inform such policies and laws.

Table 5 Nutrient contents of chestnut leaves

	N (%)	C (%)
Control	2.81 ± 0.04^b	46.48 ± 0.41
P compost	3.16 ± 0.05^a	46.62 ± 0.23
T compost	3.12 ± 0.04^a	46.62 ± 0.23

The data presented are means of three replicates (mean ± SE). Means within a column followed by the same superscript letter are not significantly different at $p = 0.05$
P Phragmites australis, T Typha angustifolia

Acknowledgements
This work was supported by the National Research Foundation of Korea Grant funded by the Korean Government (NRF-2015S1A5B8046155).

Funding
This work was supported by the National Research Foundation of Korea Grant funded by the Korean Government (NRF-2015S1A5B8046155).

Authors' contributions
SU conducted the experiment and partially wrote the manuscript. HP designed the experiment and partially wrote the manuscript. Both authors read and approved the final manuscript.

Competing interests
The authors declare that they have no competing interests.

Author details
[1]Department of Biology and Research Institute for Basic Sciences, Jeju National University, Jeju 63243, South Korea. [2]Institute for Legal Studies, Yonsei University, Seoul 03722, South Korea.

References

Abma, W. R., Driessen, W., Haarhuis, R., & van Loosdrecht, M. C. M. (2010). Upgrading of sewage treatment plant by sustainable and cost-effective separate treatment of industrial wastewater. *Water Science and Technology, 61,* 1715–1722.

Asaeda, T., Trung, V. K., & Manatunge, J. (2000). Modeling the effects of macrophyte growth and decomposition on the nutrient budget in shallow lakes. *Aquatic Botany, 68,* 217–237.

Banuelos, G. S. (2006). Phyto-products may be essential for sustainability and implementation of phytoremediation. *Environmental Pollution, 144,* 19–23.

Barac, T., Taghavi, S., Borremans, B., Provoost, A., Oeyen, L., Colpaert, J. V., Vangronsveld, J., & van der Lelie, D. (2004). Engineered endophytic bacteria improve phytoremediation of water-soluble, volatile, organic pollutants. *Nature Biotechnology, 22,* 583–588.

Brooks, R. R., Chambers, M. F., Nicks, L. J., & Robinson, B. H. (1998). Phytomining. *Trends in Plant Science, 3,* 359–362.

Cheng, S., Grosse, W., Karrenbrock, F., & Thoennessen, M. (2002). Efficiency of constructed wetlands in decontamination of water polluted by heavy metals. *Ecological Engineering, 18,* 317–325.

Chimney, M. J., & Pietro, K. C. (2006). Decomposition of macrophyte litter in a subtropical constructed wetland in south Florida (USA). *Ecological Engineering, 27,* 301–321.

Choi, I.-W., Seo, D.-C., Kang, S.-W., Lee, S.-G., Seo, Y.-J., Lim, B.-J., Park, J.-H., Kim, K.-S., Heo, J.-S., & Cho, J.-S. (2012). Evaluation of treatment efficiencies of pollutants in Juksancheon constructed wetlands for treating non-point source pollution. *Korean Journal of Soil Science and Fertilizer, 45,* 642–648.

Choi, M J, Byeon, M S, Park, H K, Jeon, N H, Yoon, S H, Kong, D S (2007). The growth anti nutrient removal efficiency of hydrophytes at an artificial vegetation island, Lake Paldang J Kor Soc Water Environ, 23:348–355

Eapen, S., & D'Souza, S. F. (2005). Prospects of genetic engineering of plants for phytoremediation of toxic metals. *Biotechnology Advances, 23,* 97–114.

EM-center (2003). Teaching material of environmental agriculture. Effective microorganisms Center of Korea: Jeju Island, Korea

Fendall, L. S., & Sewell, M. A. (2009). Contributing to marine pollution by washing your face: microplastics in facial cleansers. *Marine Pollution Bulletin, 58,* 1225–1228.

Helfield, J. M., & Diamond, M. L. (1997). Use of constructed wetlands for urban stream restoration: a critical analysis. *Environmental Management, 21,* 329–341.

Justin, M. Z., & Zupancic, M. (2009). Combined purification and reuse of landfill leachate by constructed wetland and irrigation of grass and willows. *Desalination, 246,* 157–168.

Korea Environmental Institute (KEI) (2003). Management protocols for sewage sludge. KEI, Incheon

Kim, H.-S., & Ihm, B.-S. (1998). Studies on the application plan and purification capacity of the hydrophytes for improvement of water quality of effluent from agricultural land. *Korean Journal of Environmental Management, 4,* 1–8.

Kim, H. H., & Yun, Y. H. (2012). Analysis of patents artificial floating island for maximizing the development of water purification. *Journal of Environmental Science International, 21,* 825–835.

Kim, K D (2001). Vegetation structure and ecological restoration of the waste landfills in Seoul metropolitan area. Ph.D thesis of Seoul National University.

Kim, K. J., Kim, J. S., Kim, Y. H., & Yang, G. C. (2012). Characteristics of nutrient uptake by aquatic plant in constructed wetlands for treating livestock wastewate. *Journal of Wetlands Research, 14,* 121–130.

Korea Ministry of Environment (KME) (2010). Waste management act for bio solids.Available at: https://www.srf-info.or.kr/srfEneIntro/srfEneIntroR.do

Lee, G. S., Jang, J. R., Kim, Y. K., & Park, B. H. (1999). A study on the floating island for water quality improvement of a reservoir. *Korean Journal of Environmental Agriculture, 18,* 77–82.

Lee, J. S. (2005). A study on the water quality purification effect of aquatic plants in field work. *Journal of Environmental Science International, 14,* 937–944.

Lee, S. C., Lee, S. Y., Jung, B. Y., Jo, Y. S., Kim, G. S., Kim, W. S., & Lee, J. S. (2009). Reduction of cadmium (Cd) in water using yellow flag (Iris pseudacorus L.) and sweet flag (Acorus calamus L.). *Journal of the Korean Society for Horticultural Science, 5,* 169–169.

Lu, Y., Song, S., Wang, R., Liu, Z., Meng, J., Sweetman, A. J., Jenkins, A., Ferrier, R. C., Li, H., Luo, W., & Wang, T. (2015). Impacts of soil and water pollution on food safety and health risks in China. *Environment International, 77,* 5–15.

Mohan, S., & Gandhimathi, R. (2009). Removal of heavy metal ions from municipal solid waste leachate using coal fly ash as an adsorbent. *Journal of Hazardous Materials, 169,* 351–359.

No, H. M., Choi, W. J., Lee, E. J., Youn, S. I., & Choi, Y. D. (2002). Uptake patterns of N and P by reeds (Phragmites australis) of newly constructed Shihwa tidal freshwater marshes. *Journal of Ecology and Environment, 25,* 359–364.

Paerl, H. W., & Whitall, D. R. (1999). Anthropogenically-derived atmospheric nitrogen deposition, marine eutrophication and harmful algal bloom expansion. *Ambio, 28,* 307–311.

Salt, D. E., Blaylock, M., Kumar, N. P., Dushenkov, V., Ensley, B. D., Chet, I., & Raskin, I. (1995). Phytoremediation: a novel strategy for the removal of toxic metals from the environment using plants. *Nature Biotechnology, 13,* 468–474.

Sas-Nowosielska, A., Kucharski, R., Małkowski, E., Pogrzeba, M., Kuperberg, J. M., & Kryński, K. (2004). Phytoextraction crop disposal—an unsolved problem. *Environmental Pollution, 128,* 373–379.

Smith, L. E. D, & Siciliano, G. (2015). A comprehensive review of constraints to improved management of fertilizers in China and mitigation of diffuse water pollution from agriculture. *Agriculture, Ecosystems and Environment, 209,* 15–25.

Song, G. Y., & Kang, H. J. (2006). Application of constructed wetland for water purification. *Korean Journal of Geotechnical Engineering, 7,* 6–10.

Song, U (2010). Ecological monitoring and management of plant, soil and leachate channel in the Sudokwon landfill, Korea. Ph.D thesis of Seoul National University:pp. 95–111.

Song, U., Kim, D. W., Waldman, B., & Lee, E. J. (2016). From phytoaccumulation to post-harvest use of water fern for landfill management. *Journal of Environmental Management, 182,* 13–20.

Song, U, Lee, E (2010a). Ecophysiological responses of plants after sewage sludge compost applications. Journal of Plant Biology, 53:259–267

Song, U, Lee, E J (2010b). Environmental and economical assessment of sewage sludge compost application on soil and plants in a landfill. Resources, Conservation and Recycling 54:1109–1116

Song, U, Shin, M, Lee, G, Roh, J, Kim, Y, Lee, E (2013a). Functional analysis of TiO_2 nanoparticle toxicity in three plant species. Biological Trace Element Research, 155:93–103

Song, U, Waldman, B, Lee, E (2013b). Ameliorating topsoil conditions by biosolid application for a waste landfill landscape. International Journal of Environmental Research, 7:1–10

Witters, N., Mendelsohn, R. O., Van Slycken, S., Weyens, N., Schreurs, E., Meers, E., Tack, F., Carleer, R., & Vangronsveld, J. (2012). Phytoremediation, a sustainable remediation technology? Conclusions from a case study. I: Energy production and carbon dioxide abatement. *Biomass and Bioenergy, 39,* 454–469.

You, G H (2016). Evaluation of removal efficiencies of water pollutants by duckweed, MS Thesis of Chungbuk National University

13

Hydrodynamic fish modeling for potential-expansion evaluations of exotic species (largemouth bass) on waterway tunnel of Andong-Imha Reservoir

Ji-Woong Choi and Kwang-Guk An[*]

Abstract

Background: The objectives of this study were to establish a swimming capability model for largemouth bass using the FishXing (version 3) program, and to determine the swimming speed and feasibility of fish passage through a waterway tunnel. This modeling aimed to replicate the waterway tunnel connecting the Andong and Imha Reservoirs in South Korea, where there is a concern that largemouth bass may be able to pass through this structure. As largemouth bass are considered an invasive species, this spread could have repercussions for the local environment.

Results: Flow regime of water through the waterway tunnel was calculated via the simulation of waterway tunnel operation, and the capability of largemouth bass to pass through the waterway tunnel was then estimated. The swimming speed and distance of the largemouth bass had a positive linear function with total length and negative linear function with the flow rate of the waterway tunnel. The passing rate of small-size largemouth bass (10–30 cm) was 0% at a flow of 10 m^3/s due to rapid exhaustion from prolonged upstream swimming through the long (1.952 km) waterway tunnel.

Conclusions: The results of FishXing showed that the potential passing rate of large size largemouth bass (>40 cm) through the waterway tunnel was greater than 10%; however, the passage of largemouth bass was not possible because of the mesh size (3.4 × 6.0 cm) of the pre-screening structures at the entrance of the waterway tunnel. Overall, this study suggests that the spread of largemouth bass population in the Imha Reservoir through the waterway tunnel is most likely impossible.

Keywords: Invasive species, Largemouth bass (*Micropterus salmoides*), Waterway tunnel, Fish swimming capability, Inflow rate

Background

Largemouth bass (*Micropterus salmoides*) were introduced into South Korea in 1973 from Louisiana, USA, with the intent of providing a sustainable food source. However, their population has rapidly increased in the ecosystems of rivers and reservoirs across the country and consequently has exerted great influence on freshwater food webs (Lee et al. 2008; Kim et al. 2013). Largemouth bass prey upon native fishes, amphibians, crustaceans, and other organisms, and this high predation pressure, coupled with the absence of any known natural enemies, has enabled largemouth bass to occupy the position of top predator in freshwater environments. Furthermore, largemouth bass inhabit large reservoirs and rivers in high densities because they can spawn up to 100,000 eggs and the males protect their eggs and young (Wheeler and Allen 2003; Almeida et al. 2012). Because of the threat to food webs posed by the feeding habits of largemouth bass, their rapidly growing population, and the lack of natural enemies, in 1998 the Ministry of Environment designated largemouth bass as an invasive alien species that disrupts local ecosystems.

Cases of ecological disturbances due to high predation pressure and rapid population growth of largemouth bass have been reported in many countries across the world. In Guatemala, the majority of indigenous fish species were reported to have disappeared from Atitlan

* Correspondence: kgan@cnu.ac.kr
Department of Biological Science, College of Biosciences and Biotechnology, Chungnam National University, Daejeon 305-764, South Korea

Reservoir after the introduction of largemouth bass (Zaret and Paine 1973); in South Africa, three indigenous species were reported to have been extirpated by largemouth bass (Hickley et al. 1994); and Cuba also witnessed a considerable decrease in the number of individuals of indigenous fish species after the introduction of largemouth bass, which increased the number of malaria mosquitoes and consequently raised malaria infection rates among residents (Lasenby and Kerr 2000). Furthermore, such ecological disturbances caused by largemouth bass have been reported at a global level, including Canada, Japan, and Europe (Lasenby and Kerr 2000; Yasunori and Tadashi 2003; Wasserman et al. 2011; Almeida et al. 2012). In fact, largemouth bass were originally endemic only to the eastern half of the United States, and their spread to other regions within the country has also resulted in ecological disturbances (Findlay et al. 2000; Brown et al. 2009). The detrimental spread of largemouth bass directly affects freshwater food webs, thereby reducing species richness and biodiversity, and their potential for damaging the integrity of aquatic ecosystems has led to widespread discussion (and disagreement) with regard to the best methods to manage this ecological challenge.

In general, categories of fish swimming are classified into three types: sustained swimming, prolonged swimming, and burst swimming (Beamish 1978). Sustained swimming is defined as having the capability of swimming over 200 min without muscle fatigue, an adequate energy supply for metabolism, and the ability to excrete metabolic wastes before they accumulate. Prolonged swimming refers to swimming for a time ranging between 20 s and 200 min with metabolic waste accumulating in the muscles. In this type of swimming, muscle fibers turn pink (fast oxidative glycolytic when aerobic) or white (fast glycolytic when anaerobic) due to low energy supply. In particular, white muscle fibers have a high-energy efficiency but rapidly fatigue after all energy is consumed (Webb 1975). Burst swimming involves swimming for a short length of time (≤20 s) when maximum speed is required. After burst swimming under excessive energy consumption, a large amount of intracellular waste accumulates (Colavecchia et al. 1998). Most fish species perform burst swimming when entering or exiting a waterway tunnel by stiffening the caudal fin and giving strong horizontal thrusts to achieve the greatest propulsive force (Nursall 1962). Burst swimming requires a long recovery time, which varies depending on the fish species, ranging from several hours to several days, and can even be fatal to some fish (Black et al. 1962). For example, when trout are forced to swim intensively for 6 min, 40% die, and almost 100% die after additional forced swimming of 4–6 h (Wood et al. 1983). Largemouth bass have a carangiform body that minimizes friction with water, enabling them to swim rapidly. They can swim most efficiently in water temperatures of 10–25 °C, which corresponds to the water temperature of

summer season in South Korea, and their swimming capability is rapidly reduced at temperatures ≥30 or ≤10 °C (Beamish 1978, Hammer 1995; Brown et al. 2009).

In South Korea, over two thirds of the annual precipitation occurs during the summer monsoon period and because of these concentrated rainfalls, a large portion of national water resources flows into the ocean without being efficiently processed through stable water resource management (Ahn and Kim 2010; Yum 2010). To ensure secure water resource management and address fundamental water problems such as floods and droughts, growing attention has been paid to the integrated management of neighboring dams (Lee 2005; Kang et al. 2007). As part of such nationwide projects, the Andong-Imha Integrated Project was implemented in the Andong and Imha Reservoirs. The purpose of the Andong-Imha Integrated Project was to connect the Andong and Imha Reservoirs with a waterway tunnel to secure additional water resources by reducing spillway discharge during flood events and improve downstream water quality. However, this waterway tunnel could also be used as a passage by largemouth bass, an invasive species threatening native species and disrupting local ecosystems, thereby accelerating their spread. According to the study *Monitoring of Invasive Alien Species Designated by the Wildlife Protection Act* by the National Institute of Environmental Research (Ministry of Environment, Korea 2013), the Andong Reservoir has a large population of largemouth bass and the Imha Reservoir has not been invaded by this species. If these two reservoirs are connected, there is a concern that largemouth bass could be introduced into the Imha Reservoir, which would result in a disturbance of its aquatic ecosystems.

This study was conducted to establish a swimming capacity model of largemouth bass using the computer program FishXing (Love et al. 1999; Furniss et al. 2006), which was developed to assess the potential for fish passage through an artificial structure depending on water flow rate and velocity, and to calculate the swimming speed of largemouth bass based on body length and inflow rate into the waterway tunnel. Additionally, swimming distance was calculated using these swimming speeds to determine the feasibility of largemouth bass passing through the waterway tunnel in order to derive the optimal waterway tunnel management strategy. The rate of water flow at the inlet that would inhibit their entrance into the waterway tunnel was also determined, with the overall goal of finding ways to inhibit the spread of largemouth bass and conserve local ecosystems.

Methods
Overview of the survey sites and waterway tunnel
The Andong Reservoir study site is located upstream at the northernmost extent of the Nakdong River, approximately 340 km from its estuary. The Nakdong River

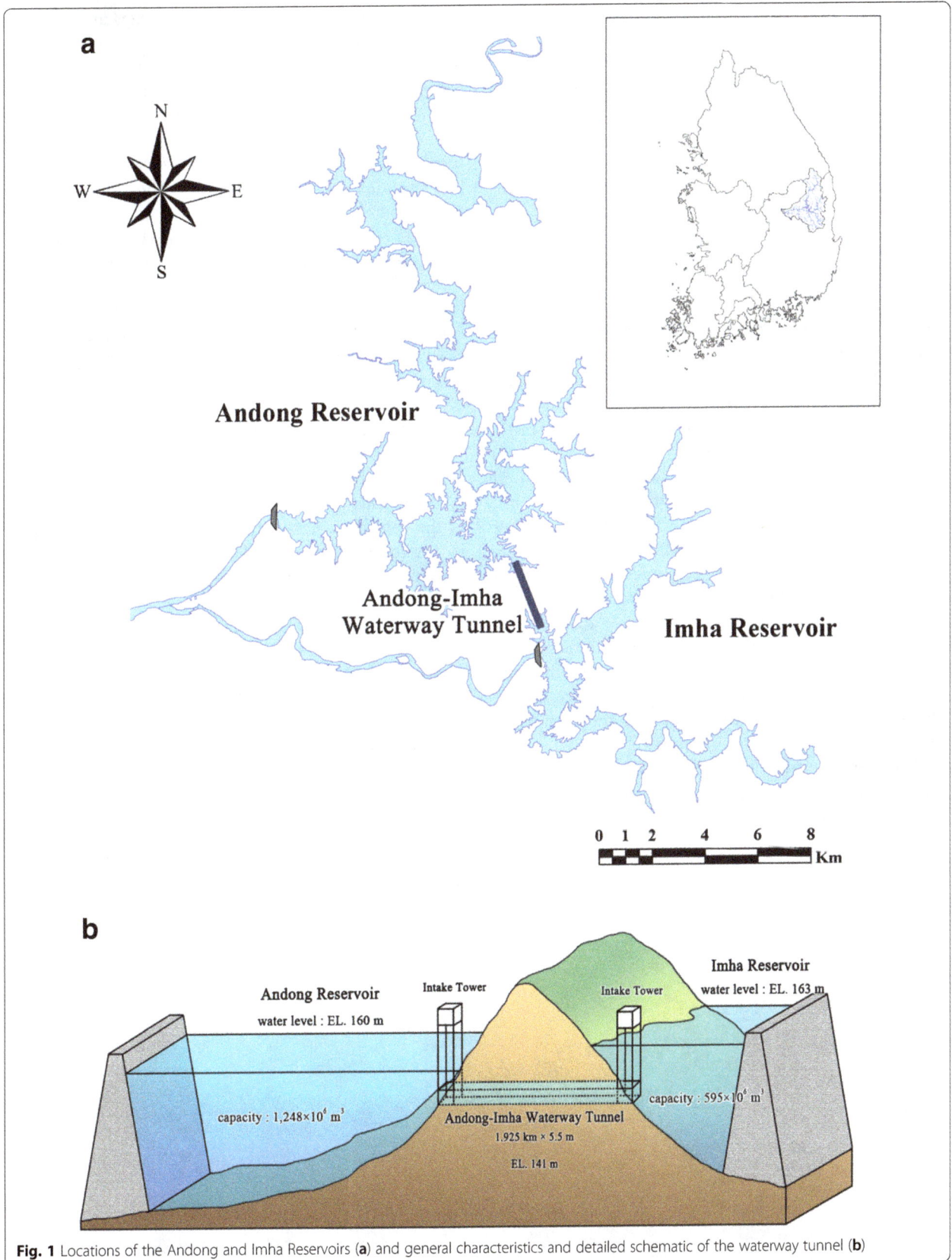

Fig. 1 Locations of the Andong and Imha Reservoirs (**a**) and general characteristics and detailed schematic of the waterway tunnel (**b**)

Table 1 Empirical model (regression) for the swimming speed of largemouth bass (B_{SS}, m/s) using water temperature and body length (T_L, cm) (Beamish, 1970)

Habitat water temperature (°C)	Largemouth bass swimming speed (B_{SS}, m/s)	Period of study sites
10	Log B_{SS} = 0.9342 + 0.0303 T_L	April/December
15	Log B_{SS} = 1.2068 + 0.0210 T_L	May/November
20	Log B_{SS} = 1.4465 + 0.0137 T_L	June–July/October
25	Log B_{SS} = 1.5023 + 0.0117 T_L	August–September
30	Log B_{SS} = 1.5008 + 0.0120 T_L	Expected water temperature
35	Log B_{SS} = 1.3968 + 0.0139 T_L	

plays a vital role as a source of household and industrial water for major industrial cities located in the southern part of the East Sea, including Busan, Pohang, and Ulsan. The Andong Reservoir occupies 6.6% of the total Nakdong drainage area and has maximum and mean water depths of 60 and 19.4 m, respectively, a total drainage area of 1584 km², full-water surface area of 51.5 km², and reservoir capacity of 1,248,000,000 m³. The Imha Reservoir is a multipurpose artificial reservoir constructed next to the Andong Reservoir to facilitate efficient development of water resources, reduce flood-induced damages in downstream areas, improve water quality, and provide an alternative source of water for the industries in the mid- and downstream areas of the Nakdong River. The reservoir occupies 5.7% of the total Nakdong drainage area and has a total drainage area of 1361 km², full-water surface area of 26.4 km², and reservoir capacity of 595,000,000 m³ (Fig. 1).

While the two reservoirs have drainage areas of similar size, the water storage capacity of the Imha Reservoir is less than half that of the Andong Reservoir. In other words, the low ratio of water storage capacity to drainage area of the Imha Reservoir makes it vulnerable to water release through spillways due to rapidly rising water levels whenever its drainage area is inundated. The Andong-Imha waterway tunnel was installed to connect the two reservoirs in order to ensure stable procurement of water resources and their efficient management by preventing spillway losses (Fig. 1). The waterway tunnel is a round tunnel with a total length of 1.925 km, inner diameter of 5.5 m, and invert elevation (EL) of 141 m. The structure procures additional water resources by inducing the water released from the Imha Reservoir to flow into the Andong Reservoir. Also, pre-screening structure (mesh size 3.4 × 6.0 cm) at the entrance of the waterway tunnel was installed to block waste and local fish over 40 cm into the waterway tunnel.

Calculation of the water flow through the waterway tunnel

Water flow from one reservoir to the other through the Andong-Imha waterway tunnel depends on the water level of the two reservoirs, with water flowing from the higher to the lower water level. The quantity of water flowing through a waterway tunnel widely varies depending on the physical and hydrological characteristics of the waterway tunnel, including length, head loss, difference of water head, and flow rate. For calculating the inflow rate, surface water levels of both reservoirs are taken into consideration, because water level determines whether the water flowing within the waterway tunnel has the characteristics of pipe flow or open channel flow. If the reservoir water level is higher than the invert elevation of the waterway tunnel (i.e., in the case of pipe flow), then the Darcy–Weisbach formula can be applied as follows:

Table 2 Inflow rate (m³/s) into the waterway tunnel depending on water head difference and inner diameter of the waterway tunnel

Water head difference (m)	Inner diameter of the waterway tunnel									
	1 m	2 m	3 m	4 m	5 m	6 m	7 m	8 m	9 m	10 m
1.0	15.1	23.4	29.9	35.3	39.8	43.7	47.0	50.0	52.5	54.8
2.0	21.4	33.1	42.3	49.9	56.3	61.7	66.5	70.6	74.3	77.6
3.0	26.2	40.6	51.8	61.1	68.9	75.6	81.4	86.5	91.0	95.0
4.0	30.2	46.9	59.9	70.5	79.6	87.3	94.0	99.9	105.1	109.7
5.0	33.8	52.4	66.9	78.9	89.0	97.6	105.1	111.7	117.5	122.6
6.0	37.0	57.4	73.3	86.4	97.4	106.9	115.2	122.4	128.7	134.3
7.0	40.0	62.0	79.2	93.3	105.2	115.5	124.4	132.2	139.0	145.1
8.0	42.7	66.3	84.6	99.8	112.5	123.5	133.0	141.3	148.6	155.1
9.0	45.3	70.3	89.8	105.8	119.3	131.0	141.0	149.9	157.6	164.5
10.0	47.8	74.1	94.6	111.5	125.8	138.0	148.7	158.0	166.2	173.4
15.0	58.5	90.8	115.9	136.6	154.1	169.1	182.1	193.5	203.5	212.4
20.0	67.6	104.8	133.8	157.7	177.9	195.2	210.2	223.4	235.0	245.3

Table 3 Number of days and quantity of water flow through the waterway tunnel estimated via simulation

	Number of days of water flow			Quantity of water flow ($\times 10^6$ m³)		
	Total	Flood period	Dry period	Total	Flood period	Dry period
Flow through the waterway tunnel	9349 (85.3%)	2351	6998	8214 (100.0%)	4152	4062
Transferring from I_R to A_R	7818 (71.3%)	2016	5802	7552 (91.9%)	3859	3693
Transferring from A_R to I_R	1531 (14.0%)	335	1196	661 (8.1%)	293	368
Non-flow through the waterway tunnel	1609 (14.7%)	–	–	–	–	–

Inner diameter of the waterway tunnel: 5.5 m; invert elevation: EL 141 m
A_R Andong Reservoir, I_R Imha Reservoir

$$H_f = f \times \frac{L}{D} \times \frac{V^2}{2g} \qquad (1)$$

where H_f is head loss (m); f is friction factor; L is length of pipe work (m); D is inner diameter of pipe work (m); V is average velocity of water flow within the waterway tunnel (m/s); and g is acceleration due to gravity (m/s²).

In the case of open channel flow in which the reservoir water level is lower than the invert elevation, the Manning formula was applied as follows:

$$V = \frac{1}{n} R_h^{2/3} S^{1/2} \qquad (2)$$

$$Q = A \times V = (\theta - \sin\theta) \frac{D^2}{8} \times \frac{1}{n} R_h^{2/3} S^{1/2} \qquad (3)$$

where V is average velocity of water flow within the waterway tunnel (m/s); R_h is hydraulic radius (m); S is energy slope (m/m); Q is inflow rate (m³/s); A is cross-sectional area of pipe work (m²); and n is Manning's roughness coefficient.

Hydrodynamic modeling of the swimming capability of largemouth bass

The FishXing program is a hydrodynamic model for dam engineers, hydrologists, and fish biologists that evaluates the design of waterway tunnels for fish passage. The FishXing program demonstrates the complexities of waterway tunnel hydraulics and fish swimming performance for a certain fish species and is frequently used to identify waterway tunnels that impede fish passage, which can lead

to the removal of dams or fish passage barriers. Fish swimming capabilities against waterway tunnel hydraulics across a range of expected stream discharges are estimated, in addition to comparing flow regime, velocities, and leap conditions according to the specific swimming abilities of a fish species.

Fish swimming capability is determined by many variables, including body length, weight, time to exhaustion, water temperature, and water flow rate. The following fish swimming speed formula used in the FishXing program was developed by Hunter and Mayor (1986):

$$V = aL^b t^{-c} \qquad (4)$$

where V (m/s) is swim speed of fish relative to the water; L (cm) is body length of the fish; t (min) is time to exhaustion of the fish; and a, b, and c are regression constants.

Fish swimming capability analysis and modeling involve the calculation of the migration potential of fish through an artificial structure based on the flow regime of water and the range of swimmable water. If an artificial structure cannot be traversed, factors contributing to the failure to pass are analyzed. In addition, the swimming distance, swimming type, and time it takes for fish to pass through the artificial structure are calculated. Based on these results, we compared fish swimming capability based on the size of individuals and flow rate and analyzed the comparison results.

Variables and characteristics of hydrodynamic model

Variables used for the calculation of the swimming capability of largemouth bass can be grouped into 3 categories: (1) fish-related data such as body length, weight, swimming speed, and time to exhaustion; (2) waterway tunnel-related data such as dimensions, materials, and slope; and (3) hydraulic-hydrological data such as minimum and maximum flow rate and water depths of the waterway tunnel inlet and outlet.

The swimming capability of individuals by size can be analyzed using fish-related data such as body length, weight, sustained and burst swimming speeds, and optimal habitat water temperature. We entered the waterway data necessary (length, diameter, material, shape, and gradient of the waterway) for fish to migrate upstream or move from pool to pool, in order to analyze the factors influencing the upstream migration of fish.

Table 4 Number and percentage of days by inflow rate into the waterway tunnel estimated via simulation

	Inflow rate (m³/s)							Flow	Non-flow	Maximum inflow rate (m³/s)
	>100	>80	>60	>40	>20	>10	≤10			
Number of days of water flow	37	83	98	280	717	853	7281	9349	1609	135.0
Ratio (%)	0.3	0.8	0.9	2.6	6.5	7.8	66.4	85.3	14.7	

Finally, by analyzing the hydraulic-hydrological data, such as minimum and maximum flow rate and water depths in the inlet and outlet, we determined the effects of varying flow velocity and rate on the migration of fish.

In order to calculate the swimming capability and maximum swimming speed of largemouth bass, we first calculated the swimming speed by size of individuals using the empirical model (regression) proposed by Beamish (1970) (Table 1). Swimming distance and maximum possible speed were then calculated using the FishXing program based on the pre-calculated swimming speed.

Results and discussion

Inflow rate of waterway tunnel via simulation

The inflow rate depended on the inner diameter of the waterway tunnel and the difference in water head between the two reservoirs (Table 2). The water inflow rate into tunnel tended to increase proportionally as the waterway tunnel inner diameter and water level difference between two reservoirs. The peak inflow rate into the Andong-Imha waterway tunnel (ø 5.5 m) was shown to be 135.0 m^3/s at the water head difference of 10.92 m.

The results of the simulation run of waterway tunnel operation are as follows. The number of days that water flowed from the Imha Reservoir to the Andong Reservoir and in the reverse direction were 7818 (71.3%) and 1531 days (14.0%), respectively (Table 3), and water flow during the flood period (June 21 to September 20) and dry period were 21.5 and 63.9%, respectively. In terms of the quantity of water that passed through the waterway tunnel, water flow from the Imha Reservoir to the Andong Reservoir and in the opposite direction accounted for 91.9 and 8.1%, and the total water flow during the flood and dry periods accounted for 50.5 and 49.5%, which was not a significant difference. Comparing the number of water flow days by inflow rate, water flowed at an inflow rate of ≥100 m^3/s on 37 days (0.3%), with the highest inflow rate being 135.0 m^3/s. The inflow rate ranged from 10 to 20 m^3/s on 853 days (7.8%), and the inflow rate of ≤10 m^3/s accounted for the majority of time (7281 days, 66.4%) (Table 4).

Swimming speed of largemouth bass

The swimming speed increased as individuals grew in size. At a water temperature of 25 °C, which is the optimal habitat water temperature for largemouth bass, the swimming speed ranged between 0.42 (individuals with 10 cm body length) and 1.60 m/s (60 cm), demonstrating remarkable size-dependent differences (Table 5). According to a study conducted by Hocutt (1973), at water temperatures of 15–30 °C, juveniles of 5–10 cm body length swam at 0.3–0.5 m/s and adult fish of 20–25 cm body length swam at

Table 5 Swimming speed of largemouth bass (m/s) by total length (cm) and water temperature (°C)

Fish growth stage	Total length	Water temperature (°C)					
		10 °C	15 °C	20 °C	25 °C	30 °C	35 °C
Juvenile to young stage	10 cm	0.17	0.26	0.38	0.42	0.42	0.34
	15 cm	0.24	0.33	0.45	0.48	0.48	0.40
	20 cm	0.31	0.42	0.53	0.54	0.55	0.47
Adult stage of medium size	30 cm	0.44	0.69	0.72	0.71	0.73	0.65
	35 cm	0.56	0.87	0.84	0.82	0.83	0.76
	40 cm	0.68	1.11	0.99	0.93	0.96	0.90
Adult stage of large size	50 cm	0.84	1.81	1.35	1.22	1.26	1.24
	55 cm	0.96	2.30	1.58	1.40	1.45	1.45
	60 cm	1.03	2.93	1.86	1.60	1.66	1.70

0.5–0.6 m/s, concordant with the results of the present study. Additionally, Bell (1991) reported that the prolonged swimming speed corresponded to 50–70% of the burst swimming speed. When this percentage was applied to the results of the present study, the burst swimming speed of 10-cm juveniles was 0.8 m/s, and that of 20–25-cm

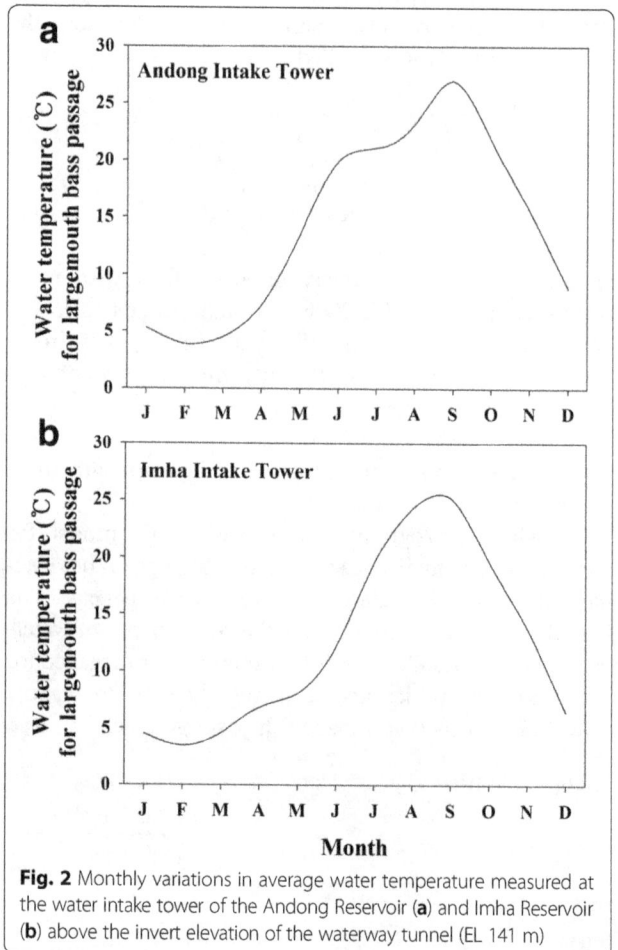

Fig. 2 Monthly variations in average water temperature measured at the water intake tower of the Andong Reservoir (**a**) and Imha Reservoir (**b**) above the invert elevation of the waterway tunnel (EL 141 m)

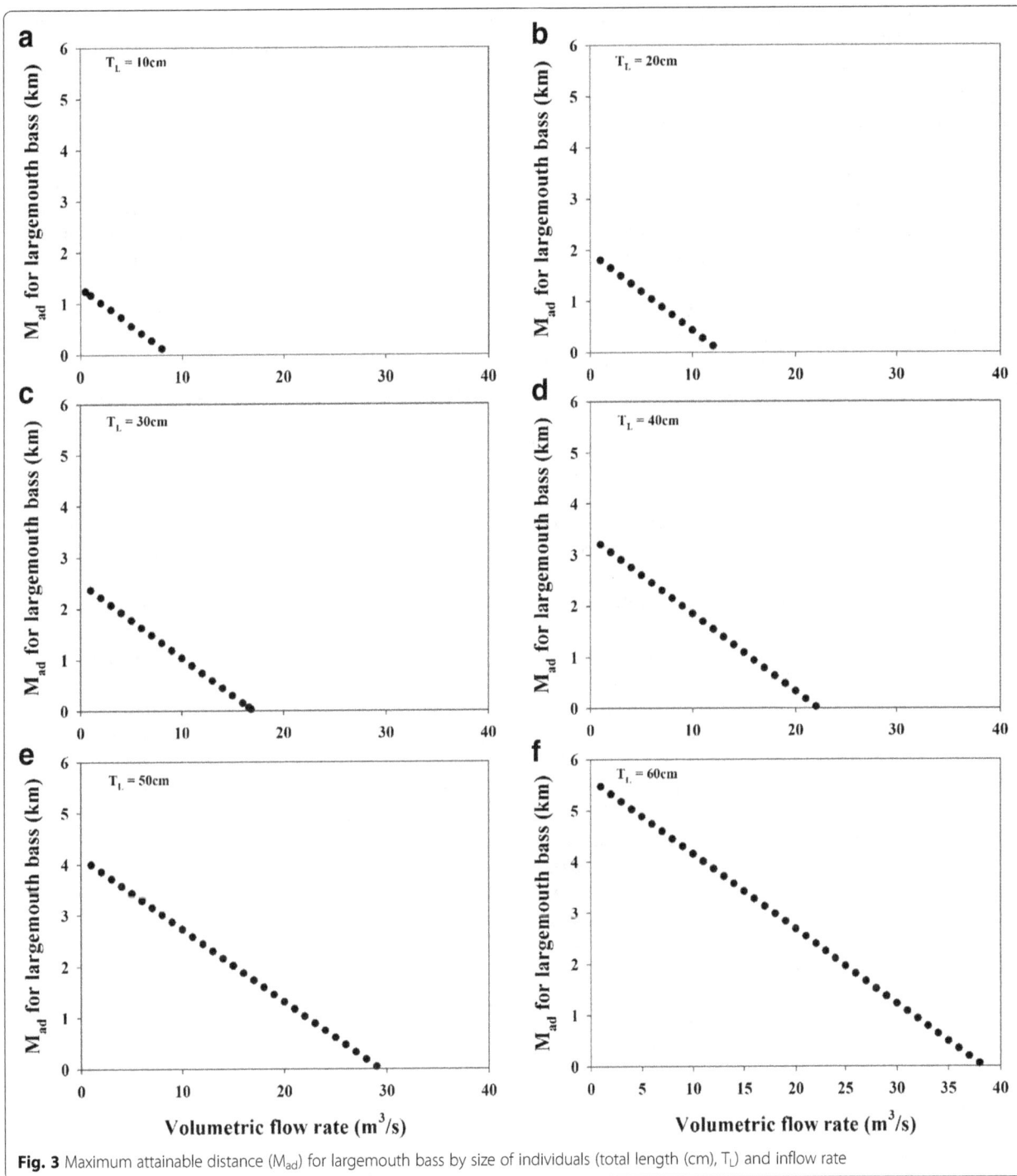

Fig. 3 Maximum attainable distance (M_{ad}) for largemouth bass by size of individuals (total length (cm), T_L) and inflow rate

individuals was 1.1–1.3 m/s, which was consistent with the results of Bell (1991).

In addition to the size of individuals and the inflow rate into the waterway tunnel, the distribution pattern of water temperature was also a key factor influencing fish swimming capability. At the Andong and Imha Reservoirs study sites, the drinking water source had a high content of dissolved oxygen (DO) and low carbon dioxide concentration

and salinity and was thus considered to have no to negligible toxic components (An et al. 2006; Park and Chung 2014). Therefore, of all factors presented by Beamish (1978) and physical constraint factors, including water temperature distribution, solar radiation, nutrient composition, DO concentration, carbon dioxide concentration, and salinity, water temperature was considered to exert the greatest influence on fish swimming performance.

Largemouth bass showed widely varying swimming performance at different water temperatures. For example, 10-cm juveniles swam at 0.17 m/s at 10 °C and 0.42 m/s at 25 °C. The swimming capability and maximum swimming speed of largemouth bass were found to increase linearly in the temperature range of 10–30 °C and decrease in the range of 30–40 °C (Beamish 1970). Large individuals more markedly demonstrated the temperature-dependent difference, with some individuals showing the highest swimming speed at temperatures lower than 25 °C (Table 5).

The average water depth measured at the water intake tower of the Andong Reservoir was 7.7 m above the invert elevation of the Andong-Imha waterway tunnel (EL 141 m), and the average, lowest, and highest water temperatures were measured at 13.8, 3.7, and 28.3 °C, respectively. Equivalent values for the Imha Reservoir were 9.6, 3.0, and 29.11 °C (Fig. 2). The results presented in Table 5 revealed that fish swimming speed rapidly declined by over 30% at water temperatures of 10 °C or less and consequently, the potential swim distance also decreased.

Swim distance of largemouth bass into the waterway tunnel

The distance covered by largemouth bass for 60 min was calculated based on the swimming speed of largemouth bass at the lowest inflow rate of 1 m³/s. The results showed that the 10-cm largemouth bass swam 1.17 km upstream, which linearly increased to 1.80, 2.36, 3.21, 3.99, and 5.47 km, as the size of individuals grew to 20, 30, 40, 50, and 60 cm, respectively.

According to the results of swim distance based on the inflow rate into the waterway tunnel, the 10-cm largemouth bass that swam 1.17 km at an inflow rate of 1 m³/s tended to cover short distances as the inflow rate increased and could not pass through the waterway tunnel at inflow rates of 9 m³/s or more. The same result was found for the 20-cm largemouth bass that swam 1.8 km at inflow rates of 13 m³/s or more. The length of the Andong-Imha waterway tunnel is 1.925 km, which was too long for the 10- to 20-cm largemouth bass to traverse, even at the lowest inflow rate of 1 m³/s. Smaller individuals reached a state of exhaustion more rapidly than larger individuals, making it impossible to pass the waterway tunnel. The minimum size of individuals capable of swimming through the waterway tunnel was 25 cm (Fig. 3).

At inflow rates of ≤3 m³/s, the 30-cm largemouth bass were capable of swimming 2 km or more (i.e., passing the waterway tunnel), with a time requirement of 49–57 min. However, they could not pass the waterway tunnel at inflow rates of 4 m³/s or more. At inflow rates of 17 m³/s or higher, migration via the waterway tunnel was determined to be nearly impossible.

According to the results regarding the time requirement for the passage of the waterway tunnel, the 40-cm largemouth bass took 37 min to traverse the waterway tunnel at an inflow rate of 1 m³/s, the 50-cm largemouth bass took 28 min, and the 60-cm largemouth bass took 21 min, which demonstrated that the larger the individuals, the less time it took them to cross the waterway tunnel. The time requirement for passage increased as the inflow rate increased. The 40-cm largemouth bass took 37–60 min at an inflow rate of 9 m³/s or less, the 50-cm largemouth bass took 28–56 min at 15 m³/s or less, and the 60-cm largemouth bass took 21–56 min at 24 m³/s or less (Fig. 4).

Modeling analysis of the swimming capability of largemouth bass

According to the results of FishXing, the success potential for 10-cm largemouth bass migrating via the waterway tunnel was analyzed to be 0%, and the lowest and highest inflow rates allowing them to enter and swim along the waterway tunnel were estimated at 1 and 9 m³/s, respectively (Fig. 5). The success potential for the 20-cm largemouth bass was also analyzed to be 0%, with the highest inflow rate allowing them to enter and swim along the waterway tunnel being 13 m³/s. Largemouth bass with 30-cm body length could migrate via the waterway tunnel at the inflow rate of 3 m³/s or less, with a success rate of 7%. The success rates for the 40-, 50-, and 60-cm largemouth bass were 10.9% at 8 m³/s or less, 15% at 15 m³/s or less, and 20% at 24 m³/s or less, respectively. At higher inflow rates (25–38 m³/s), larger largemouth bass were found to be capable of entering the waterway tunnel and swimming, but migration from the Andong Reservoir to the Imha Reservoir and vice versa was found to be impossible. Analysis also

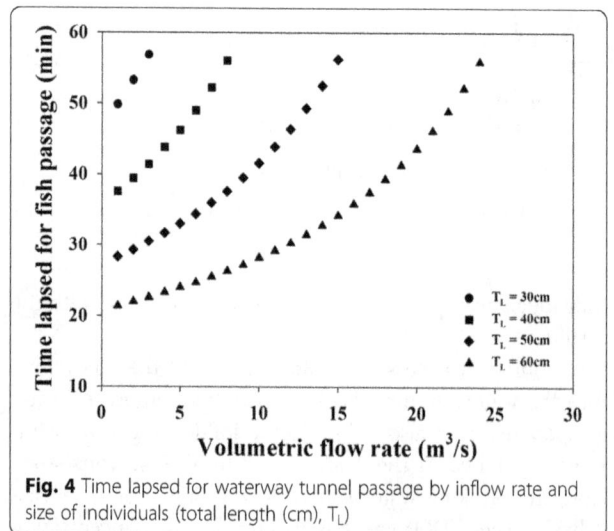

Fig. 4 Time lapsed for waterway tunnel passage by inflow rate and size of individuals (total length (cm), T_L)

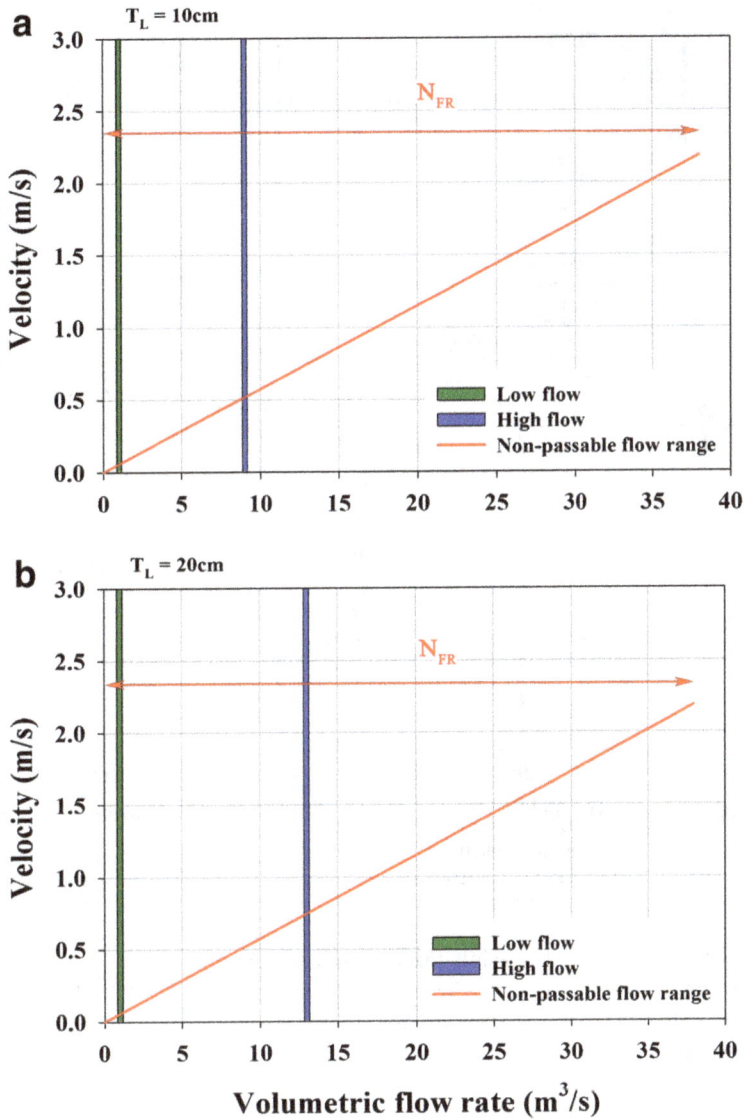

Fig. 5 Passable (P_FR) or non-passable (N_FR) flow range of the calculated inflow rates for juvenile/young largemouth bass based on size of individuals (total length (cm), T_L)

showed that at inflow rates of 38 m³/s or higher, the use of the waterway tunnel by largemouth bass was next to impossible (Fig. 6).

Conclusions

Swimming speed and distance increased as the size of largemouth bass increased, and largemouth bass of the same body length swam slower and covered less distance as the inflow rate of the waterway tunnel increased. The frequency of transferring the waterway tunnel was highest (66.4%) at an inflow rate of 10 m³/s or less. For smaller individuals (10–30 cm), the migration rate was 0% at 10 m³/s. The success rate of migration was 10.9% at 9 m³/s or less for the 40-cm largemouth bass, 15% at 15 m³/s or less for

the 50-cm largemouth bass, and 20% at 24 m³/s or less for the 60-cm largemouth bass. However, these migration rates were estimated at an assumed water temperature of 25 °C, the optimal habitat condition for largemouth bass. At water temperatures of 10 °C or lower, their swimming speed rapidly decreased by more than 30%, and the migration rate and swimming distance in the waterway tunnel were also assumed to decrease (Mitchell 1989; Bell 1991).

During the dry period in which water levels do not fluctuate significantly in the Andong and Imha Reservoirs, larger largemouth bass (≥30 cm) are expected to have higher migration rates than those during the flood period because of low water head differences between the two reservoirs and consequently, low inflow rates. However,

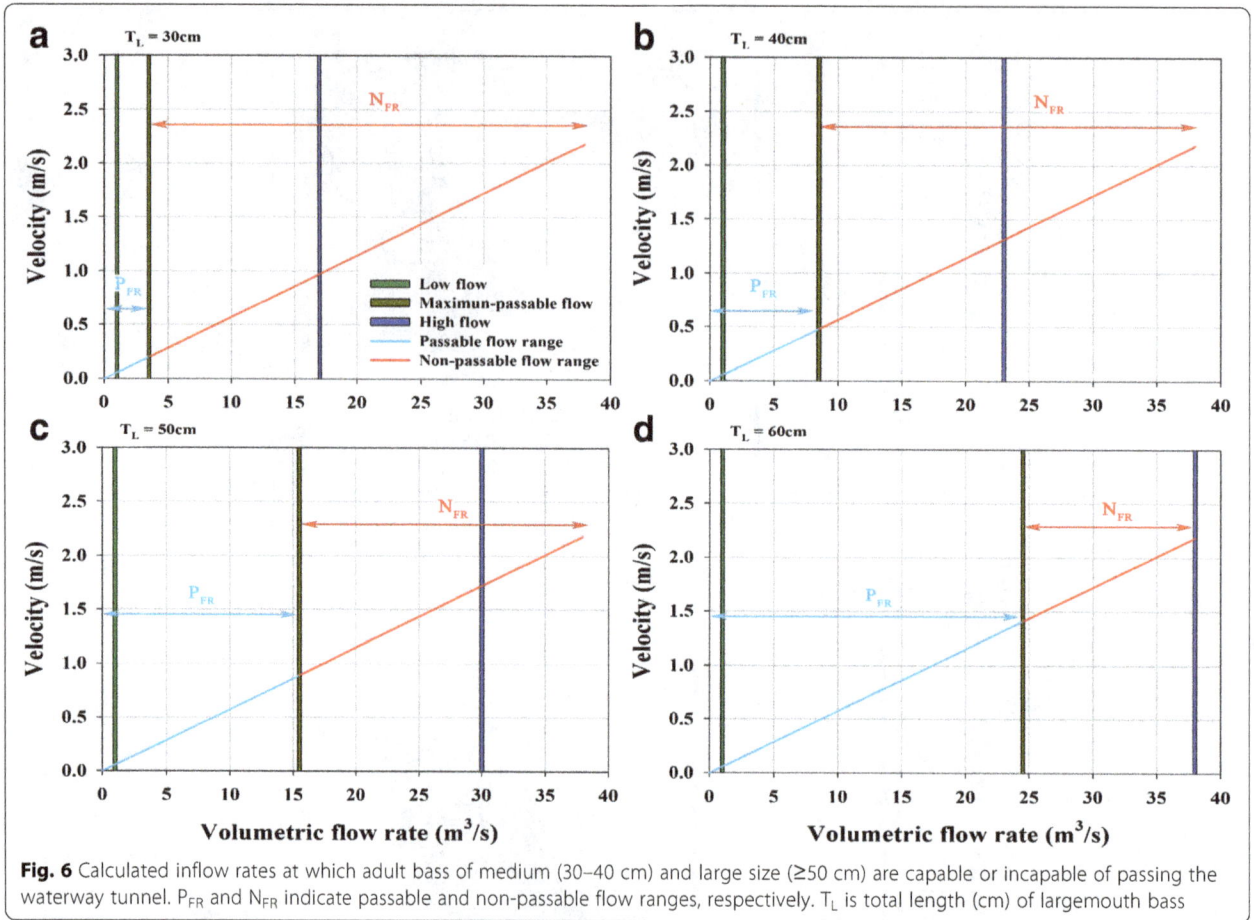

Fig. 6 Calculated inflow rates at which adult bass of medium (30–40 cm) and large size (≥50 cm) are capable or incapable of passing the waterway tunnel. P_{FR} and N_{FR} indicate passable and non-passable flow ranges, respectively. T_L is total length (cm) of largemouth bass

during the flood period with concentrated precipitation, increased inflow rates (≥40 m³/s) due to the rapid water level rise of the Imha Reservoir and subsequent increase in water head difference (≥1 m) would impede the migration of largemouth bass via the waterway tunnel. In particular, smaller individuals (10–30 cm) are assumed to be unable to pass through the long waterway tunnel because they quickly reach a state of exhaustion during prolonged upstream swimming against rapidly flowing water (Farlinger and Beamish 1977). In summary, the potential for the spread of largemouth bass to the Imha Reservoir via the waterway tunnel is not remarkable enough to raise any concern.

Abbreviations
EL: Elevation; M_{ad}: Maximum attainable distance; N_{FR}: Non-passable flow range; P_{FR}: Passable flow range; T_L: Total length

Acknowledgements
This research was supported by "Daejeon Green Environment Center under the Research Development Program (Yr 2009)", so the authors would like to acknowledge for the assistance.

Authors' contributions
Prof. KGA got a research project for the topic, and Dr. JWC analyzed the data with Prof. KGA and Dr. JWC and KGA wrote the manuscript and then edited together. Both authors read and approved the final manuscript.

Competing interests
The authors declare that they have no competing interests.

References
Ahn, J. B., & Kim, H. J. (2010). Characteristics of monsoon climate in the Korean peninsula. *Journal Climate Research, 5*, 91–99 (in Korean).
Almeida, D., Almodóvar, A., Nicola, G. G., Elvira, B., & Grossman, G. D. (2012). Trophic plasticity of invasive juvenile largemouth bass *Micropterus salmoides* in Iberian streams. *Fisheries Research, 113*, 153–158.
An, K. G., Park, S. J., Choi, S. M., & Park, J. S. (2006). Comparative analysis of long-term water quality data monitored in Andong and Imha reservoirs. *Korean Journal of Limnology, 39*, 21–31 (in Korean).
Beamish, F. W. H. (1970). Oxygen consumption of largemouth bass, *Micropterus salmoides*, in relation to swimming speed and temperature. *Canadian Journal of Zoology, 48*, 1221–1228.
Beamish, F. W. H. (1978). Swimming capacity. In W. H. Hoar & D. J. Randall (Eds.), *Fish physiology* (pp. 101–187). New York: Academic.
Bell MC. 1991. Fisheries handbook of engineering requirements and biological criteria. Fish Passage Development and Evaluation Program, U.S. Army Corps of Engineers
Black, E. C., Connor, A. R., Lam, K., & Chiu, W. (1962). Changes in glycogen, pyruvate and lactate in rainbow trout (*Salmo gairdneri*) and following muscular activity. *Journal of the Fisheries Research Board of Canada, 19*, 409–436.
Brown, T. G., Runciman, B., Pollard, S., & Grant, A. D. A. (2009). Biological synopsis of largemouth bass (*Micropterus salmoides*). *Canadian Manuscript Report of Fisheries and Aquatic Sciences, 2884*.

Colavecchia, M., Katopodis, C., Goosney, R., Scruton, D. A., & Mckinley, R. S. (1998). Measurement of burst swimming performance in wild Atlantic salmon (*Salmo salar* L.) using digital telemetry. *Regulated River, 14*, 41–51.

Farlinger, S., & Beamish, F. W. H. (1977). Effects of time and velocity increments on the critical swimming speed of largemouth bass (*Micropterus salmoides*). *Transactions of the American Fisheries Society, 106*, 436–439.

Findlay, C. S., Bert, D. G., Zheng, L. (2000). Effect of introduced piscivores on native minnow communities in Adirondack lakes. *Canadian Journal of Fisheries and Aquatic Sciences, 57*, 570–580.

Furniss, M., Love, M., Firor, S., Moynan, K., Llanos, A., Guntle, J., & Gubernick, R. (2006). *FishXing version 3.0.* Corvallis: US Forest Service.

Hammer, C. (1995). Fatigue and exercise tests with fish. *Comparative Biochemistry and Physiology Part A;, 112*, 1–20.

Hickley, P., North, R., Muchiri, S. M., & Harper, D. M. (1994). The diet of largemouth bass (*Micropterus salmoides*) in Lake Naivasha, Kenya. *Journal of Fish Biology, 44*, 607–619.

Hocutt, C. H. (1973). Swimming performance of three warmwater fishes exposed to a rapid temperature change. *Chesapeake Science, 14*, 11–16.

Hunter LA, Mayor L. 1986. Analysis of fish swimming performance data. Volume I. North-south consultants for the Department of Fisheries and Oceans, Government of Canada, and Alberta Department of Transportation

Kang, M. G., Lee, G. M., & Cha, H. S. (2007). Evaluation of effects of real joint-operation of multi-purpose dams. *Journal of Korea Water Resources Association, 40*, 101–112 (in Korean).

Kim, H. M., Kil, J. H., Lee, E. H., & An, K. G. (2013). Distribution characteristics of largemouth bass (*Micropterus salmoides*) as an exotic species, in some medium-to-large size Korean reservoirs and physico-chemical water quality in the habitats. *Korean Journal of Ecology and Environment, 46*, 541–550 (in Korean).

Lasenby, T. A., & Kerr, S. J. (2000). *Bass stocking and transfers: an annotated bibliography and literature review.* Peterborough: Fish and Wildlife Branch, Ontario Ministry of Natural resources.

Lee, J. U. (2005). Development of optimal operation rule for multipurpose reservoirs systems. *Journal of Korea Water Resources Association, 37*, 487–497 (in Korean).

Lee, W. O., Yang, H., Yoon, S. W., & Park, J. Y. (2008). Study on the feeding habits of *Micropterus salmoides* in Lake Okjeong and Lake Yongdam. *Korean Journal of Ichthyology, 21*, 200–207 (in Korean).

Love, M. S., Firor, S., Furniss, M., Gubernick, R., Dunklin, T., & Quarles, R. (1999). *FishXing version 2.2. USDA forest service.* San Dimas: San Dimas Technology and Development Center.

Ministry of Environment, Korea. (2013). *Monitoring of invasive alien species designated by the wildlife protection act (VII).* Incheon: National Institute of Environmental Research (NIER) (in Korean).

Mitchell, C. P. (1989). Swimming performances of some native freshwater fishes. *New Zealand Journal of Marine and Freshwater, 23*, 181–187.

Nursall, J. R. (1962). Swimming and the origin of paired appendages. *American Zoologist, 2*, 127–141.

Park, H. S., & Chung, S. W. (2014). Water transportation and stratification modification in the Andong-Imha linked reservoirs system. *Journal of Korean Society on Water Environment, 30*, 31–43 (in Korean).

Wasserman, R. J., Strydom, N. A., & Weyl, O. L. F. (2011). Diet of largemouth bass, *Micropterus salmoides* (Centrarchidae), an invasive alien in the lower reaches of an Eastern Cape river, South Africa. *African Zoology, 46*, 378–386.

Webb, P. W. (1975). Hydrodynamics and energetics of fish propulsion. *Journal of the Fisheries Research Board of Canada, 190*, 1–158.

Wheeler, A. P., & Allen, M. S. (2003). Habitat and diet partitioning between shoal bass and largemouth bass in the Chipola River, Florida. *Transactions of the American Fisheries Society, 132*, 438–449.

Wood, C. M., Turner, J. D., & Graham, M. S. (1983). Why do fish die after severe exercise? *Journal of Fish Biology, 22*, 189–201.

Yasunori, M., & Tadashi, M. (2003). Community-level impacts induced by introduced largemouth bass and bluegill in farm ponds in Japan. *Biological Conservation, 109*, 111–121.

Yum, K. (2010). The roles and effects of the four rivers restoration project to cope with climate change. *Journal of the Korean Society of Civil Engineers, 58*, 10–15 (in Korean).

Zaret, T. M., & Paine, R. T. (1973). Species introduction in a tropical lake. *Science, 182*, 449–455.

14

Distribution and attachment characteristics of *Sida crystallina* (O.F. Müller, 1776) in lentic freshwater ecosystems of South Korea

Jong-Yun Choi[1*], Kwang-Seuk Jeong[2,3], Seong-Ki Kim[4], Se-Hwan Son[1] and Gea-Jae Joo[2]

Abstract

Background: Macrophytes are commonly utilised as habitat by epiphytic species; thus, complex macrophyte structures can support high diversities and abundances of epiphytic species. We tested the hypothesis that the presence of aquatic macrophytes is an important factor determining *Sida crystallina* (O.F. Müller, 1776) distribution.

Results: An ecological survey was conducted in 147 lentic freshwater bodies. *S. crystallina* was frequently observed, and its density was strongly associated with macrophyte abundance. *S. crystallina* was found on emergent plant species such as *Phragmites australis* and *Paspalum distichum*, attached to the stem surfaces by adhesive substances secreted by the nuchal organ. Thus, *S. crystallina* was more strongly attached to macrophytes than to other epiphytic cladoceran species. We found higher densities of *S. crystallina* in filtered water with increased macrophyte shaking effort (i.e. 10, 20, 40, or 80 times). *S. crystallina* attachment was not related to fish predation. Stable isotope analysis showed that *S. crystallina* utilises epiphytic organic matter (EOM) on macrophytes as a food source.

Conclusions: Consequently, *S. crystallina* seems to have a strong association with species-specific macrophyte biomass than with other cladoceran species, which may contribute to this species' predominance in various freshwater ecosystems where macrophytes are abundant.

Keywords: *Sida crystallina*, Aquatic macrophytes, Attachment characteristics, Stable isotope analysis, Lentic freshwater ecosystems

Background

Aquatic macrophytes commonly occur in shallow aquatic ecosystems and can have dramatic effects on their physical structure (O'Hare et al. 2006; Smokorowski and Pratt 2007). Aquatic macrophytes give rise to heterogeneous spaces with varying degrees of structural complexity (Denny 1994; Findlay and Bourdages 2000). Some studies have suggested that vegetated beds with high structural heterogeneity provide small animals with refuges from predators and suitable spawning and foraging substrates, mediating trophic interactions among diverse organisms (Vieira et al. 2007; Thomaz et al. 2008). The effectiveness

of macrophytes as a refuges and/or habitat varies with their life form, density, and species (see review in Burks et al. 2002). In particular, macrophyte morphological characteristics have a significant bearing upon the availability of small animals as a food source, mediated via detritus trapping (Rooke 1984) and the growth of periphytic algae (Cattaneo et al. 1998). Thus, the presence of aquatic macrophytes facilitates increased abundance and diversity in aquatic animal communities.

Among the many animals utilising macrophyte habitats, freshwater cladocerans are well known to exploit macrophytes as habitats and/or refuges (Kuczyńska-Kippen and Nagengast 2006; Choi et al. 2014a). The majority of studies that have focused on the interactions between macrophytes and cladocerans have considered the influence of macrophytes on pelagic cladoceran species (Jeppesen et al.

* Correspondence: jyc311@nie.re.kr
[1]National Institute of Ecology, Seo-Cheon Gun, Chungcheongnam province 325-813, South Korea
Full list of author information is available at the end of the article

1998; Meerhoff et al. 2007). These studies argued that aquatic macrophytes are capable of providing suitable habitats for mainly pelagic cladocerans. However, pelagic species are continuously exposed to predators such as fish due to their frequent movement; thus, it would be relatively difficult for such pelagic species to develop higher abundances in freshwater ecosystems. In comparison with other aquatic systems, shallow wetlands are characterised by abundant aquatic macrophytes, and this abundance of habitat tends to attract more cladoceran species. These are known as epiphytic cladocerans (i.e. plant-attached species; Castilho-Noll et al. 2010; Gyllström et al. 2005; Kuczyńska-Kippen and Nagengast 2006) and are strongly affected by the abundance, morphology, and arrangement of plant species (Choi et al. 2014b). Unfortunately, epiphytic cladoceran distribution patterns and modes of macrophyte utilisation are still unclear, and their abundances are usually underestimated.

Sida crystallina (Cladocera: Sididae, O.F. Müller 1776) is a typical epiphytic cladoceran species occurring in temperate and tropical waters. They attach to aquatic macrophytes by means of maxillary glands and thus filter feed in fixed positions (Fairchild 1981). Compared with other cladoceran species, *S. crystallina* occurs at relatively high water temperatures (approximately 21 to 22 °C; Kotov and Boikova 1998) and is prevalent in temperate zones during summer (Balayla and Moss 2003). They are found in most shallow freshwater ecosystems in South Korea during summer (Choi et al. 2014c). Unfortunately, although the overall distribution of the species has been reported (Downing and Peters 1980; Lauridsen et al. 1996), its distribution patterns, attachment characteristics, and feeding habits have been insufficiently studied.

In this study, we investigated the distribution patterns of *S. crystallina* in lentic freshwater ecosystems in South Korea. We hypothesised that *S. crystallina* may prefer shallow wetland microhabitats in which macrophytes dominate and may utilise epiphytic algae growing on macrophyte stands as their food source. To test these hypotheses, we investigated (i) the influence of diverse physicochemical parameters and macrophytes on *S. crystallina* distribution and (ii) the nature of attachment of *S. crystallina* in relation to fish predation and food availability. We surveyed 147 lentic ecosystems in South Korea and recorded physicochemical parameters of water, macrophyte occurrence, and *S. crystallina* densities.

Methods
Study sites
South Korea is located in East Asia and has a temperate climate. Four distinct seasons lead to the dynamic succession of biological communities in the freshwater ecosystems of South Korea. Annual mean rainfall is ca. 1150 mm, more than 60 % of which occurs from June to

early September (Choi et al. 2011; Jeong et al. 2007). Our study sites were located in south-eastern South Korea, along the middle and lower reaches of the Nakdong River. Historically, there were numerous riverine wetlands in this river basin (Son and Jeon 2002); however, large areas of wetland have vanished due to the expansion of human society (Burkett and Kusler 2000). The dominant land cover surrounding reservoirs is agricultural, and non-point source pollution continuously influences the study sites (Korean Ministry of Environment 2006).

We investigated 147 lentic freshwater ecosystems in the river basin (wetlands, ponds, and reservoirs; see Fig. 1). The wetlands and shallow lakes are dominated by various macrophyte species; however, the development and growth of macrophytes is inhibited in reservoirs and some lakes due to their impermeable floors. In addition, some wetlands support only a few plant species because of high water levels and low nutrient concentrations. Therefore, the study sites encompassed a wide range of microhabitat characteristics (i.e. different types of lentic systems and different patterns in their constituent plant communities).

Monitoring strategy
The target species *S. crystallina* is known to prefer relatively high water temperatures (approximately 21 to 22 °C; Kotov and Boikova 1998) and is prevalent in temperate zones during summer (Balayla and Moss 2003). Based on this information, we monitored study sites in summer (June to July 2012). At each site, three sampling points were established in the littoral zone.

Physicochemical parameters were measured and *S. crystalline* collected at each sampling point. Water temperature, dissolved oxygen, conductivity, pH, chlorophyll a, and turbidity were measured at each site. Water samples were collected at a depth of 0.5 m. We used a DO meter (YSI DO meter; Model 58, YSI Research Inc., OH, USA) to measure water temperature and dissolved oxygen. Conductivity and pH were measured using a conductivity meter (YSI Model 152; Yellow Springs Instruments, Yellow Springs, OH, USA) and pH meter (Orion Model 250A; Orion Research Inc., Boston, MA, USA). Turbidity and chlorophyll a concentration were measured in the laboratory. Turbidity was measured using a turbidimeter (Model 100B; Scientific Inc., Ft. Myers, FL, USA). The water samples were filtered through mixed cellulose ester (MCE) membrane filters (Advantech; Model No., A045A047A; pore size, 0.45 μm), and chlorophyll a concentration was ascertained based on the methodology of Wetzel and Likens (2000).

At each sampling point, we took an additional 10 L of water for zooplankton collection from the surface layer (to a depth of 0.5 m), using a 10-L column sampler. This water was filtered through a plankton net (68-μm mesh size), and the filtrate was preserved in sugar formalin

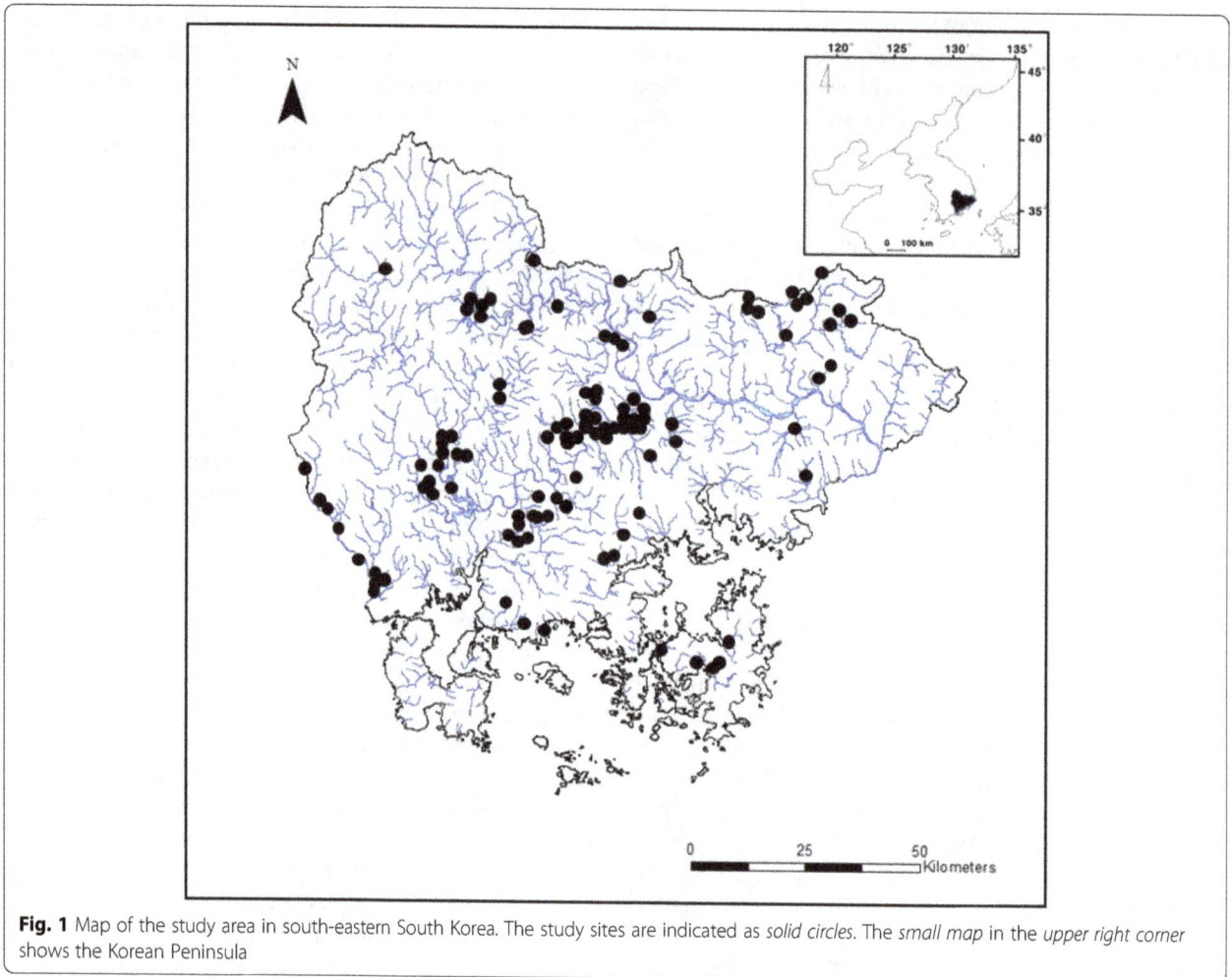

Fig. 1 Map of the study area in south-eastern South Korea. The study sites are indicated as *solid circles*. The *small map* in the *upper right corner* shows the Korean Peninsula

(final concentration 4 % in the form of aldehyde). *S. crystallina* and other zooplankton species were identified and counted using a microscope (ZEISS, Model Axioskop 40; ×200 magnification) using the classification key of Mizuno and Takahashi (1991).

To investigate attachment characteristics of *S. crystallina*, we additionally collected *S. crystallina* from stems and leaves of macrophytes at six sites where high abundances of *S. crystallina* were observed. We established five quadrats (0.5 m × 0.5 m) along the littoral zone at each of these sites and counted all *S. crystallina* within each quadrat. We did not include emergent organs of macrophytes above the water surface (i.e. stalks and flowers) because *S. crystallina* inhabit underwater environments. The submerged parts were handled carefully to prevent *S. crystallina* from accidentally detaching. *S. crystallina* were kept alive using filtered wetland water. Small animals including zooplankton were removed from 2 L of water using a plankton net (32 μm mesh size), and the filtered water was stored in 5-L tanks. This water was used as temporary storage for epiphytic species including *S. crystallina*. Collected macrophytes were shaken in the tank to detach *S. crystallina*

(for the detaching process, see Sakuma et al. 2002). *S. crystallina* on the plants were detached by shaking 10, 20, 40, and 80 times. After collection of *S. crystallina* at the study sites, macrophyte samples were carried to the laboratory and dried at 60 °C for 2 days. Epiphytic species, including *S. crystallina*, were filtered from the water using a 68 μm mesh net and immediately fixed with sugar formalin (final concentration 4 % in the form of aldehyde). We counted numbers of *S. crystallina* using a microscope (ZEISS, Model Axioskop 40; ×200 magnification). Densities of *S. crystallina* attached to plants were expressed as number of individuals per gramme dry weight of macrophyte (ind. g^{-1} dw).

Microcosm experiment

To understand how the attachment characteristics of *S. crystallina* influence fish predation, we conducted additional microcosm experiments. Approximately 200 *S. crystallina* adult individuals with similar life-history traits (body size and condition of clutch) were selected. These *S. crystallina* individuals were acclimatised for approximately 48 h in a stock culture environment

(Elendt M4 medium; Elendt 1990). To simulate the fish predation, we used fish chemical cues (De Meester and Cousyn 1997) in this microcosm experiment. A total of five fish species were considered, *Micropterus salmoides* (Lacepéde 1802), *Lepomis macrochirus* (Rafinesque 1819), *Pseudorasbora parva* (Temminck and Schlegel 1846), *Rhinogobius brunneus* (Temminck and Schlegel 1845), and *Misgurnus anguillicaudatus* (Cantor 1842). We collected samples of these fish at the study sites where we obtained *S. crystallina* using a 7 mm × 7 mm cast net and a 5 mm × 5 mm scoop net. To obtain fish chemical cues from the collected fish, we allocated each fish species to one of five tanks, and five individuals of each fish species were acclimatised in each tank for 24 h (tanks filled with 10 L Elendt M4 medium).

The experiment was designed as follows: a total of six groups—control (no predation) and five experimental groups by using each of the fish species. First, we prepared 60 500-mL beakers (10 beakers per experimental group) and filled each with 300 mL of clear M4 medium. We put five *S. crystallina* individuals in each beaker and allowed them to acclimatise to the new environment for 30 min. During the acclimatisation period, we prepared the fish chemical cues used in the experiment: the fish-exposed M4 medium was filtered using a 30 μm mesh net (used only in this experiment and not used for field plankton collection) to remove particulate matter. Before introducing the resulting fish chemical cues to the experimental beakers, we very carefully marked the position of each *S. crystallina* individual on the outer surface of every beaker. Then, we injected 200 mL of fish chemical cue into each beaker. For the control group, unexposed, clear M4 medium was added. The *S. crystallina* individuals were allowed to respond to the changed environment for 30 min, and then, we investigated the number of moved individuals.

Data analysis

We used two-way ANOVA ($\alpha = 0.05$) to analyse how the density of *S. crystallina* varied with shaking procedure (i.e. 10, 20, 40, and 80 times) and collection site. Differences in *S. crystallina* movement among fish predation treatments were analysed statistically using one-way ANOVA. Furthermore, the relationships between *S. crystallina* density and environmental variables were tested using stepwise multiple regression. All statistical analyses, including stepwise multiple regression and ANOVA, were conducted using the statistical package SPSS for Windows ver. 14.

Stable isotope analysis

Stable isotope analysis was conducted to identify the food sources of *S. crystallina*. Particulate organic matter (POM), epiphytic organic matter (EOM), and *S. crystallina* individuals were sampled at six different sites where *S. crystallina*

was abundant. To process the POM samples, any small animals were first removed using a plankton net (32 μm mesh size), and then, the water samples were filtered through GF/F glass-fibre (pre-combusted at 500 °C for 2 h). The surfaces of submerged parts of macrophytes from each study point were gently brushed in a tank filled with distilled water, in order to obtain the EOM. Similar to the processing of POM, micro- and macroinvertebrates were removed using the plankton net (32 μm mesh size). *S. crystallina* individuals were isolated using a micropipette.

POM and EOM samples were treated with 1 N HCl to remove inorganic carbon and rinsed with distilled water to remove the acid. *S. crystallina* samples were not acidified to remove inorganic carbon, because acidification affects nitrogen values (Pinnegar and Polunin, 1999). All samples were freeze-dried and homogenised with a mortar and pestle, and the powdered samples were kept frozen (-70 °C) until analysis. Carbon and nitrogen isotope ratios were determined using continuous-flow isotope mass spectrometry. Dried samples (ca. 1 mg for *S. crystallina* samples and 1.5 mg for POM and EOM) were combusted in an elemental analyser (Euro EA 3000 Elemental Analyzer, Eurovector SPA., Milano, Italy), and the resultant gas (CO_2 and N_2) was introduced to an isotope ratio mass spectrometer (CF-IRMS, IsoPrime) in a continuous flow using a helium carrier. Data were expressed as the relative concentration (‰) difference between sample and conventional standards of Pee Dee Belemnite carbonate (PDB) for carbon and atmospheric N_2 for nitrogen according to the following equation:

$$\delta X(‰) = \left[(R_{sample}/R_{standard}) - 1 \right] \times 1000 \qquad (1)$$

where X is ^{13}C or ^{15}N and R is the $^{13}C{:}^{12}C$ or $^{15}N{:}^{14}N$ ratio. A secondary standard (Peptone) of known relation to the international standard was used as a reference material. Standard deviations of $\delta^{13}C$ and $\delta^{15}N$ for analyses with 20 replicates of Peptone standard were ±0.1 and ±0.2 (‰), respectively.

To determine which of the two food sources (POM and EOM) was assimilated more readily by *S. crystallina*, we used two-source isotope mixing models. The carbon isotope values of POM and EOM significantly differed among sites (see 'Results'). The model was defined as:

$$\begin{aligned} \delta^{15}C_M &= f_X \left(\delta^{13}C_X + \Delta^{13}N \right) \\ &+ f_Y \left(\delta^{13}C_Y + \Delta^{13}C \right); 1 \\ &= f_X + f_Y \end{aligned} \qquad (2)$$

where X, Y and M represent the two food sources and a mixture of the two, respectively; f represents the proportion of N from each food source in the consumer's diet; and $\Delta^{15}C$ is the assumed trophic fractionation (i.e. the change in $\delta^{15}C$ over one trophic step from prey to predator; Phillips and Gregg 2001). Trophic fractionation was assumed to be constant at either 3.4 or 2.4 % (Minagawa and Wada 1984).

Results

Physicochemical parameters, macrophytes, and zooplankton

There were few differences in physicochemical characteristics of water among the study sites (Table 1). Although some study sites had exceptionally high or low values, the coefficients of variation (CV; standard deviation/mean × 100 %) were lower than 100 %. Conductivity had the highest CV, but this was only approximately 66.9 %.

Macrophyte species composition and dry weight differed among study sites. *Paspalum distichum* L. dominated most of the study sites; a total of 10 species of macrophyte were found (*Phragmites australis* Trin. (Cav.), *P. distichum*, *Zizania latifolia* Griseb., *Scirpus tabernaemontani* Gmel., *Spirodela polyrhiza* L., *Salvinia natans* L., *Trapa japonica* Flerov., *Ceratophyllum demersum* L., *Hydrilla verticillata* (L.F.) Royle, and *Nymphoides indica* (L.) Kuntze).

A total of 122 species of zooplankton were identified (86 rotifers, 27 cladocerans, and 9 copepods). The highest abundance of zooplankton was 4135 ind. L^{-1}, followed by 3736 ind. L^{-1}. *Lecane hamata* (Stokes 1897), *Polyarthra vulgaris* (Carlin 1943), *Chydorus sphaericus* (O.F. Müller 1785), and *Diaphanosoma brachyurum* (Lievin 1848) were recorded frequently.

S. crystallina was observed at 81 out of 147 sites. From the observation, *S. crystallina* was mostly found in shallow wetlands where macrophytes were present. Stepwise multiple regression showed that density and distribution of *S. crystallina* were clearly related to macrophyte biomass (d.f. for regression, residuals, and total = 2, 143, 146, respectively; $F = 94.32$, $P = 0.001$, see Table 2 for independent variables). In addition, *S. crystallina* density was related to water depth and chlorophyll a concentration. However, other environmental parameters did not have significant influence on *S. crystallina* distribution or abundance.

Table 1 Mean macrophyte dry weights and physicochemical parameters measured at the study sites

Variable	Units	Max	Min	Mean ± SD	CV (%)
Macrophyte biomass	gdw	114.1	0	42.8 ± 23.1	68.4
Water depth	m	231	14	78.4 ± 40.6	51.7
Water temperature	°C	27.1	22.5	25.6 ± 2.5	11.2
Dissolved oxygen	%	217.2	21.6	87.9 ± 34.2	38.2
Conductivity	μs cm^{-1}	746.7	83.6	187.8 ± 125.6	66.9
pH	–	8.8	6.3	7.8 ± 0.6	9.2
Chlorophyll a	μg L^{-1}	42.8	2.1	25.8 ± 9.0	35.2
Turbidity	NTU	24.0	1.33	9.4 ± 6.0	64.5
Total nitrogen	mg L^{-1}	13.2	1.4	7.3 ± 2.6	25.4
Total phosphorous	mg L^{-1}	123.2	23.6	76.7 ± 32.5	36.7

Table 2 Summary of stepwise multiple regression between *Sida crystallina* abundance (response variable) and physicochemical parameters (explanatory variables)

Response variable	Explanatory variables	B_j	t	P
Sida crystallina	Constant	27.201	−2.488	0.014
	Macrophyte biomass (g)	2.469	12.287	0.000
	Water depth (mm)	0.748	3.735	0.005
	TP (mg L^{-1})	−1.794	−2.185	0.031

Data were transformed prior to analysis using either the arcsine-square root (proportion of agricultural land) or log (all other variables) transformation

Interestingly, *S. crystallina* was frequently observed in sites where emergent plants such as *P. australis* and *P. distichum* dominated (Fig. 2). In particular, sites dominated by *P. distichum* supported higher densities of *S. crystallina*. By contrast, *S. crystallina* densities were low in sites where free-floating (*S. natans*), floating-leaved (*T. Japonica*), and submerged (*C. demersum*) macrophyte species dominated.

Attachment characteristics of *S. crystallina*

Numbers of *S. crystallina* detached from macrophytes differed with shaking effort at each site (Fig. 3). Two-way ANOVA revealed that *S. crystallina* counts were significantly affected by both shaking effort (10, 20, 40, or 80 times; d.f. = 3, $F = 265.98$, $P < 0.05$) and sites (between total 6 sites; d.f. = 5, $F = 101.43$, $P < 0.05$). *S. crystallina* counts were proportional to the shaking effort, but the increase in counts tailed off as the efforts increased. This phenomenon was observed at all six study sites.

Interestingly, *S. crystallina* attachment was not affected by simulated fish predation (one-way ANOVA, d.f. = 5, $F = 0.236$, $P = 0.945$; Fig. 4). We counted the number of moved individuals of *S. crystallina* following exposure to fish chemical cues, but almost no individuals moved in any experimental group, including the control.

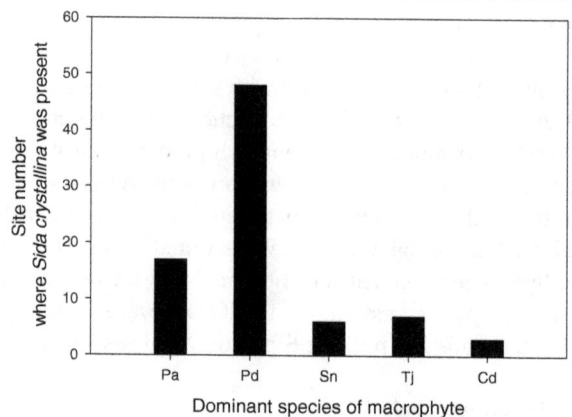

Fig. 2 Number of sites where *S. crystallina* was present classed by dominant species of macrophyte. *Pa Phragmites australis* Trin. (Cav.), *Pd Paspalum distichum* L., *Sn Salvinia natans* L., *Tj Trapa japonica* Flerov., *Cd Ceratophyllum demersum* L

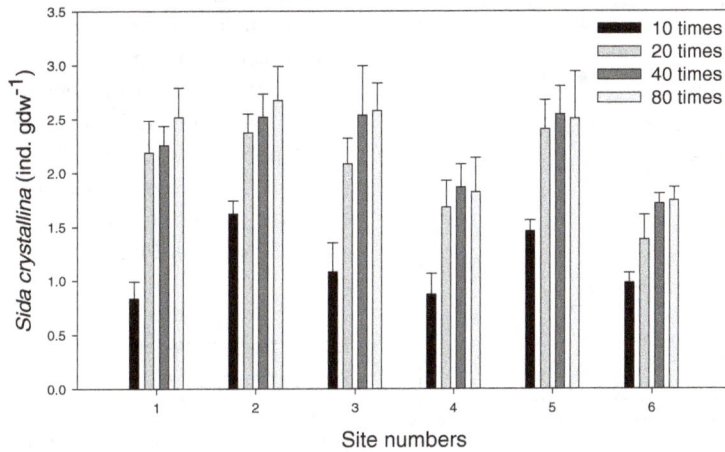

Fig. 3 Numbers of *Sida crystallina* detached from macrophyte stands based on the shaking effort (10, 20, 40, and 80 times) at each study sites

Figure 5 shows the organ used by *S. crystallina* for attachment. They attached to substrata using adhesive anchors that are connected to the cuticle by anchor threads.

S. crystallina food sources

The results of stable isotope analysis indicated potential food sources of *S. crystallina*. *S. crystallina* $\delta^{13}C$ values indicated a bias in the composition of their food (between SPOM and EPOM) and reflected differences in the $\delta^{13}C$ values of SPOM and EPOM (Fig. 6). At all sites, *S. crystallina* was more dependent on EPOM on macrophyte surfaces than SPOM. Although $\delta^{13}C$ and $\delta^{15}N$ values of *S. crystallina* and EPOM differed among sites, a common relationship between the grazer and its diet could be assumed for all sites (a fractionation coefficient of 1 ‰ was used per trophic step for carbon isotopes

and 2~3 ‰ per trophic step for nitrogen isotopes). Moreover, when the contributions of the two potential food sources to *S. crystallina* diet were calculated from isotope analyses, the contribution of EPOM (average, 83 %) was higher than that of SPOM (average, 17 %), according to the two-source mixing model.

Discussion

In this study, *S. crystallina* distribution showed a strong relationship with macrophyte biomass. Among cladoceran species, *S. crystallina* is well known to have an epiphytic character and mainly attaches to stem and leaf surfaces of

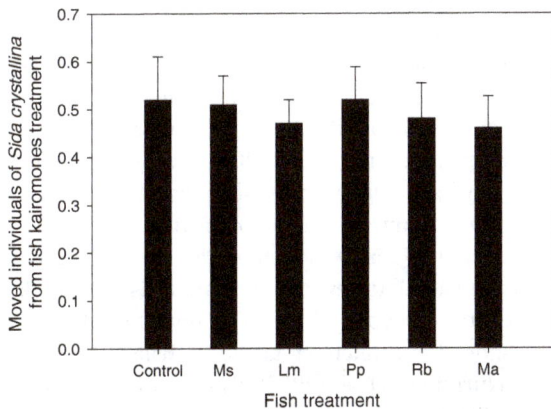

Fig. 4 Number of *Sida crystallina* individuals to have moved in response to simulated predation by five fish species. *Ms Micropterus salmoides* (Lacepéde, 1802), *Lm Lepomis macrochirus* (Rafinesque, 1819), *Pp Pseudorasbora parva* (Temminck and Schlegel, 1846), *Rb Rhinogobius brunneus* (Temminck and Schlegel, 1845), *Ma Misgurnus anguillicaudatus* (Cantor, 1842)

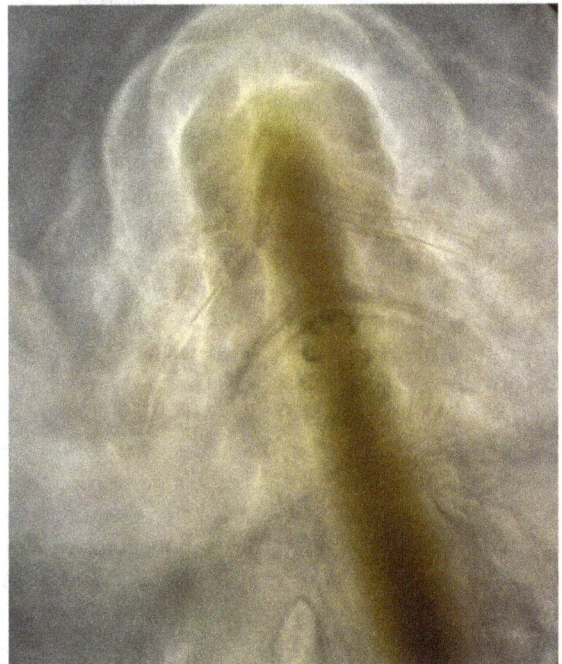

Fig. 5 Adhesive anchors of *Sida crystallina*

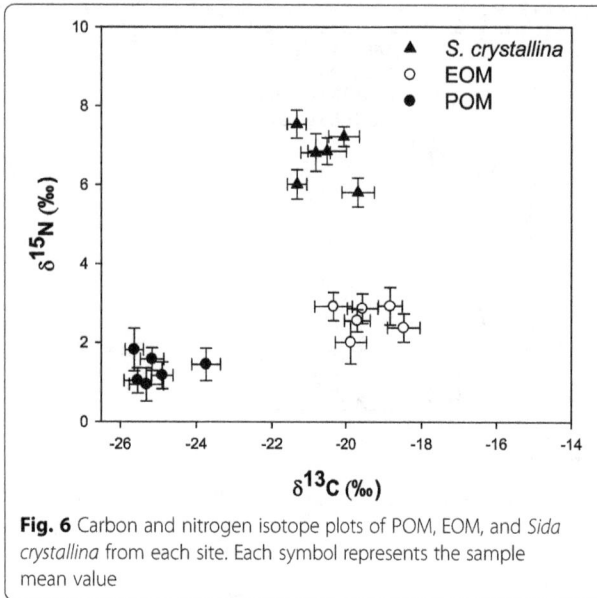

Fig. 6 Carbon and nitrogen isotope plots of POM, EOM, and *Sida crystallina* from each site. Each symbol represents the sample mean value

macrophytes. Thus, *S. crystallina* was not frequently observed in ecosystems where macrophytes were absent or not abundant. We found distribution patterns of *S. crystallina* varied with dominant macrophyte species. Moss et al. (1998) suggested that some epiphytic zooplankton species could attain high biomasses in free-floating and floating-leaved macrophyte beds. However, macrophyte species occupy limited space in the water (mostly at the water surface) and are mainly utilised as habitat by small species (i.e. rotifers; Choi et al. 2014c). Although submerged macrophytes make a large contribution to aquatic habitat complexity, the leaves and stems of submerged macrophytes are more easily agitated by wind and water currents than those of other plant species (Vermaat et al. 2000). As a result, they are less suitable for the attachment of epiphytic species. Some reports suggest that submerged macrophytes are mainly used by pelagic zooplankton, such as daphnids, as daytime refuges (Lauridsen and Lodge 1996; Burks et al. 2002). In contrast, emergent macrophyte species are tightly fixed in place and thus are suitable for attachment of large epiphytic species such as *S. crystallina*. Although emergent macrophytes are known to have simpler structure than other aquatic plant species (Choi et al. 2014b), it seems emergent plants are important as habitat for *S. crystallina*. In particular, *S. crystallina* was more abundant in ecosystems where *P. distichum* dominated than those dominated by other emergent macrophyte species (e.g. *P. australis*). We suggest that *S. crystallina* prefers *P. distichum* because it has a more complex structure (more diversified arrangement of stems and leaves) than *P. australis*. In addition, *S. crystallina* distribution was related to water depth and total phosphorus. Deeper water may increase the volume of water available to be occupied by macrophytes and thus support greater densities of *S.*

crystallina. In addition, *S. crystallina* mostly occurred in aquatic environments where nutrient levels were relatively low. In such wetland, macrophytes are thought to make a large contribution to nutrient removal and improve water quality (Sooknah and Wilkie 2004). Thus, the positive relationship between *S. crystallina* and macrophyte biomass can explain the negative relationship with total phosphorus. However, further investigation is needed to better understand the relationships between *S. crystallina* and nutrient status.

From the results of an additional experiment on the interactions between *S. crystallina* and macrophytes, we found densities of *S. crystallina* on macrophyte varied with shaking effort (10, 20, 40, or 80 times) in filtered wetland water. We collected more *S. crystallina* individual as the number of shakes, suggesting that *S. crystallina* was strongly attached to the macrophytes. Similarly, Sakuma et al. (2002) compared densities of some epiphytic rotifer and cladoceran species on plants after shaking for different numbers of times and found large numbers of *Lecane* and *Collotheca* remained on plants even after shaking 50 times. However, numbers of the epiphytic cladoceran genus *Alona* did not vary with shaking effort. Some crustaceans, such as conchostracans or cladocerans, have nuchal organs named maxillary glands, which secrete excreta (Thorp and Covich 2001) that are used to attach to substrate surfaces. In particular, *S. crystallina* has well developed nuchal organs, allowing it to firmly attach to leaf and stem surfaces of macrophytes. Therefore, we considered that this strong attachment of *S. crystallina* to macrophytes caused the differences in observed densities with shaking effort. Moreover, *S. crystallina* populations had statistically different demographic structures among sites.

Attachment of *S. crystallina* was not affected by fish predation. Pelagic species such as daphnids actively move to avoid predators, but epiphytic species are relatively less influenced by predation. Epiphytic cladoceran species are more sensitive to food availability than predation, and their distribution is related to food sources. For example, Sakuma et al. (2004) suggested that epiphytic chydorid cladocerans such as *Alona* migrated from the reed zone to the submerged macrophyte zone in summer and may select food-rich habitats and migrate seasonally. However, *S. crystallina* is vulnerable to fish predation and periods of high population density are often short (Nurminen et al. 2001; Balayla and Moss 2003). Nurminen et al. (2007) also reported that predation by fish has a strong effect on the migration of *S. crystallina*. They have relatively larger than other epiphytic cladocerans (*Alona*, *Chydorus*, and *Pleuroxus*) and thus can be easily captured by predators. However, the lack of response of *S. crystallina* to simulated predation may indicate adaption to long-term movement patterns, such as diurnal migration (vertical and horizontal migration; Zaret and Suffern 1976; Burks et al. 2002). Therefore, we conclude that *S. crystallina* was not influenced by fish chemical cues in the short term.

The stable isotope analysis indicated that *S. crystallina* are affected more by availability of EPOM on macrophyte stands than that of SPOM in the water. Their habit is to attach to aquatic macrophytes by means of the maxillary gland and thus filter feed from a fixed position (Fairchild 1981). Therefore, we suggest that *S. crystallina* consumes more EPOM than SPOM. Consumption of EPOM by *S. crystallina* plays an important role in the freshwater food web. Epiphytic materials on macrophyte stands are transported across food webs to epiphytic species such as *S. crystallina* (Schindler and Scheuerell 2002, Takai et al. 2002), and the trophic interactions can be interpreted as another route in food webs of freshwater ecosystems with developed macrophytes. Multiple trophic interactions can contribute not only to increased biodiversity in wetlands but also to sustaining an ecologically healthy food web.

In this study, we found that *S. crystallina* was more abundant in relatively less eutrophicated ecosystems where macrophytes were dominant. However, agricultural land surrounds most lentic freshwater ecosystems in South Korea, and non-point pollution sources continuously influence these ecosystems. Even though these ecosystems were frequently dominated by macrophytes, they had high nitrogen and phosphorus concentrations. Moreover, the development and growth of macrophytes is inhibited in some reservoirs due to their impermeable floors. We suggest that it is for these reasons that *S. crystallina* was not frequently observed in lentic freshwater ecosystems of South Korea. *Diaphanosoma* and *Simocephalus* have similar habitat requirements to *S. crystallina*, and they actively utilise macrophytes as habitat and refuges, resulting in greater abundances when macrophytes with complex structures, such as free-floating and submerged macrophytes, coexist (Lauridsen et al. 1996; Stansfield et al. 1997). *S. crystallina* was less dependent on such complexity and coexistence, but rather macrophyte biomass, water quality, and food availability. In addition, *S. crystallina* preferred emergent macrophytes that have relatively simple structures and thus is more vulnerable to predation by fish. Even though *S. crystallina* utilises emergent macrophytes as habitat, such plants did not play the role of refuges from fish predation. Moreover, *S. crystallina* did not show any avoidance behaviour to simulated fish presence. These habitat utilisation characteristics of *S. crystallina* may underlie its restricted distribution in lentic freshwater ecosystems of South Korea. Moreover, the species' preference for high water temperatures may make a significant contribution to their restricted distribution. Further investigation is needed to better understand the importance of macrophytes, fish predation, food availability, and water temperature in sustaining growth and development of *S. crystallina* populations.

Conclusions

S. crystallina was most frequently observed in shallow wetlands, and density was strongly associated with macrophyte biomass. Emergent macrophytes (*P. australis* and *P. distichum*) were preferred by *S. crystallina* to other plant species. Although free-floating or submerged macrophytes make a large contribution to aquatic habitat complexity, because they are more easily agitated by wind and water currents, they were not suitable for attachment of *S. crystallina*. Empirical studies report that some epiphytic species utilise stem and leaves surfaces of free-floating macrophytes as habitat, but this is true of most small species (e.g. those belonging to *Alona*, *Chydorus*, and *Pleuroxus*). The space occupied by free-floating macrophytes in the water is relatively small; thus, it is difficult for the large *S. crystallina* to use free-floating macrophytes. They attach to stem and leaf surfaces of macrophytes by means of secreted glues from the nuchal organ. Therefore, *S. crystallina* can more firmly attach to macrophytes than other epiphytic cladoceran species. We found greater densities of *S. crystallina* as the shaking effort increased. However, attachment of *S. crystallina* to substrate surfaces was not affected by simulated predation. Some studies suggest that predation threats by fish have strong effects on *S. crystallina*, but *S. crystallina* did not immediately respond to such threats in this study. The results of stable isotope analysis showed that *S. crystallina* primarily utilises EPOM as its food source. *S. crystallina* consumed food source through filter feeding from a fixed position on macrophyte stands, attached by means of the maxillary gland. Consequently, *S. crystallina* abundance seems to be more strongly correlated with macrophyte biomass than other cladoceran species. This may contribute to this species' predominance in various freshwater ecosystems where macrophytes dominate.

Acknowledgements

This research was fully supported by Basic Science Research Program through the National Research Foundation of Korea (NRF) funded by the Ministry of Education (grant number: NRF-2012-R1A6A3A04040793; http://www.nrf.re.kr). The funders had no role in study design, data collection and analysis, decision to publish, or preparation of the manuscript.

Authors' contributions

JYC carried out the field studies (collection and experiment), participated in the sequence alignment, and drafted the manuscript. KSJ participated in the design of the study and performed the statistical analysis. SKK and SHS participated in the sequence alignment. GJJ conceived the study, participated in its design and coordination, and helped to draft the manuscript. All authors read and approved the final manuscript.

Competing interests

The authors declare that they have no competing interests.

Author details

[1]National Institute of Ecology, Seo-Cheon Gun, Chungcheongnam province 325-813, South Korea. [2]Department of Biological Sciences, Pusan National University, Busan 609-735, South Korea. [3]Institute of Environmental Technology and Industry, Pusan National University, Busan 609-735, South Korea. [4]Nakdong River Environment Research Center, Goryeong-Gun, Gyeongsangbuk-do, South Korea.

References

Balayla, D. J., & Moss, B. (2003). Spatial patterns and population dynamics of plant-associated microcrustacea (cladocera) in an English shallow lake (Little Mere, Cheshire). *Aquatic Ecology, 37*, 417–435.

Burkett, V., & Kusler, J. (2000). Climate change: potential impacts and interactions in wetlands of the United States. *Journal of the American Water Resources Association, 36*, 313–320.

Burks, R., Lodge, D. M., Jeppesen, E., & Lauridsen, T. L. (2002). Diel horizontal migration of zooplankton: costs and benefits of inhabiting littoral zones. *Freshwater Biology, 47*, 343–365.

Castilho-Noll, M. S. M., Câmara, C. F., Chicone, M. F., & Shibata, E. H. (2010). Pelagic and littoral cladocerans (Crustacea, Anomopoda and Ctenopoda) from reservoirs of the Northwest of São Paulo State, Brazil. *Biota Neotropica, 10*, 21–30.

Cattaneo, A., Galanti, G. G., Gentinetta, S., & Romo, S. (1998). Epiphytic algae and macroinvertebrates on submerged and floating-leaved macrophytes in an Italian lake. *Freshwater Biology, 39*, 725–740.

Choi, J. Y., Jeong, K. S., La, G. H., Kim, H. W., Chang, K. H., & Joo, G. J. (2011). Inter-annual variability of a zooplankton community: the importance of summer concentrated rainfall in a regulated river ecosystem. *Journal of Ecology and Field Biologyl, 34*, 49–58.

Choi, J. Y., Jeong, K. S., La, G. H., & Joo, G. J. (2014). Effect of removal of free-floating macrophytes on zooplankton habitat in shallow wetland. *Knowledge Management Aquatic Ecosystems, 414*, 11.

Choi, J. Y., Jeong, K. S., La, G. H., Kim, S. K., & Joo, G. J. (2014). Sustainment of epiphytic microinvertebrate assemblage in relation with different aquatic plant microhabitats in freshwater wetlands (South Korea). *Journal of Limnology, 73*, 197–202.

Choi, J. Y., Jeong, K. S., Kim, S. K., La, G. H., Chang, K. H., & Joo, G. J. (2014). Role of macrophytes as microhabitats for zooplankton community in lentic freshwater ecosystems of South Korea. *Ecology Information, 24*, 177–185.

De Meester, L., & Cousyn, C. (1997). The change in phototactic behaviour of a *Daphnia magna* clone in the presence of fish kairomones: the effect of exposure time. In *Cladocera: the Biology of Model Organisms* (pp. 169–175). Netherlands: Springer.

Denny, P. (1994). Biodiversity and wetlands. *Wetland Ecology and Management, 3*, 55–61.

Downing, J. A., & Peters, R. H. (1980). The effect of body size and food concentration on the in situ filtering rate of *Sida crytallina*. *Limnology and Oceanography, 25*, 883–895.

Fairchild, G. W. (1981). Movement and microdistribution of *Sida crystallina* and other littoral microcrustacea. *Ecology, 62*, 1341–1354.

Findlay, C. S. T., & Bourdages, J. (2000). Response time of wetland biodiversity to road construction on adjacent lands. *Conservation Biology, 14*, 86–94.

Gyllström, M., Hansson, L. A., Jeppesen, E., Garcia-Criado, F., Gross, E., Irvine, K., Kairesalo, T., Kornijow, R., Miracle, M., Nykänen, M., Nõges, T., Romo, S., Stephen, D., Van Donk, E., & Moss, B. (2005). The role of climate in shaping zooplankton communities of shallow lakes. *Limnology and Oceanography, 50*, 2008–2021.

Jeong, K. S., Kim, D. K., & Joo, G. J. (2007). Delayed influence of dam storage and discharge on the determination of seasonal proliferations of *Microcystis aeruginosa* and *Stephanodiscus hantzschii* in a regulated river system of the lower Nakdong River (South Korea). *Water Research, 41*, 1269–1279.

Jeppesen, E., Lauridsen, T. L., Kairesalo, T., & Perrow, M. R. (1998). *Impact of submerged macrophytes on fish-zooplankton interactions in lakes. In The structuring role of submerged macrophytes in lakes* (pp. 91–114). New York: Springer.

Korean Ministry of Environment. (2006). *[Inland wetlands investigation: Sandle Wetland, Hwapo Wetland, Jangcheok Wetland and Gumgang Wetland]. [Report*

in Korean] (p. 348). Seoul: Korean Ministry of Environment & National Wetlands Center.

Kotov, A. A., & Boikova, O. (1998). Comparative analysis of the late embryogenesis of *Sida crystallina* (O.F. Müller, 1776) and *Diaphanosoma brachyurum* (Lievin, 1848) (Crustacea: Branchiopoda: Ctenopoda). *Hydrobiologia, 380*, 103–125.

Kuczyńska-Kippen, N. M., & Nagengast, B. (2006). The influence of the spatial structure of hydromacrophytes and differentiating habitat on the structure of rotifer and cladoceran communities. *Hydrobiologia, 559*, 203–212.

Lauridsen, T. L., & Lodge, D. M. (1996). Avoidance by *Daphnia magna* of fish and macrophytes: chemical cues and predator-mediated use of macrophyte habitat. *Limnology and Oceanography, 41*, 794–798.

Lauridsen, T., Pedersen, L. J., Jeppesen, E., & Sønergaard, M. (1996). The importance of macrophyte bed size for cladoceran composition and horizontal migration in a shallow lake. *Journal of Plankton Research, 18*, 2283–2294.

Meerhoff, M., Iglesias, C., De Mello, F. T., Clemente, J. M., Jensen, E., Lauridsen, T. L., & Jeppesen, E. (2007). Effects of habitat complexity on community structure and predator avoidance behaviour of littoral zooplankton in temperate versus subtropical shallow lakes. *Freshwater Biology, 52*, 1009–1021.

Minagawa, M., & Wada, E. (1984). Stepwise enrichment of δ ^{15}N along food chains: further evidence and the relation between d15N and animal age. *Geochem Cosmochim Acta, 48*, 1135–1140.

Mizuno, T., & Takahashi, E. (1991). *An illustrated guide to freshwater zooplankton in japan*. Tokyo: Tokai University Press.

Moss, B., Kornijow, R., & Measey, G. (1998). The effect of nymphaeid (Nuphar lutea) density and predation by perch (Perca fluviatilis) on the zooplankton communities in a shallow lake. *Freshwater Biology, 39*, 689–697.

Nurminen, L., Horppila, J., & Tallberg, P. (2001). Seasonal development of the cladoceran assemblage in a turbid lake: role of emergent macrophytes. *Archiv für Hydrobiologie, 151*, 127–1540.

Nurminen, L., Horppila, J., & Pekcan-Hekim, Z. (2007). Effect of light and predator abundance on the habitat choice of plant-attached zooplankton. *Freshwater Biology, 52*, 539–548.

O'Hare, M. T., Baattrup-Pedersen, A., Nijboer, R., Szoszkiewicz, K., & Ferreira, T. (2006). Macrophyte communities of European streams with altered physical habitat. *Hydrobiology, 566*, 197–210.

Phillips, D. L., & Gregg, J. W. (2001). Uncertainty in source partitioning using stable isotopes. *Oecologia, 127*, 171–179.

Pinnegar, J. K., & Polunin, N. V. C. (1999). Differential fractionation of δ^{13}C and δ^{15}N among fish tissues: implications for the study of trophic interactions. *Functional Ecology, 13*, 225–231.

Sakuma, M., Hanazato, T., Nakazato, R., & Haga, H. (2002). Methods for quantitative sampling of epiphytic microinvertebrates in lake vegetation. *Limnology, 3*, 115–119.

Sakuma, M., Hanazato, T., Saji, A., & Nakazato, R. (2004). Migration from plant to plant: an important factor controlling densities of the epiphytic cladoceran *Alona* (Chydoridae, Anomopoda) on lake vegetation. *Limnology, 5*, 17–23.

Schindler, D. E., & Scheuerell, M. D. (2002). Habitat coupling in lake ecosystems. *Oikos, 98*, 177–189.

Smokorowski, K. E., & Pratt, T. C. (2007). Effect of a change in physical structure and cover on fish and fish habitat in freshwater ecosystems – a review and meta-analysis. *Environmental Review, 15*, 15–41.

Son, M. W., & Jeon, Y. G. (2002). Physical geographical characteristics of natural wetlands on the downstream reach of Nakdong River. *Journal of the Korean Association of Geographic Information Studies, 9*, 66–76.

Sooknah, R. D., & Wilkie, A. C. (2004). Nutrient removal by floating aquatic macrophytes cultured in anaerobically digested flushed dairy manure wastewater. *Ecological Engineering, 22*, 27–42.

Stansfield, J. H., Perrow, M. R., Tench, L. D., Jowitt, A. J., & Taylor, A. A. (1997). Submerged macrophytes as refuges for grazing Cladocera against fish predation: observations on seasonal changes in relation to macrophyte cover and predation pressure. In *Shallow Lakes' 95* (pp. 229–240). Netherlands: Springer.

Takai, N., Mishima, Y., Yorozu, A., & Hoshika, A. (2002). Carbon sources for demersal fish in the western Seto Inland Sea, Japan, examined by δ^{13}C and δ^{15}N analyses. *Limnology and Oceanography, 47*, 471–730.

Thomaz, S. M., Dibble, E. D., Evangelista, L. R., Higuti, J., & Bini, L. M. (2008). Influence of aquatic macrophyte habitat complexity on invertebrate abundance and richness in tropical lagoons. *Freshwater Biology, 53*, 358–367.

Thorp, J., & Covich, A. P. (2001). *Ecology and classification of North Amirican invertebrates* (2nd ed., p. 950). San Diego: Academic Press.

Vermaat, J. E., Santamaria, L., & Roos, P. J. (2000). Water flow across and sediment trapping in submerged macrophyte beds of contrasting growth form. *Archives of Hydrobiology, 148*, 549–562.

Vieira, L. C. G., Bini, L. M., Velho, L. F. M., & Mazão, G. R. (2007). Influence of spatial complexity on the density and diversity of periphytic rotifers, microcrustaceans and testate amoebae. *Fundamental and Applied Limnology, 170*, 77–85.

Wetzel, R. G., & Likens, G. E. (2000). *Limnological analyses*. 429 pp, Springer-Vera lag New York. Berlin Heidelberg Spin springer.

Zaret, T. M., & Suffern, J. S. (1976). Vertical migration in zooplankton as a predator avoidance mechanism. *Limnology and Oceanography, 21*, 804–813.

15

Effects of vegetation structure and human impact on understory honey plant richness: implications for pollinator visitation

Yoori Cho[1], Dowon Lee[1] and SoYeon Bae[2*]

Abstract

Background: Though the biomass of floral vegetation in understory plant communities in a forested ecosystem only accounts for less than 1% of the total biomass of a forest, they contain most of the floral resources of a forest. The diversity of understory honey plants determines visitation rate of pollinators such as honey bee (*Apis mellifera*) as they provide rich food resources. Since the flower visitation and foraging activity of pollinators lead to the provision of pollination service, it also means the enhancement of plant-pollinator relationship. Therefore, an appropriate management scheme for understory vegetation is essential in order to conserve pollinator population that is decreasing due to habitat destruction and disease infection. This research examined the diversity of understory honey plant and studied how it is related to environmental variables such as (1) canopy density, (2) horizontal heterogeneity of canopy surface height, (3) slope gradient, and (4) distance from roads. Vegetation survey data of 39 plots of mixed forests in Chuncheon, Korea, were used, and possible management practices for understory vegetation were suggested.

Results: This study found that 113 species among 141 species of honey plant of the forests were classified as understory vegetation. Also, the understory honey plant diversity is significantly positively correlated with distance from the nearest road and horizontal heterogeneity of canopy surface height and negatively correlated with canopy density.

Conclusions: The diversity of understory honey plant vegetation is correlated to vegetation structure and human impact. In order to enhance the diversity of understory honey plant, management of density and height of canopy is necessary. This study suggests that improved diversity of canopy cover through thinning of overstory vegetation can increase the diversity of understory honey plant species.

Keywords: *Apis mellifera*, Forest ecology, Airborne LiDAR, Pollination service, Vegetation structure

Background

Effects of pollinators on plant community have been widely studied since pollination was recognized as a good example of an ecosystem service, providing not only an economic benefit (Ricketts et al. 2008) but also basic life-support processes (Daily 2000). As about 80% of wild plant species rely on insect pollination for fruit and their reproduction, pollinator declines can cause a great loss in pollination services, which could significantly harm the ecosystem stability (Potts et al. 2010). Especially, honey bees (*Apis mellifera*) affected by colony collapse disorder (CCD), a recent, pervasive phenomenon

in Northern Hemisphere, are of concern in terms of conservation. Although there has been little study that proves the causes for the collapse of honey bee colonies (Hadley and Betts 2012; Wratten et al. 2012), parallel declines of plant and pollinator relationship in a large scale highlight the need for management of plants as resources for pollinators (Hadley and Betts 2012).

Among a number of insect pollinators, bees are the most significant anthophiles that are well adapted to the environment (Kevan and Baker 1983). Flowers are important food resources for both larvae and adult bees, their pollen and nectar are needed for raising young and their own overwintering (Kevan and Baker 1983). Therefore, maintaining the pollination services can be achieved by the conservation and management of enough resources for pollinators (Ricketts et al. 2008). One important factor

* Correspondence: baelovejx@gmail.com
[2]Division of Ecological Survey Research, NIE, 1210 Geumgang-ro, Seocheon-gun 33657 Seoul, South Korea
Full list of author information is available at the end of the article

for the diversity and abundance of pollinator populations is the composition of floral communities (Hegland and Boeke 2006).

The understory floral vegetation in vascular and nonvascular plants accounts for most of the plant species diversity in forested ecosystems, and anthophilous insects use flowers from these plants for their food (Proctor et al. 2012). Understory vegetative layer hardly covers 1% of the biomass of the forest but has 90% or more of the plant species that represents the majority of floristic diversity of the ecosystem (Proctor et al. 2012; Gilliam 2007). For these reasons, well-planned management scheme that understands the mechanisms maintaining diversity of understory vegetation can play a key role in conserving declining pollinators. A number of environmental factors inside and outside a forest can have an influence on richness of understory vegetation. For example, highly diverse understory vegetation could be a result of intra-stand heterogeneity due to differences in availability in limiting resources such as nutrient, light, and water (Huebner et al. 1995). Therefore, a sufficient presence of stands that are dominated by overstory species through management could be an important way to sustain the diversity of understory species (Fourrier et al. 2015).

According to Ebeling et al. (2008), visitation rates of pollinators linearly increased with the degree of blooming of flowers and the number of bloomed flowers in a particular plant community. However, existence of plants that are known to be favored by pollinators in general did not have a significant effect on richness of pollinators. This result indicates that planting abundant honey plants only is not enough to protect pollinator species, but there needs to be something more complicated and policy makers and conservationists should consider environmental variables. Honey plants, in this study, are defined as plant species that produce nectar required by honey bees. Bees collect nectar and convert it into honey which is the primary food source for maintenance of their colony.

This study is to quantify and compare understory honey plant diversity among sampled plots in Chuncheon, South Korea, and to determine how this diversity was related to different environmental variables, in particular vegetation structure, topography, and human impact. Then, we suggest a possible forest management scheme for pollinator conservation as an implication of this study.

Methods
Study site
This study was conducted in temperate mixed forests in Chuncheon (37° 55′ N, 127° 52′ E), South Korea (Fig. 1). It is located in Gangwon province, approximately 90 km east of Seoul. The forests of the study area have 29% of broadleaved forests cover, 30% of coniferous forests cover, and 41% of mixed forests (Bae 2015). Mongolian oak

(*Quercus mongolica*) and cork oak (*Quercus variabilis*) are the dominant broadleaved species, while Japanese red pine (*Pinus densiflora*), pitch pine (*Pinus rigida*), and Korean pine (*Pinus koraiensis*) are the dominant coniferous species (Bae 2015). The area of the forests is 400.55 km^2, and the annual average precipitation and temperature of the study site are 1347.3 mm and 11.1 °C (information from Korea Meteorological Administration).

Environmental variables
Understory species diversity may be the results from many different environmental variables (Huebner et al. 1995). The factors selected for this study are the variables that are known to be influencing plant establishment and growth (Table 1). The environmental variables measured at each site were (1) canopy density, (2) horizontal heterogeneity of canopy surface height, (3) slope gradient, and (4) distance from roads. Airborne LiDAR (Light Detection and Ranging), a remote sensing tool for measuring three-dimensional habitat structure, was adopted to measure canopy density, horizontal heterogeneity of canopy surface height, and slope gradient. GIS (geographic information system, provided by © Naver Corp.) data was also adopted to measure distances between the sampled plots and the surrounding roads. LiDAR data are comprised of coordinates and elevation and can be classified into ground and vegetation returns, so that by using the data, it is possible to measure stand structure additionally such as vertical complexity or horizontal heterogeneity. We calculated horizontal heterogeneity of canopy surface height by using a measure of spatial autocorrelation with local Geary's c. Each environmental variable has no significant correlation with another one (Spearman's rank correlation coefficient among four variables $\rho < .05$).

Sampling
Thirty-nine plots were sampled across the study region for calculating diversity of honey plants. Plots that are at least 200 m apart from each other were only chosen in order to control potential environmental effects. The sampling was conducted at 39 sampling points in June and September, 2014 and May 2015. Each point was sampled in a 10 × 10 m quadrant. In each sample site, species in the ground vegetation (shrubs and herbaceous plants) and the canopy cover of the tree layer were recorded. The elevation survey points below 400 m were limited. In addition, sampling sites were chosen as apart from road at least 100 m. "Overstory species" in this study means woody species of the canopy stand and "understory species" describes plants growing on the forest floor whose heights are less than 2 m, including bryophytes, herbaceous, and woody species. Honey plant identification was conducted with reference to Jang (2009).

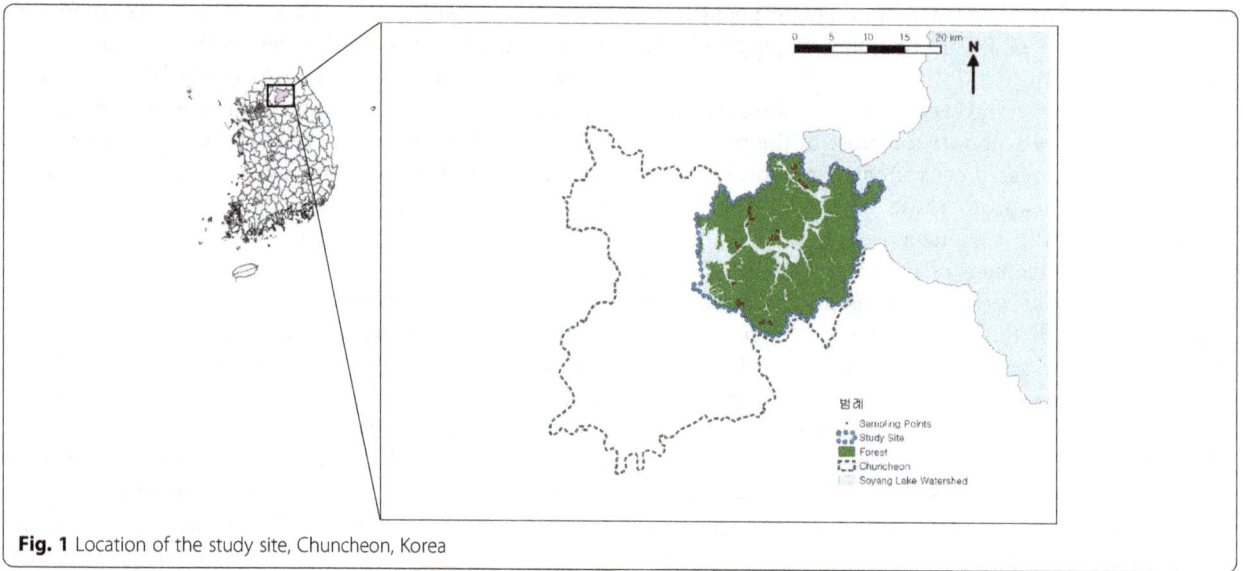

Fig. 1 Location of the study site, Chuncheon, Korea

Statistics

Statistical dependence between overall, overstory, and understory vegetation was measured with Spearman rank correlation. The generalized linear models with Poisson distribution were employed for modeling understory species diversity, since the dependent variable, species richness per plot, is count data. All the data were statistically analyzed using R (R Core Team 2015).

Results

Among 251 plant species excluding 5 unspecified species, 141 species are identified as honey plants in our study site (Table 2). The mean of occurred honey plants species number was 18.13 ± 7.48 (mean ± SD), and most of them, 15.31 ± 7.82 (mean ± SD), were understory honey plant (Table 3). Dominant honey plant species of understory communities were *Quercus* spp., *Corylus heterophylla*, *Castanea crenata*, *Styrax obassia*, and *Lindera obtusiloba*. The rank correlation between the overall honey plants richness and species richness of understory honey plant was 0.97 (Spearman rank correlation, $p < .001$) (Fig. 2). Those results indicated the overall honey plants richness were determined by the species richness of understory honey plant.

In Poisson regression, understory diversity is significantly positively correlated with horizontal heterogeneity of canopy surface height and distance from a road and negatively correlated with canopy density (Table 4).

Discussion

Canopy density and horizontal heterogeneity of canopy surface height

Canopy dynamics of overstory is a product from permanent processes of succession and disturbance, and they contribute to heterogeneity of community, meeting ecological requirements such as light, water, and nutrient (Valverde and Silvertown 1997). And it is understory vegetation that is much benefited from heterogeneity of the canopy properties. Especially, canopy gaps and the light that passed through the gaps have been considered as the most important limiting factor that affects composition and growth of understory vegetation (Lefrancois et al. 2008; Latif and Blackburn 2010). Furthermore, canopy gaps created by harvesting increase not only species diversity but also functional diversity of the understory vegetation (Kern et al. 2014). According to Walters and Stiles (1996), both floral density and pollinator visitation rates were considerably higher in patches under forest canopy gaps than patches under closed canopy.

Table 1 The environmental variables selected and description for each variable

	Environmental variables	Variable description	Mean ± SD
Vegetation structure	Canopy density (%)	Average percentage of canopy density above 2 m within 50-m radius of plot point	86.46 ± 16.34
	Horizontal heterogeneity of canopy surface height	A measure of spatial autocorrelation of maximum vegetation height with local Geary's c within 50-m radius	1.62 ± 1.19
Topography	Slope gradient (°)	Average slope gradient within 50-m radius of plot point	31.05 ± 7.30
Human impact	Distance from a road (m)	Distance to the closest road from plot point	258.79 ± 199.13

Table 2 Overall and honey plant richness of the studied sites

	Richness	Overstory	Understory	Co-occurring
Overall	251	17	210	24
Honey plant	141	9	113	19

"Overstory" and "understory" stands for the richness of plant species that were only found in each story. Among 251 plant species, 141 species are honey plants in our study site. Richness of understory honey plants accounts for 93.62% of overall honey plant richness while that of overstory accounts for only 19.86%

Canopy gaps have a positive relation with species richness of shrub layer especially (Gazol and Ibáñez 2009). In spite of some exceptions, the relationship between forest canopy gap and light transmittance are the results of complex interactions leading to denser canopy space filling and more light interception (Angelini et al. 2015). Thus, understory vegetation may receive enough amount of radiance. Different traits of forest overstory affect the creation of canopy gaps which determines light intensity in the understory. The structure of overstory is one of the characteristics mainly correlated to light availability for understory vegetation. High values of local Geary from our result are determined by significant differences between the pivotal locality and its neighbors (Sokal et al. 1998). As local Geary values were significantly positively related to understory diversity, it illustrates that sufficient canopy openings with horizontal heterogeneity of the height of overstory vegetation should be guaranteed (Fig. 3).

According to Angelini et al. (2015), when the canopy is closed and homogeneous, the passage of light through overstory vegetation relies on the characteristics of the crowns, then on overstory plant richness. When management practices open the crown cover with the same overstory richness, stand density accounts for the light availability. In our study, however, the local Geary mean value of maximum vegetation height was correlated to neither the canopy density (2–60 m, Pearson's correlation coefficient $r = -0.268$, p value $= .099$) nor species richness of overstory vegetation ($r = -0.201$, p value $= 0.219$). This result demonstrates that it is solely the heterogeneity of overstory's vertical structure, rather than its species composition or density, which plays an important role in determining understory plant's richness. It is in contrast with the examination of literatures by Angelini et al. (2015). Angelini et al. (2015) concluded that in regular mixed forests, the understory light availability such as the

Table 3 Mean and standard deviation of honey plant richness in overall, overstory, and understory layers over 39 sites

	Mean ± standard deviation
Overall honey plant richness	18.13 ± 7.48
Overstory honey plant richness	2.82 ± 1.71
Understory honey plant richness	15.31 ± 7.82

spatial and temporal distribution of radiance are controlled by tree composition of canopy covers. The variability between conclusions may come from differences in arrangement of stems in space and their size of overstory vegetation.

Distance from a road

Roads have become an important landscape feature (Watkins et al. 2003) that may influence ecological processes in a surrounding habitat. It is still unclear how distance from roads can affect the diversity of understory vegetation in a forest, and the ecological impacts of roads and traffic, however, have become a great concern (Spellerberg 1998), especially on understory plant richness (Enoki et al. 2014). Guirado et al. (2007) studying Mediterranean forests found that the total species richness of understory vegetation increased with distance to the main roads. Pollutants affecting plants and animals range from noise, light, sand, dust to other particulates, metals, and gases (Spellerberg 1998). Also, road dust could be an important factor for plant diversity. Dust that are mobilized and spread by the road traffic can restrict photosynthesis, respiration, and transpiration of plants (Trombulak and Frissell 2000). According to Wong et al. (1984, in Watkins et al. 2003), plant materials sampled from a site with a high traffic density showed a significant decrease in root growth whereas increased root growth was found for plants sampled from low-traffic-density areas. In addition to physiological stress generated from roads, the closer the habit is to a road, the more disturbances on vegetation by human influence there may be. On the other hand, Watkins et al. (2003) found higher richness of understory vegetation on the roadside edge since lower canopy cover of the edge allows more light to reach understory. Thus, road density, vehicle transportation, and distribution of roads around the area of interest may interplay, affecting the richness of understory vegetation.

Implications

From the findings of this study, honey plant richness in understory is significantly related to canopy density and distance from the closest road. Understory light and its distribution, especially, can be controlled, at least partially, by silvi-cultural and harvesting practices according to the overstory structure (Angelini et al. 2015). Therefore, in order to enhance the diversity of understory honey plant as a pollinator conservation strategy, canopy density of overstory should be managed. Canopy gaps created by human activities such as logging could change not only micro-environmental conditions of a forest increasing understory richness but also the total bee abundance. According to Jackson et al. (2014), the number of pollinator groups was more abundant in

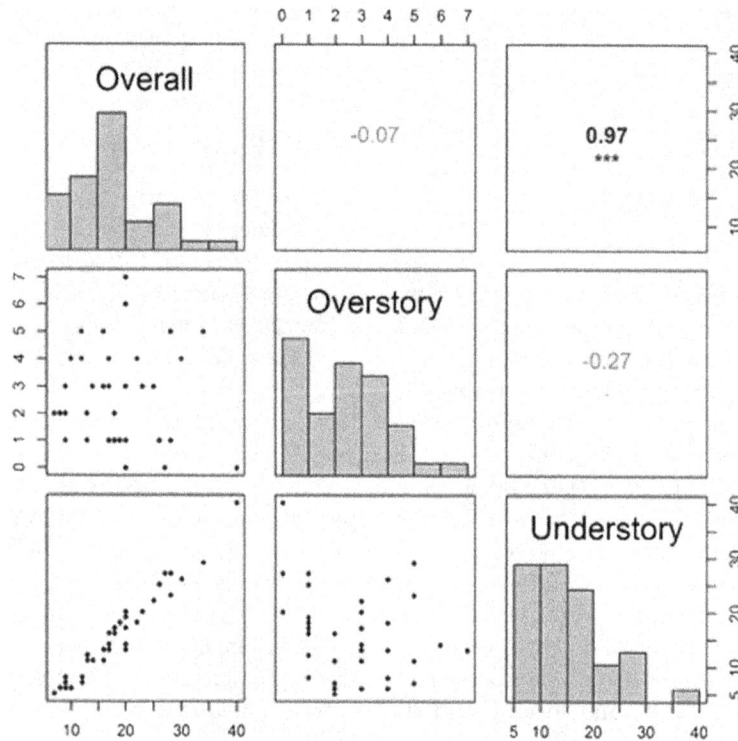

Fig. 2 Multiple displays of pairwise relationships between overall, overstory, and understory honey plant richness with Spearman rank correlation of significance levels (***p < .001)

forests that are younger and closer to logging roads, likely due to more available light and more floral resources near the roads. In a recent study by Girling et al. (2013), it was found that pollution from diesel exhaust emissions altered the constituents of a synthetic blend of floral odors. Exposure of flowers to diesel exhaust can reduce both the ability of honey bees to navigate to the flowers and efficacy of pollination services they provide.

In addition to management of canopy gaps, planting diverse honey plant species is certainly a supplementary strategy for pollinators like honey bee since existence of diverse species guarantee different flowering dates. So planting diverse species and increasing the understory richness for pollinator improves functional diversity of a forest, making it more stable and resilient in different

Table 4 Poisson regression with understory diversity (species richness) as the dependent variable and environmental variables included as independent variables. n = 39, p value ≤.05 level of significance

Variables	B	Wald χ^2	p value
Canopy density	−1.033	50.267	<.000
Horizontal heterogeneity of canopy surface height	0.188	38.785	<.000
Slope gradient	−0.404	1.568	.211
Distance from a road	0.001	8.302	.004

blooming periods of understory vegetation (McCann 2000). Blossom density, however, could be more significant than plant richness in explaining pollinator activity (Hegland and Boeke 2006). Also, a forest habitat that provides only a few floral resources, i.e., limited number of honey plants, may be inadequate to nurture healthy bees (Vaudo et al. 2013). Bees can self-select their own food to satisfy their requirements, so that they will be negatively affected if various foraging choices and diets are not guaranteed throughout their lives (Vaudo et al. 2013).

According to the Korea Beekeeping Association (Korea Beekeeping Association 2007), plant species that were selected for governmental honey plant planting scheme were all tree species such as *Quercus acutissima*, *Betula platyphylla* var. *japonica*, *Fraxinus rhynchophylla*, and *Liriodendron tulipifera* for better honey harvesting and eventually for economic purposes. However, it is still not obvious how trees can contribute to conservation of pollinators. Roubik (1993) in his research hypothesized that bees either like to forage in the canopy or specialize there. The results showed that the data obtained in a long term did not prove the hypothesis, except for the nocturnal *Megalopta* and one stingless bee, *Partamona*, for 20 species and 10 genera including *A. mellifera*. This demonstrates that bees are quite opportunistic and unpredictable (Roubik 1993).

Fig. 3 Illustration that conceptualizes local Geary mean value of maximum vegetation height. As the vertical structure of vegetation of a plot is more heterogeneous, local Geary mean is higher than less heterogeneous that the other

Overstory vegetation like understory vegetation also provides abundant floral resources for pollinators; fluctuations of floral resources, however, highly vary every year (Inari et al. 2012). This temporal discrepancy in floral resources may affect pollinator population (Inari et al. 2012). However, the flowering days of understory species is longer than that of overstory tree species, lasting from early spring to autumn (Inari et al. 2012).

Yet, there are still limitations of this study. Light availability for understory vegetation varied by canopy structures is one of the most important factors in this study. Light transmittance, however, significantly varies among tree species, and it depends on their light demands and shade tolerance (Inari et al. 2012). The composition of understory vegetation can be more related to their trait of shade tolerance than canopy gaps (Kern et al. 2014), regardless of whether or not they are the plants that pollinators favor.

In addition, the effect of slope aspect on the diversity of understory vegetation was not considered due to the lack of the number of samples. Slop aspect, however, is inherently related to available light and soil moisture, which should be discussed in detail. In general, understory vegetation shows high species richness and alpha diversity when it is shed on south-facing slope (Messier et al. 1998; Valladares and Niinemets 2008). In Northern Hemisphere, south-facing slope usually has higher temperature and light intensity and lower humidity with a more variable microclimate (Auslander et al. 2003). Still, there is no consensus agreed between authors on which aspect vegetation develops. It is more dependent on regions. According to a study conducted at Mt. Jumbong, South Korea, high productivity and species diversity on north-facing slope having less evaporation rate since water use efficiency was one of the most important limiting factors for plant growth in Korea. Therefore, the relationship between understory diversity and slope aspect in our study may be because understory vegetation has a higher evaporation rate, so that plant species on south-directed slopes could be under stress due to the higher atmospheric moisture demand (Huebner et al. 1995). South-directed slopes are expected to have six times higher solar radiation than north-directed slope (Auslander et al. 2003). Slopes facing south retain flora that are more tolerant to stress and demand more light, which is revealed by long-term monitoring (Bennie et al. 2006; Gong et al. 2008). Although located only a few hundred meters apart, the microclimatic conditions generated by different slope aspects could vary dramatically. Between slopes, a sharp microclimatic gradient can stress the plants (Auslander et al. 2003) and their pollinators. Thus, differences and diversity in species composition on different slope aspects were explained by resource availability. Another possible explanation for higher understory vegetation is different pedologic processes. The difference of vegetation composition and productivity, and erosion processes of soil are generally accentuated on slopes facing south (Rech et al. 2001), which leads to contrasting soil characteristics of north- and south-directed slopes (Gong et al. 2008).

Although the findings of this study are meaningful, understory vegetation richness in general can be explained by intermediate disturbance hypothesis. As stated above, partially removing overstory canopy can create more growing space and release more resources that are available and therefore have a positive effect on understory plant diversity. Again, this means that opening canopy gaps as much as possible by managing canopy density is not a definite solution for enhancing understory honey plant richness. Like the local Geary value illustrated, sufficient canopy openings with horizontal heterogeneity of height of overstory vegetation should be the first to be considered.

For further research, spatial and temporal variability of understory light determined by not only gaps but also other qualitative and quantitative attributes of canopy cover such as heterogeneity of stand spacing and characteristics of crowns should be studied. The causal relationship between the actual amount of pollutants from roads and the growth of understory vegetation should be further studied as well. Pollination efficiency measured by richness and abundance of pollinators in the study site should also be empirically examined to validate the findings of this study. Also, collecting and statistically examining more slope aspect data will contribute to the findings of this study.

Conclusions

The floral vegetation in understory plants community in forested ecosystems plays an important role as the main food source for pollinators such as honey bees. This research shows that vegetation structure and human impact such as distance to roads are correlated to the diversity of understory honey plant. This study suggests that management of density and height of canopy is necessary in order to improve the diversity of understory honey plant.

Acknowledgements

We sincerely thank Minjoo Lee of the National Institute of Forest Science and Dr. Engkyong Lee of Teoal for their contribution in developing this paper. We greatly acknowledge Dr. Wanmo Kang of the Korea Environment Institute, Ho Choi of Seoul National University, and Hyo-In Lim of the National Institute of Forest Science for exerting much effort for the field work of this research. We also would like to thank the National Forest Management Office at Chuncheon for their cooperation throughout the study period and two anonymous reviewers for their valuable comments.

Funding

This research was supported by the Basic Science Research Program through the National Research Foundation of Korea (NRF) funded by the Ministry of Education (NRF-2013R1A1A2012874) and through the 4th National Ecosystem Survey by the National Institute of Ecology.

Authors' contributions

YC and SYB undertook the statistical analysis and drafted the manuscript. SYB also participated in collecting the field data for the vegetation surveys. DL assisted in the data collection (vegetation surveys) and contributed in the discussion of this study. All authors read and approved the final manuscript.

Competing interests

The authors declare that they have no competing interests.

Author details

[1]Department of Environmental Planning, Seoul National University, 1 Gwanak-ro, Seoul 08826, South Korea. [2]Division of Ecological Survey Research, NIE, 1210 Geumgang-ro, Seocheon-gun 33657 Seoul, South Korea.

References

Angelini, A., Corona, P., Chianucci, F., & Portoghesi, L. (2015). Structural attributes of stand overstory and light under the canopy.

Auslander, M., Nevo, E., & Inbar, M. (2003). The effects of slope orientation on plant growth, developmental instability and susceptibility to herbivores. *Journal of Arid Environments, 55*, 405–416.

Bae, S. Y. (2015). *Modelling avian taxonomic, functional, and phylogenetic diversity in relation to 3-D forest structure.* Seoul: PhD dissertation, Seoul National Univ.

Bennie, J., Hill, M. O., Baxter, R., & Huntley, B. (2006). Influence of slope and aspect on long-term vegetation change in British chalk grasslands. *Journal of Ecology, 94*, 355–368.

Daily, G. C. (2000). Management objectives for the protection of ecosystem services. *Environmental Science & Policy, 3*, 333–339.

Ebeling, A., Klein, A. M., Schumacher, J., Weisser, W. W., & Tscharntke, T. (2008). How does plant richness affect pollinator richness and temporal stability of flower visits? *Oikos, 117*, 1808–1815.

Enoki, T., Kusumoto, B., Igarashi, S., & Tsuji, K. (2014). Stand structure and plant species occurrence in forest edge habitat along different aged roads on Okinawa Island, southwestern Japan. *Journal of Forest Research, 19*, 97–104.

Fourrier, A., Bouchard, M., & Pothier, D. (2015). Effects of canopy composition and disturbance type on understorey plant assembly in boreal forests. *Journal of Vegetation Science, 26*, 1225–1237.

Gazol, A., & Ibáñez, R. (2009). Different response to environmental factors and spatial variables of two attributes (cover and diversity) of the understorey layers. *Forest Ecology and Management, 258*, 1267–1274.

Gilliam, F. S. (2007). The ecological significance of the herbaceous layer in temperate forest ecosystems. *Bioscience, 57*, 845–858.

Girling, R. D., Lusebrink, I., Farthing, E., Newman, T. A., & Poppy, G. M. (2013). Diesel exhaust rapidly degrades floral odours used by honeybees. *Scientific Reports, 3*, 2779.

Gong, X., Brueck, H., Giese, K., Zhang, L., Sattelmacher, B., & Lin, S. (2008). Slope aspect has effects on productivity and species composition of hilly grassland in the Xilin River Basin, Inner Mongolia, China. *Journal of Arid Environments, 72*, 483–493.

Guirado, M., Pino, J., & Roda, F. (2007). Comparing the role of site disturbance and landscape properties on understory species richness in fragmented periurban Mediterranean forests. *Landscape Ecology, 22*, 117–129.

Hadley, A. S., & Betts, M. G. (2012). The effects of landscape fragmentation on pollination dynamics: absence of evidence not evidence of absence. *Biological Reviews, 87*, 526–544.

Hegland, S. J., & Boeke, L. (2006). Relationships between the density and diversity of floral resources and flower visitor activity in a temperate grassland community. *Ecological Entomology, 31*, 532–538.

Huebner, C. D., Randolph, J., & Parker, G. (1995). Environmental factors affecting understory diversity in second-growth deciduous forests. *American Midland Naturalist, 155–165*.

Inari, N., Hiura, T., Toda, M. J., & Kudo, G. (2012). Pollination linkage between canopy flowering, bumble bee abundance and seed production of understorey plants in a cool temperate forest. *Journal of Ecology, 100*, 1534–1543.

Jackson, M. M., Turner, M. G., & Pearson, S. M. (2014). Logging legacies affect insect pollinator communities in Southern Appalachian Forests. *Southeastern Naturalist, 13*, 317–336.

Jang, J. W. (2009). *A study on honey plants in Korea.* Daegu: PhD Dissertation, Daegu Univ.

Kern, C. C., Montgomery, R. A., Reich, P. B., & Strong, T. F. (2014). Harvest-created canopy gaps increase species and functional trait diversity of the forest ground-layer community. *Forest Science, 60*, 335–344.

Kevan, P., & Baker, H. (1983). Insects as flower visitors and pollinators. *Annual Review of Entomology, 28*, 407–453.

Korea Beekeeping Association. (2007). Current status of honey plant afforestation. *Korea Beekeeping Bulletin, 331*, 52–23.

Latif, Z. A., & Blackburn, G. A. (2010). The effects of gap size on some microclimate variables during late summer and autumn in a temperate broadleaved deciduous forest. *International Journal of Biometeorology, 54*, 119–129.

Lefrancois, M.-L., Beaudet, M., & Messier, C. (2008). Crown openness as influenced by tree and site characteristics for yellow birch, sugar maple, and eastern hemlock. *Canadian Journal of Forest Research, 38*, 488–497.

McCann, K. S. (2000). The diversity–stability debate. *Nature, 405*, 228–233.

Messier, C., Parent, S., & Bergeron, Y. (1998). Effects of overstory and understory vegetation on the understory light environment in mixed boreal forests. *Journal of Vegetation Science, 9*, 511–520.

Potts, S. G., Biesmeijer, J. C., Kremen, C., Neumann, P., Schweiger, O., & Kunin, W. E. (2010). Global pollinator declines: trends, impacts and drivers. *Trends in Ecology & Evolution, 25*, 345–353.

Proctor, E., Nol, E., Burke, D., & Crins, W. J. (2012). Responses of insect pollinators and understory plants to silviculture in northern hardwood forests. *Biodiversity and Conservation, 21*, 1703–1740.

R Core Team. (2015). *R: a language and environment for statistical computing.* Vienna, Austria: R Foundation for Statistical Computing. https://www.R-project.org/.

Rech, J. A., Reeves, R. W., & Hendricks, D. M. (2001). The influence of slope aspect on soil weathering processes in the Springerville volcanic field, Arizona. *Catena, 43*, 49–62.

Ricketts, T. H., Regetz, J., Steffan-Dewenter, I., Cunningham, S. A., Kremen, C., Bogdanski, A., Gemmill-Herren, B., Greenleaf, S. S., Klein, A. M., & Mayfield, M. M.

(2008). Landscape effects on crop pollination services: are there general patterns? *Ecology Letters, 11*, 499–515.

Roubik, D. W. (1993). Tropical pollinators in the canopy and understory: field data and theory for stratum "preferences". *Journal of Insect Behavior, 6*, 659–673.

Sokal, R. R., Oden, N. L., & Thomson, B. A. (1998). Local spatial autocorrelation in a biological model. *Geographical Analysis, 30*, 331–354.

Spellerberg, I. (1998). Ecological effects of roads and traffic: a literature review. *Global Ecology and Biogeography, 7*, 317–333.

Trombulak, S. C., & Frissell, C. A. (2000). Review of ecological effects of roads on terrestrial and aquatic communities. *Conservation Biology, 14*, 18–30.

Valladares, F., & Niinemets, U. (2008). Shade tolerance, a key plant feature of complex nature and consequences. *Annual Review of Ecology, Evolution, and Systematics, 39*, 237–257.

Valverde, T., & Silvertown, J. (1997). Canopy closure rate and forest structure. *Ecology, 78*, 1555–1562.

Vaudo, A. D., Tooker, J. F., Grozinger, C. M., & Patch, H. M. (2013). Bee nutrition and floral resource restoration. *Current Opinion in Insect Science, 10*, 133–141.

Walters, B. B., & Stiles, E. W. (1996). Effect of canopy gaps and flower patch size on pollinator visitation of Impatiens capensis. *Bulletin of the Torrey Botanical Club, 184–188*.

Watkins, R. Z., Chen, J., Pickens, J., & Brosofske, K. D. (2003). Effects of forest roads on understory plants in a managed hardwood landscape. *Conservation Biology, 17*, 411–419.

Wong, M., Cheung, L., & Wong, W. (1984). Effects of roadside dust on seed germination and root growth of Brassica chinensis and B. parachinensis. *Science of the Total Environment, 33*, 87–102.

Wratten, S. D., Gillespie, M., Decourtye, A., Mader, E., & Desneux, N. (2012). Pollinator habitat enhancement: benefits to other ecosystem services. *Agriculture Ecosystems & Environment, 159*, 112–122.

16

Effects of thinning intensity on nutrient concentration and enzyme activity in *Larix kaempferi* forest soils

Seongjun Kim[1], Seung Hyun Han[1], Guanlin Li[1], Tae Kyung Yoon[2], Sang-Tae Lee[3], Choonsig Kim[4] and Yowhan Son[1*]

Abstract

Background: As the decomposition of lignocellulosic compounds is a rate-limiting stage in the nutrient mineralization from organic matters, elucidation of the changes in soil enzyme activity can provide insight into the nutrient dynamics and ecosystem functioning. The current study aimed to assess the effect of thinning intensities on soil conditions. Un-thinned control, 20 % thinning, and 30 % thinning treatments were applied to a *Larix kaempferi* forest, and total carbon and nitrogen, total carbon to total nitrogen ratio, extractable nutrients (inorganic nitrogen, phosphorus, calcium, magnesium, potassium), and enzyme activities (acid phosphatase, β-glucosidase, β-xylosidase, β-glucosaminidase) were investigated.

Results: Total carbon and nitrogen concentrations were significantly increased in the 30 % thinning treatment, whereas both the 20 and 30 % thinning treatments did not change total carbon to total nitrogen ratio. Inorganic nitrogen and extractable calcium and magnesium concentrations were significantly increased in the 20 % thinning treatment; however, no significant changes were found for extractable phosphorus and potassium concentrations either in the 20 or the 30 % thinning treatment. However, the applied thinning intensities had no significant influences on acid phosphatase, β-glucosidase, β-xylosidase, and β-glucosaminidase activities.

Conclusions: These results indicated that thinning can elevate soil organic matter quantity and nutrient availability, and different thinning intensities may affect extractable soil nutrients inconsistently. The results also demonstrated that such inconsistent patterns in extractable nutrient concentrations after thinning might not be fully explained by the shifts in the enzyme-mediated nutrient mineralization.

Keywords: Extracellular enzyme, Japanese larch, Nutrient availability, Soil organic matter, Thinning intensity

Background

Forest soil is a key component in ecosystems as a major storage and source of plant-available nutrients. In addition, it can be affected by the changes in other living and non-living components of forests since they interact with each other (Attiwill and Adams 1993). Accordingly, anthropogenic activities on forests may directly or indirectly shift soil conditions, disturb nutrient cycle, and consequently hinder future forest productivity if they are inappropriately conducted (Grigal 2000). In this context,

the influences of common forest management practices on soil conditions have been extensively assessed to find optimal strategies for sustaining forest productivity in the long term (Boerner et al. 2008; Achat et al. 2015).

Thinning, the selective tree cutting to achieve various management purposes, is one of the most frequently applied forest management practices. It can affect various processes related to soil nutrient dynamics. For example, tree density reduction after thinning is known to alter microclimate by increasing soil temperature and moisture (Son et al. 2004; Masyagina et al. 2006). Thinning can also influence soil conditions such as the quantity and quality of soil organic matter (Son et al. 2004; Smolander et al. 2013) and extractable nutrient

* Correspondence: yson@korea.ac.kr
[1]Department of Environmental Science and Ecological Engineering, Graduate School, Korea University, Seoul 02841, South Korea
Full list of author information is available at the end of the article

concentration (Hwang and Son 2006; Kim et al. 2015). These changes in microclimate and soil conditions are known to impact the activity of extracellular enzymes mediating the decomposition of lignocellulosic compounds (Giai and Boerner 2007; Adamczyk et al. 2015). As the decomposition of lignocellulosic compounds is a rate-limiting stage in the nutrient mineralization from organic matters, elucidation of the changes in soil enzyme activity can provide insight into the nutrient dynamics and ecosystem functioning (Sinsabaugh et al. 2008).

Japanese larch (*Larix kaempferi* Lamb.) is a major deciduous coniferous species in South Korea. This species accounts for approximately 16.5 % of the coniferous forest area in South Korea and has been widely used for reforestation and large log production (Korea Forest Service 2015). As large areas of *L. kaempferi* forests were established by national reforestation programs since the 1960s, thinning of *L. kaempferi* forests has become one of the principal challenges in the Korean forestry (Son et al. 2004; Hwang and Son 2006; Hwang et al. 2007; Lee et al. 2010; Ko et al. 2012).

The present study aimed to test the effects of thinning intensity on nutrient status and enzyme activity in *L. kaempferi* forest soils. To assess the effect on soil nutrient status, we measured concentrations of total carbon (TC) and nitrogen (TN), total carbon to total nitrogen ratio (CN ratio), inorganic N, and extractable phosphorous (P) and cations, including calcium (Ca), magnesium (Mg), and potassium (K). Moreover, we investigated activities of acid phosphatase (AP), β-glucosidase (BG), β-xylosidase (BX), and β-glucosaminidase (NAG) that are involved in decompositions of the phosphomonoesters, celluloses, hemicelluloses, and chitins, respectively (Baldrian and Štursová 2011). These enzymes are known to mediate nitrogen mineralization (NAG) and litter decomposition (BG, BX) and act as significant indices of nitrogen and phosphorus availability (AP and NAG) (Baldrian and Štursová 2011; Olander and Vitousek 2000; Tabatabai et al. 2010).

Methods
Study site
The study site is a *L. kaempferi* forest in Muju, South Korea ($35°\ 52'$ N, $127°\ 50'$ E). The overstory is comprised of *L. kaempferi* trees planted in 1968, whereas the understory is dominated by broad-leaved species such as *Acer pseudosieboldianum, Cornus controversa, Lindera obtusiloba, Styrax obassia, Quercus mongolica*, and *Q. serrata*. The site is an experimental forest, designed to establish a sustainable forest management system as well as to produce large logs. The average altitude and slope are 910 m and 21° (Kim et al. 2016). The soil is an acidic soil with a loamy sand texture (79 % sand, 19 % silt, 2 % clay), and the gravimetric water content, pH, and cation

exchange capacity are 29.0 %, 4.8, and 16.2 $cmol_c\ kg^{-1}$, respectively (Kim et al. 2016). Three thinning intensities, including un-thinned control, 20 % thinning (T20), and 30 % thinning (T30) according to the removed basal area, were conducted. Each thinning intensity included three circular plots (314 m^2) that were established on the same slope to minimize unexpected heterogeneity of initial conditions. The thinning treatments were applied in October, 2011, and the thinning residues were left in the site.

Soil sampling and processing
The mineral soil samples were collected using a 10-cm-long cylindrical corer (7.2 cm in diameter) at 0–10-cm depth within three random points of each plot in May, 2014. This soil depth is known to have concentrated soil microbial biomass and nutrients (Hwang and Son 2006; Park et al. 2011). The collected soil samples were sealed in the plastic bags and brought to the laboratory. Some field-moist subsamples were separated from each soil sample and were stored in a refrigerator at 4 °C until measurements of the enzyme activities and inorganic N. Other soil samples were air dried and sieved through a 2-mm mesh screen before quantifications of TC, TN, CN ratio, and extractable P and cations including Ca, Mg, and K.

Laboratory analysis
TC, TN, and CN ratio were determined by the dry combustion method using an elemental analyzer (vario Macro, Elementar Analysensysteme GmbH, Germany). Inorganic nitrogen (Inorganic N) including ammonium ion (NH_4^+) and nitrate ion (NO_3^-) were extracted with 6 g of the field-moist subsamples and 30 ml of 2 M KCl. NH_4^+ was quantified by the salicylate method (Mulvaney 1996), and NO_3^- was determined using the griess and vanadium chloride (VCl_3) reagents (Miranda et al. 2001). P was extracted by Bray No. 1 method (Bray and Kurtz 1945), and Ca, Mg, and K were extracted with 1 N ammonium acetate buffer (pH = 7). P, Ca, Mg, and K in the extracts were quantified by using an inductively coupled plasma optical emission spectrometer (ICP-OES) (730 series, Agilent, USA).

Activities of AP, BG, BX, and NAG were analyzed by the fluorogenic substrate method (DeForest 2009). Soil suspensions were made by mixing 2 g of the field-moist soil sample and 125 ml of sodium acetate buffer (pH = 5). The homogenized soil suspensions were incorporated into 96-well microplates with the 4-MUB or the 4-MUB-linked substrates, and were incubated at 29 °C for 2 h. The fluorescence were measured at 355-nm excitation and 460-nm emission levels using a Multilable Plate Reader (Victor 3, Perkin-Elmer, USA).

The enzyme activities were expressed as nmol 4-MUB g^{-1} dry soil h^{-1}.

Statistical analysis

Differences in TC, TN, CN ratio, extractable nutrient concentrations, and enzyme activities among the three thinning intensities were assessed by the analysis of variance (ANOVA) with post hoc Tukey's test ($P < 0.05$). In addition, a simple linear regression was carried out to describe relationships of the enzyme activities to TC, TN, CN ratio, and the extractable nutrients ($P < 0.05$). If a simple linear regression model was not significant, a quadratic regression was applied ($P < 0.05$). Mean values of three soil samples from each plot were treated as replicates ($n = 3$ for ANOVA; $n = 9$ for regression tests). Proc GLM procedure of the SAS 9.4 software was used for statistical analysis.

Results and discussion

TC, TN, and CN ratio

The mean TC ($g kg^{-1}$) and TN ($g kg^{-1}$) were 61.67 and 2.65 in the control, 78.48 and 4.08 in T20, and 97.70 and 5.14 in T30, respectively (Fig. 1a, b). TC and TN of T30 were 58.4 and 94.2 % higher compared to those of the control, and these differences were statistically significant. In contrast, no significant difference in TC and TN was found between T20 and the control. Meanwhile, CN ratio was 24.11 in the control, 19.46 in T20, and 19.29 in T30 (Fig. 1c). Even though CN ratio tended to be higher in the control than in T20 and T30, there was no significant difference in CN ratio among the three thinning intensities.

Our results demonstrate that TC and TN might be increased by thinning after 3 years. The increasing patterns in TC and TN are in agreement with the results from other studies on *L. kaempferi* forests (Masyagina et al. 2006; Hwang and Son 2006; Lee et al. 2010; Ko et al. 2012), which reported marginal or remarkable increase in TC and TN due to thinning. These patterns might result from the incorporation of thinning residues

and root mortality following thinning (Hwang and Son 2006). As the C and N in forest soils are known to mainly originate from soil organic matters (Attiwill and Adams 1993), TC and TN can be relevant measures for the soil organic matter quantity. Accordingly, the obtained results imply that T30 might substantially increase the quantity of soil organic matter whereas T20 might have only marginal effect on it.

CN ratio is generally considered an important index for soil organic matter quality because it controls microbial potentials for the nutrient mineralization and immobilization (Hodge et al. 2000). In the present study, the applied thinning intensities showed no significant influences on soil CN ratio. The result is consistent with the other study on a *L. kaempferi* forest, showing that the effect of thinning on soil CN ratio was not significant (Masyagina et al. 2006). These findings are unexpected since thinning is known to accelerate the mixing of low quality plant materials (i.e., thinning residues) into soils and could consequently increase soil CN ratio (Giai and Boerner 2007). We speculate that T20 and T30 might have not substantially shifted soil CN ratio in the short term because of the heterogenic distribution of thinning residues (Boerner et al. 2008).

Extractable nutrients

The mean concentrations of inorganic N and extractable P tended to be highest in T20, followed by T30 and the control (Table 1). The average inorganic N concentration in T20 was 102.5 % higher than that in the control and the difference was statistically significant. Although extractable P concentration was also higher in T20, there was no significant difference in extractable P between T20 and the control. Concentrations of inorganic N and extractable P in T30 were not significantly different compared to those in the control. Meanwhile, the mean concentrations of extractable cations were highest in T20 (Table 1). Extractable Ca and Mg concentrations were significantly higher in T20 than in the control, but no significant difference was found between T30 and the

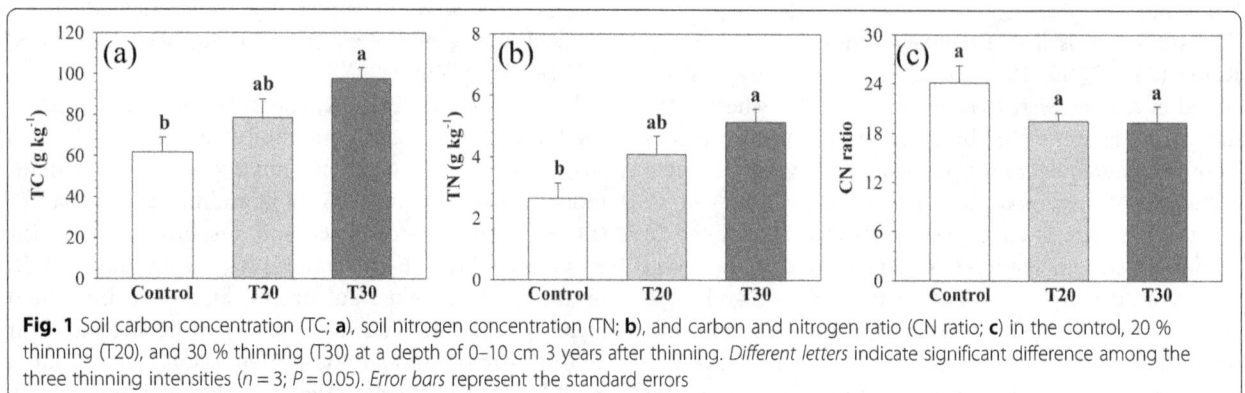

Fig. 1 Soil carbon concentration (TC; **a**), soil nitrogen concentration (TN; **b**), and carbon and nitrogen ratio (CN ratio; **c**) in the control, 20 % thinning (T20), and 30 % thinning (T30) at a depth of 0–10 cm 3 years after thinning. *Different letters* indicate significant difference among the three thinning intensities ($n = 3$; $P = 0.05$). *Error bars* represent the standard errors

Table 1 Concentrations of the inorganic nitrogen (N) and the extractable phosphorus (P), calcium (Ca), magnesium (Mg), and potassium (K) in the control, 20 % thinning (T20), and 30 % thinning (T30) at a depth of 0–10 cm 3 years following thinning

Thinning intensity	Inorganic N (mg kg^{-1})	Extractable P (mg kg^{-1})	Extractable Ca (cmol$_c$ kg^{-1})	Extractable Mg (cmol$_c$ kg^{-1})	Extractable K (cmol$_c$ kg^{-1})
Control	8.63 (0.26)b	17.50 (0.51)a	0.83 (0.08)b	0.27 (0.02)b	0.24 (0.04)a
T20	17.48 (1.91)a	20.03 (1.40)a	2.57 (0.56)a	0.57 (0.07)a	0.29 (0.02)a
T30	10.64 (1.37)b	18.14 (1.05)a	1.50 (0.17)ab	0.44 (0.05)ab	0.22 (0.01)a

Different letters indicate significant difference among the three thinning intensities ($n = 3$; $P = 0.05$). Values in parenthesis denote the standard errors

control. Extractable K concentration did not significantly differ among the three thinning intensities.

The obtained results indicate that T20 might increase extractable nutrient concentrations in soils. Significant increases in concentrations of several extractable nutrients might be attributed to the increase in total quantity of soil nutrient substrates following the initial addition of dead roots, leaves, and twigs of thinned trees (Smolander et al. 2013; Hwang et al. 2007). Moreover, decreases in tree density and nutrient demand of remaining trees might also result in the observed increases in the nutrient concentrations. In contrast, T30 had no significant effects on soil extractable nutrient concentrations. This pattern implies that different thinning intensities might affect extractable soil nutrient concentrations inconsistently. Several processes might be involved in this inconsistent pattern. Particularly, increase in thinning intensity could elevate the hydrological nutrient loss because lower canopy cover following thinning may decelerate direct rainfall interception (Kim et al. 2015; Siemion et al. 2011). The observed inconsistent effects of thinning might also be results of differences in the microbial immobilization since soil microbes may accumulate more extractable nutrients as microbial biomass when heavy intensity of thinning is applied (Chen et al. 2015). Thus, it is likely that these decreasing effects of thinning intensity might cancel out the increases in soil extractable nutrients following thinning.

Enzyme activities

The average activities of AP, BG, BX, and NAG were highest in T30, and the enzyme activities of T20 and T30 were 2.7–22.9 % and 16.9–118.5 % higher than those in the control, respectively (Table 2). Nevertheless, the enzyme activities were not significantly different among the three thinning intensities. BG, BX, and NAG activities were positively correlated to TC (Fig. 2), while only BG activity showed a positive correlation with TN ($P < 0.05$, $R^2 = 0.54$). Though any simple linear regressions were not significant for AP activity, quadratic regression was significant between AP activity and TC or TN (Fig. 2a). Other soil properties including CN ratio and extractable nutrient concentrations exhibited no significant simple linear or quadratic regression with the enzyme activities ($P > 0.05$, data not shown).

As soil organic matter could stabilize the structure of soil enzymes and provide the substrates for soil microbes (Karaca et al. 2011; Brzostek et al. 2013), incorporation of dead plant materials following thinning could enhance the soil enzyme activities (Adamczyk et al. 2015). Although BG, BX, and NAG activities tended to be promoted with significant increases in TC due to thinning, these changes in enzyme activities were not significant in the present study. Similarly, previous studies on needle-leaved forests reported no significant influences of thinning on AP, BG, BX, and NAG activities in the short term (Boyle et al. 2005; Maassen et al. 2006; Geng et al. 2012). No significant differences in AP, BG, BX, and NAG activities might be due to the negative feedback between soil enzymes and nutrient availability. Generally, microbial decomposition for assimilating nutrients is accelerated when nutrient availability acts as a limiting factor for microbial growth and activity in soils; otherwise, it can be depressed (Olander and Vitousek 2000). In the present study, soil nutrient availability seemed to be, at least, not limited by thinning and T20

Table 2 Activities of acid phosphatase (AP), β-glucosidase (BG), β-xylosidase (BX), and β-glucosaminidase (NAG) in the control, 20 % thinning (T20), and 30 % thinning (T30) at a depth of 0–10 cm 3 years following thinning

Thinning intensity	AP (nmol g^{-1} h^{-1})	BG (nmol g^{-1} h^{-1})	BX (nmol g^{-1} h^{-1})	NAG (nmol g^{-1} h^{-1})
Control	290.16 (26.54)a	143.35 (3.87)a	49.62 (7.00)a	32.08 (7.00)a
T20	298.05 (25.23)a	176.14 (16.43)a	53.48 (12.34)a	39.44 (13.71)a
T30	339.45 (27.93)a	180.62 (1.17)a	71.60 (6.85)a	70.11 (20.38)a

Different letters indicate significant difference among the three thinning intensities ($n = 3$; $P = 0.05$). Values in parenthesis denote the standard errors

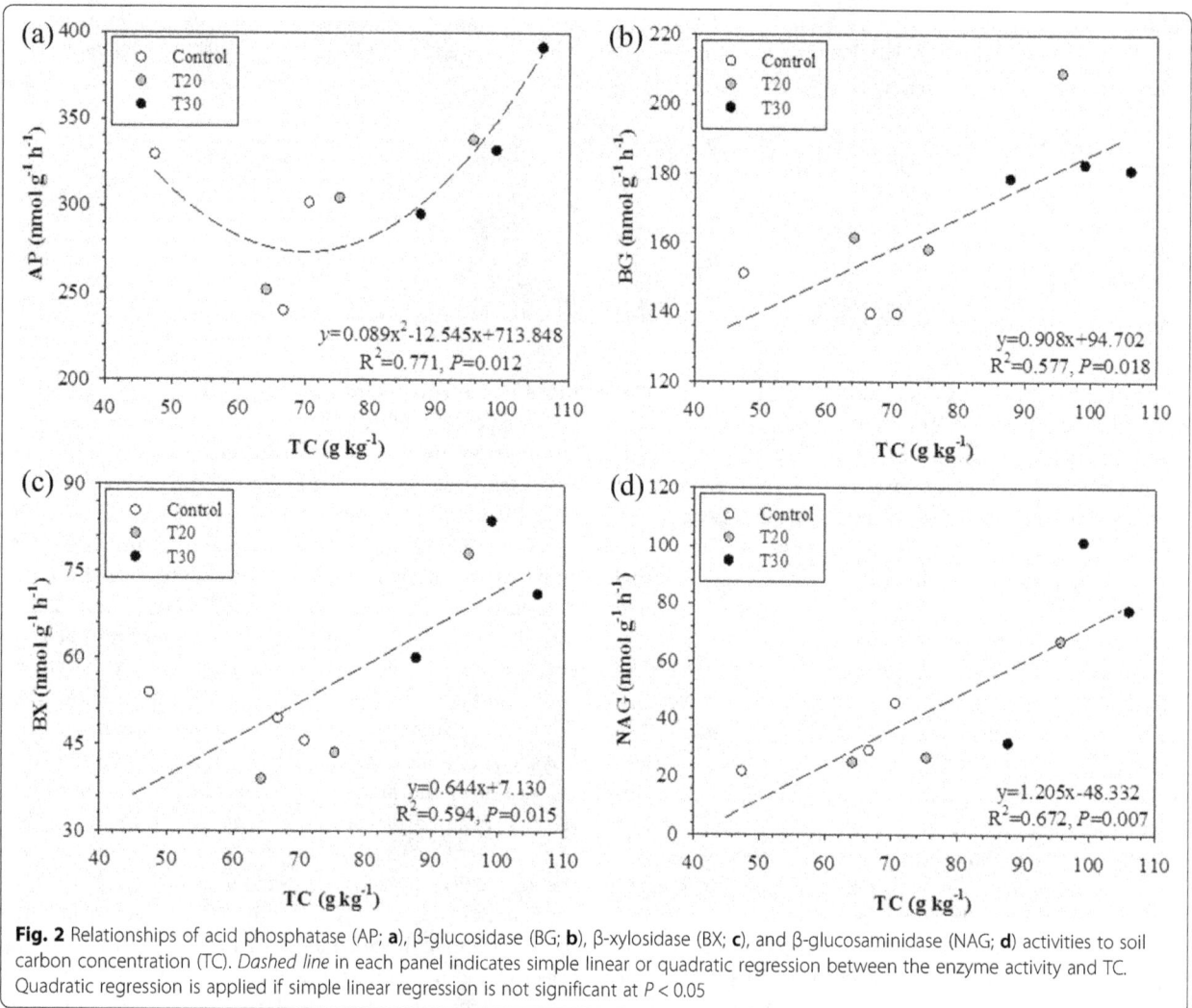

Fig. 2 Relationships of acid phosphatase (AP; **a**), β-glucosidase (BG; **b**), β-xylosidase (BX; **c**), and β-glucosaminidase (NAG; **d**) activities to soil carbon concentration (TC). *Dashed line* in each panel indicates simple linear or quadratic regression between the enzyme activity and TC. Quadratic regression is applied if simple linear regression is not significant at $P < 0.05$

significantly increased some extractable nutrient concentrations (Table 1). Hence, we speculate that no nutrient shortage for soil microbes might have restricted increases in the enzyme activities following the organic matter incorporation due to thinning.

AP, BG, BX, and NAG are soil microbe-originated enzymes participating in the soil organic matter decomposition, and play a crucial role in the nutrient mineralization (Baldrian and Štursová 2011; Olander and Vitousek 2000). Their increase or decrease is generally expected to accompany stimulation of the mineralization process (Tabatabai et al. 2010). Thus, no significant shifts in AP, BG, BX, and NAG activities due to T20 and T30 indicate that application of these thinning intensities might result in only marginal changes in the enzyme-induced nutrient mineralization 3 years after the treatments. Our results are in line with a previous study finding that 10, 20, and 40 % thinning intensities had no significant influences on litter decomposition and N mineralization in

a *L. kaempferi* forest after 4 years (Son et al. 2004). In this context, increased extractable nutrient concentrations in T20 but no significant changes in T30 might not be fully explained by the consistently non-significant effects on the enzyme activities following thinning. From this, it is speculated that this inconsistent pattern might be closely related to other mechanisms such as reduced nutrient demand of trees, accelerated hydrological loss, and increased microbial immobilization (please refer to the "Extractable nutrient" section above). Elucidation of these processes remains further research priorities in order to clarify the nutrient status in thinned forests.

Conclusions

TC, TN, CN ratio, extractable nutrients, and enzyme activities were investigated in a *L. kaempferi* forest where different thinning intensities were carried out. The obtained results showed that TC and TN were highest in T30 while no significant pattern was found for CN ratio.

Meanwhile, inorganic N, extractable Ca, and extractable Mg concentrations were significantly increased only in T20, and the effects of T20 and T30 appeared to be inconsistent for these extractable nutrients. As the activities of AP, BG, BX, and NAG were not significantly changed due to either T20 or T30, the enzyme-induced mineralization might not be enough to explain the inconsistent changes in extractable nutrient status in T20 and T30. Therefore, further studies on the inconsistent effects of thinning intensities on soil extractable nutrients are necessary for revealing the soil nutrient dynamics in thinned forests. We expect that the present study will contribute to the understanding of the nutrient status in *L. kaempferi* forests after thinning.

Abbreviations
ANOVA: Analysis of variance; AP: Acid phosphatase; BG: β-glucosidase; BX: β-xylosidase; Ca: Calcium; CN ratio: Total carbon to total nitrogen ratio; ICP-OES: Inductively coupled plasma optical emission spectrometer; Inorganic N: Inorganic nitrogen; K: Potassium; KCl: Potassium chloride; Mg: Magnesium; 4-MUB: 4-Methylumbelliferone; NAG: β-glucosaminidase; NH_4^+: Ammonium ion; NO_3^-: Nitrate ion; P: Phosphorus; T20: 20 % thinning treatment; T30: 30 % thinning treatment; TC: Total carbon; TN: Total nitrogen; VCl_3: Vanadium chloride

Acknowledgements
We would like to thank for the assistance of Soon Jin Yun, Jongyeol Lee, and Jiae An during the field and laboratory works.

Funding
This study was supported by the Forest Practice Research Center, National Institute of Forest Science (FM0101-2009-01) and the Korean Ministry of Environment (2016001300004).

Authors' contributions
SK led the field and laboratory works, data analysis and discussion, and manuscript writing. SHH, GL, and TKY participated in the field and laboratory works, discussion of the data, and manuscript writing. STL and CK contributed to the experimental design and manuscript writing. YS supervised whole process including the experimental design, data analysis and discussion, and manuscript writing. All authors read and approved the final manuscript.

Competing interests
The authors declare that they have no competing interests.

Author details
[1]Department of Environmental Science and Ecological Engineering, Graduate School, Korea University, Seoul 02841, South Korea. [2]Environmental Planning Institute, Seoul National University, Seoul 08826, South Korea. [3]Forest Practice Research Center, National Institute of Forest Science, Pocheon 11186, South Korea. [4]Department of Forest Resources, Gyeongnam National University of Science and Technology, Jinju 52725, South Korea.

References
Attiwill, P. M., & Adams, M. A. (1993). Nutrient cycling in forests. *New Phytologist, 124*, 561–582.

Grigal, D. F. (2000). Effects of extensive forest management on soil productivity. *Forest Ecology and Management, 138*, 167–185.

Boerner, R. E. J., Giai, C., Huang, J., & Miesel, J. R. (2008). Initial effects of fire and mechanical thinning on soil enzyme activity and nitrogen transformations in eight North American forest ecosystems. *Soil Biology and Biochemistry, 40*, 3076–3085.

Achat, D. L., Deleuze, C., Landmann, G., Pousse, N., Ranger, J., & Augusto, L. (2015). Quantifying consequences of removing harvesting residues on forest soils and tree growth—a meta-analysis. *Forest Ecology and Management, 348*, 124–141.

Son, Y., Jun, Y. C., Lee, Y. Y., Kim, R. H., & Yang, S. Y. (2004). Soil carbon dioxide evolution, litter decomposition, and nitrogen availability four years after thinning in a Japanese larch plantation. *Communications in Soil Science and Plant Analysis, 35*, 1111–1122.

Masyagina, O. V., Hirano, T., Ji, D. H., Choi, D. S., Qu, L., Fujinuma, Y., Sasa, K., Matsuura, Y., Prokushkin, S. G., & Koike, T. (2006). Effect of spatial variation of soil respiration rates following disturbance by timber harvesting in a larch plantation in northern Japan. *Forest Science and Technology, 2*, 80–91.

Smolander, A., Kitunen, V., Kukkola, M., & Tamminen, P. (2013). Response of soil organic matter layer characteristics to logging residues in three Scots pine thinning stands. *Soil Biology and Biochemistry, 66*, 51–59.

Hwang, J., & Son, Y. (2006). Short-term effects of thinning and liming on forest soils of pitch pine and Japanese larch plantations in central Korea. *Ecological Research, 21*, 671–680.

Kim, S., Yoon, T. K., Han, S., Han, S. H., Lee, J., Kim, C., Lee, S.-T., Seo, K. W., Yang, A.-R., & Son, Y. (2015). Initial effects of thinning on soil carbon storage and base cations in a naturally regenerated *Quercus* spp. forest in Hongcheon, Korea. *Forest Science and Technology, 11*, 172–176.

Giai, C., & Boerner, R. E. J. (2007). Effects of ecological restoration on microbial activity, microbial functional diversity, and soil organic matter in mixed-oak forests of southern Ohio, USA. *Applied Soil Ecology, 35*, 281–290.

Adamczyk, B., Adamczyk, S., Kukkola, M., Tamminen, P., & Smolander, A. (2015). Logging residue harvest may decrease enzymatic activity of boreal forest soils. *Soil Biology and Biochemistry, 82*, 74–80.

Sinsabaugh, R. L., Lauber, C. L., Weintraub, M. N., Ahmed, B., Allison, S. D., Crenshaw, C., Contosta, A. R., Cusack, D., Frey, S., Gallo, M. E., Gartner, T. B., Hobbie, S. E., Holland, K., Keeler, B. L., Powers, J. S., Stursova, M., Takacs-Vesbach, C., Waldrop, M. P., Wallenstein, M. D., Zak, D. R., & Zeglin, L. H. (2008). Stoichiometry of soil enzyme activity at global scale. *Ecology Letters, 11*, 1252–1264.

Korea Forest Service. (2015). Statistical yearbook of forestry. Daejeon: Korea Forest Service. (In Korean).

Hwang, J., Son, Y., Kim, C., Yi, M.-J., Kim, Z.-S., Lee, W.-K., & Hong, S.-K. (2007). Fine root dynamics in thinned and limed pitch pine and Japanese larch plantations. *Journal of Plant Nutrition, 30*, 1821–1839.

Lee, S. K., Son, Y., Lee, W. K., Yang, A.-R., Noh, N. J., & Byun, J.-G. (2010). Influence of thinning on carbon storage in a Japanese larch (*Larix kaempferi*) plantation in Yangpyeong, central Korea. *Forest Science and Technology, 6*, 35–40.

Ko, S., Son, Y., Noh, N. J., Yoon, T. K., Kim, C., Bae, S.-W., Hwang, J., Lee, S.-T., & Kim, H.-S. (2012). Influence of thinning on carbon storage in soil, forest floor and coarse woody debris of *Larix kaempferi* stands in Korea. *Forest Science and Technology, 8*, 116–121.

Baldrian, P., & Štursová, M. (2011). Enzymes in forest soils. In G. Shukla & A. Varma (Eds.), *Soil enzymology* (pp. 61–73). Heidelberg: Springer.

Olander, L. P., & Vitousek, P. M. (2000). Regulation of soil phosphatase and chitinase activity by N and P availability. *Biogeochemistry, 49*, 175–190.

Tabatabai, M. A., Ekenler, M., & Senwo, Z. N. (2010). Significance of enzyme activities in soil nitrogen mineralization. *Communications in Soil Science and Plant Analysis, 41*, 595–605.

Kim, S., Han, S. H., Lee, J., Kim, C., Lee, S.-T., & Son, Y. (2016). Impact of thinning on carbon storage of dead organic matter across larch and oak stands in South Korea. *iForest, 9*, 593–598.

Park, C.-W., Ko, S., Yoon, T. K., Han, S., Yi, K., Jo, W., Jin, L., Lee, S. J., Noh, N. J., Chung, H., & Son, Y. (2011). Differences in soil aggregate, microbial biomass carbon concentration, and soil carbon between *Pinus rigida* and *Larix kaempferi* plantations in Yangpyeong, central Korea. *Forest Science and Technology, 8*, 38–46.

Mulvaney, R. L. (1996). Nitrogen-inorganic forms. In D. L. Sparks, P. A. Helmke, R. H. Loeppert, P. N. Soltanpour, M. A. Tabatabai, C. T. Johnston, & M. E. Sumner

(Eds.), *Methods of soil analysis. Part 3-chemical methods* (pp. 1146–1155). Madison: Soil Sci Soc Am.

Miranda, K. M., Espey, M. G., & Wink, D. A. (2001). A rapid, simple spectrophotometric method for simultaneous detection of nitrate and nitrite. *Nitric Oxide, 5,* 62–71.

Bray, R. H., & Kurtz, L. T. (1945). Determination of total, organic, and available forms of phosphorus in soils. *Soil Science, 59,* 39–46.

DeForest, J. L. (2009). The influence of time, storage temperature, and substrate age on potential soil enzyme activity in acidic forest soils using MUB-linked substrates and L-DOPA. *Soil Biology and Biochemistry, 41,* 1180–1186.

Hodge, A., Robinson, D., & Fitter, A. (2000). Are microorganisms more effective than plants at competing for nitrogen? *Trends in Plant Science, 5,* 304–308.

Siemion, J., Burns, D. A., Murdoch, P. S., & Germain, R. H. (2011). The relation of harvesting intensity to changes in soil, soil water, and stream chemistry in a northern hardwood forest, Catskill Mountains, USA. *Forest Ecology and Management, 261,* 1510–1519.

Chen, X.-L., Wang, D., Chen, X., Wang, J., Diao, J.-J., Zhang, J.-Y., & Guan, Q.-W. (2015). Soil microbial functional diversity and biomass as affected by different thinning intensities in a Chinese fir plantation. *Applied Soil Ecology, 92,* 35–44.

Karaca, A., Cetin, S. C., Turgay, O. C., & Kizilkaya, R. (2011). Soil enzymes as indication of soil quality. In G. Shukla & A. Varma (Eds.), *Soil enzymology* (pp. 119–148). Heidelberg: Springer.

Brzostek, E. R., Greco, A., Drake, J. E., & Finzi, A. C. (2013). Root carbon inputs to the rhizosphere stimulate extracellular enzyme activity and increase nitrogen availability in temperate forest soils. *Biogeochemistry, 115,* 65–76.

Boyle, S. I., Hart, S. C., Kaye, J. P., & Waldrop, M. P. (2005). Restoration and canopy type influence soil microflora in a ponderosa pine forest. *Soil Science Society of America Journal, 69,* 1627–1638.

Maassen, S., Fritze, H., & Wirth, S. (2006). Response of soil microbial biomass, activities, and community structure at a pine stand in northeastern Germany 5 years after thinning. *Can J of For Res., 36,* 1427–1434.

Geng, Y., Dighton, J., & Gray, D. (2012). The effects of thinning and soil disturbance on enzyme activities under pitch pine soil in New Jersey pinelands. *Applied Soil Ecology, 62,* 1–7.

17

Soil CO$_2$ efflux in a warm-temperature and sub-alpine forest in Jeju, South Korea

Heon-Mo Jeong[1], Rae-Ha Jang[2], Hae-Ran Kim[3] and Young-Han You[2*]

Abstract

Background: This study investigated the temporal variation in soil CO$_2$ efflux and its relationship with soil temperature and precipitation in the *Quercus glauca* and *Abies koreana* forests in Jeju Island, South Korea, from August 2010 to December 2012. *Q. glauca* and *A. koreana* forests are typical vegetation of warm-temperate evergreen forest zone and sub-alpine coniferous forest zone, respectively, in Jeju island.

Results: The mean soil CO$_2$ efflux of *Q. glauca* forest was 0.7 g CO$_2$ m^{-2} h^{-1} at 14.3 °C and that of *A. koreana* forest was 0.4 g CO$_2$ m^{-2} h^{-1} at 6.8 °C. The cumulative annual soil CO$_2$ efflux of *Q. glauca* and *A. koreana* forests was 54.2 and 34.2 t CO$_2$ ha^{-1}, respectively. Total accumulated soil carbon efflux in *Q. glauca* and *A. koreana* forests was 29.5 and 18.7 t C ha^{-1} for 2 years, respectively. The relationship between soil CO$_2$ efflux and soil temperate at 10 cm depth was highly significant in the *Q. glauca* ($r^2 = 0.853$) and *A. koreana* forests ($r^2 = 0.842$). Soil temperature was the main controlling factor over CO$_2$ efflux during most of the study period. Also, precipitation may affect soil CO$_2$ efflux that appeared to be an important factor controlling the efflux rate.

Conclusions: Soil CO$_2$ efflux was affected by soil temperature as the dominant control and moisture as the limiting factor. The difference of soil CO$_2$ efflux between of *Q. glauca* and *A. koreana* forests was induced by soil temperature to altitude and regional precipitation.

Keywords: CO$_2$ efflux, Soil temperature, Precipitation, *Abies koreana*, *Quercus glauca*, Jeju island

Background

Soil respiration is one of the processes in the ecosystem that comprises root respiration, decomposition of soil organic matters by microorganisms, and efflux of CO$_2$ from the animals (Luo and Zhou 2006). It plays an important role in the regulation of carbon cycle in regional and global scale. Carbon cycle in global scale consists of exchange of CO$_2$ between biome on land, atmosphere, and ground surface. The terrestrial plants absorb about 120 Pg of carbon each year through photosynthesis, and it is recycled into the atmosphere through respiration in the ecosystem. The soil around the world contains 3150 Pg of carbon; 450 Pg C in wetlands, 400 Pg C in tundra, and 2300 Pg in other ecosystem (Sabine et al. 2004). It is known that other ecosystems contain 1500 Pg C up to 1 m and 800 Pg C up to 3 m in soil depth (Jobbagy and Jackson 2000). The sum of carbon in plants and soil, which is 3800 Pg C, is five times more than the carbon distributed in the atmosphere (750 Pg C) because the plants also contain 650 Pg C carbon. Furthermore, carbon is also in every living organism as the primary component (Kim et al. 2014). The plants make organic compounds using CO$_2$, water, and sunlight where light energy is stored in organic compounds through photosynthesis. The organic compounds are used by plants themselves in respiration and some become part of the plant. They ultimately carry out crucial roles in the ecosystem, for example, in net production and keeping the carbon balance.

Net ecosystem production (NEP), which shows net gain or loss of carbon in the ecosystem, should be analyzed in order to accurately estimate the carbon balance in the forest ecosystem. NEP can be obtained by subtracting the amount of organic matter consumed by heterotrophic respiration from net primary productivity (NPP), and studies are being carried out recently to accurately forecast the amount of CO$_2$ released from heterotrophic respiration of soil (Nakane 1995, Raich and Tufekcioglu 2000, Lee and Mun 2001, Lee et al. 2012).

* Correspondence: youeco21@kongju.ac.kr
[2]Department of Biology, Kongju National University, Gongju, South Korea
Full list of author information is available at the end of the article

The subject of this study, *Quercus glauca*, an evergreen broad-leaved tree, and *Abies koreana*, an evergreen coniferous tree, are typical vegetation distributed in warm-temperate forest and sub-alpine forest found in Korea. Warm-temperate forest is a narrow vegetation belt found between tropical and temperate forest zones. It is mainly located along southern coastlines and island regions that have average annual temperature above 14 °C and following vegetation are found: *Castanopsis cuspidata, Machilus thunbergii, Q. glauca, Camellia japonica, Cinnamomum japonicum, Euonymus japonicus, Trachelospermum asiaticum, Stauntonia hexaphylla*, etc. (Kong 2007). The distribution of warm-temperate evergreen broad-leaved forests in Korea has increased by approximately 2.7 times over the past 20 years, and it is expected to move up to northern regions in the future (Park et al. 2010). On the other hand, evergreen coniferous forests are found in boreal zones, such as plateau or alpine region that has an average annual temperature below 5 °C with average temperature of −12 °C during January. Trees that have adapted to cold winter and short growing period, such as *Abies holophylla, Picea jezoensis, A. nephrolepis, Pinus koraiensis, A. koreana, Larix gmelinii, Betula costata*, and *B. platyphylla*, are usually found in this region (Kong 2007). The study result on the community structure of *A. koreana* distributed around Mt. Halla revealed low vitality of *A. koreana* in the area with up to 8.11% frequency of dead trees (Kim et al. 1998). In fact, there are studies that claim plants that belong to genus *Abies*, which are main species of sub-alpine and sub-polar zone, will slowly become decline due to global warming (Kim 2002, Koo et al. 2001, Kong 1998, Kim and Kil 1996).

The study area is a forest ecosystem that is expected to respond sensitively to climate change caused by global warming. *Q. glauca* community is expected to expand,

but *A. koreana* community is expected to decline. The sub-alpine vegetation of Mt. Halla, where *A. koreana* community is located, especially is vulnerable to high winter temperatures and water stress caused by global warming, and thus affects plant productivity and soil CO_2 efflux, which results in a change in the NEP of sub-alpine vegetation. Soil CO_2 efflux plays a crucial role in controlling atmospheric CO_2 concentrations and climate change in the global ecosystem.

In addition, the importance of soil respiration should be emphasized for the accurate measurement of carbon balance in the changing forest ecosystem. The purpose of this study is to provide basic data and to analyze the features of soil respiration according to temperature in *Q. glauca* community and *A. koreana* community which are the main forest ecosystem in warm-temperate and sub-alpine zone.

Methods
Site description
This study was conducted on *Q. glauca* community and *A. koreana* community located in Jeju island (Fig. 1). *Q. glauca* community (33° 31′ 09″ N, 126° 42′ 57″ E) was distributed around Mt. Dongbaek situated in Sunheul-ri, Jocheon-eup, Jeju island. More than 10% of evergreen broad-leaved forests in this island are found in this area which is a Gotjawal terrain (Kwak et al. 2013). The stand age in *Q. glauca* community was about 38 years, and evergreen vegetation, such as *Camellia japonica, Eurya japonica*, and *Ardisia japonica*, inhabited the herb layer of the community (Table 1). The average annual temperature and precipitation of Sunheul-ri in Jocheon-eup, where *Q. glauca* community is distributed, were 13.2 °C and 2447 mm respectively during the study period. *A. koreana* community (33° 21′ 31″ N, 126° 30′ 27″ E)

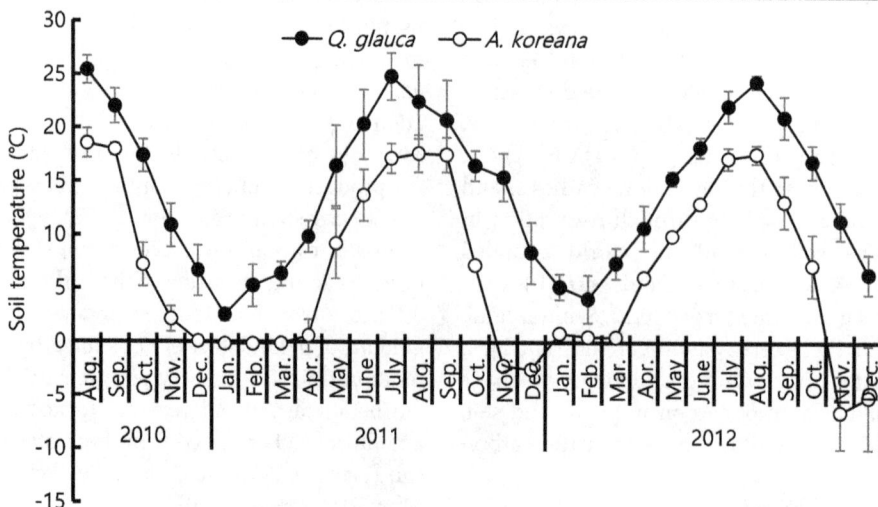

Fig. 1 Monthly variation of soil temperature at 10 cm depth below ground in *Q. glauca* and *A. koreana* communities

Table 1 Habitat characteristics of the *Q. glauca* and *A. koreana* communities

Study forest Characteristics	*Q. glauca*	*A. koreana*
Altitude (m)	126	1660
Density (tree/ha)	16,000	3700
Forest age (year)	38	90
Tree layer		
Dominant species	*Q. glauca*	*A. koreana*
Tree height (m)	16	7
Mean DBH (cm)	7.4	9.5
Coverage (%)	90	85
Shrub layer		
Dominant species	*Eurya japonica*	*Taxus cuspidata*
	Camellia japonica	*Rhododendron mucronulatum*
	Q. glauca	*Symplocos coreana*
Shrub height (m)	2	0.8
Coverage (%)	15	10
Herb layer		
Dominant species	*Ardisia japonica*	*Empetrum nigrum*
	Q. glauca	*Sasa quelpaertensis*
Herb height (m)	0.45	0.3
Coverage (%)	45	50

was distributed in Youngsil region with altitude of 1400 m above sea level around Mt. Halla. The stand age in *A. koreana* community was about 90 years, and sub-alpine vegetation, such as *Taxus cuspidate, Rhododendron mucronulatum, Empetrum nigrum*, and *Sasa quelpaertensis*, were found in the herb layer of the community. The average annual temperature and precipitation of Witseorum of Youngsil in Mt. Halla, where *A. koreana* community is distributed, were 6.1 °C and 5882 mm respectively. The difference in altitude and average annual temperature of *Q. glauca* community and *A. koreana* community were 1500 m and 7.1 °C representing warm-temperate and sub-alpine forest that has contrasting climate and ecosystem.

Measurement of soil temperature and respiration

Measurements of soil respiration efflux were taken in all seasons to take into account its characteristics at various temperature ranges, and measurement was conducted between 10 am and 4 pm on a clear day. For each vegetation community, soil respiration efflux was measured ten times at three random sites in the quadrate, and the maximum and minimum values were excluded from the analysis.

The soil respiration was measured by using IRGA portable gas analyzer (EGM4, PP System, UK). The soil

temperature at 10 cm depth after removing litterfall and CO_2 efflux was measured over the whole year throughout four seasons in order to clearly understand the relationship between soil temperature, respiration, and CO_2 efflux. Digital thermometers (Thermo recorder TR-71U, T&D Co., Japan), respectively, were placed 10 cm below ground, and the temperature was recorded every hour in order to measure the soil temperature of *Q. glauca* community and *A. koreana* community from August 2010 to December 2012.

Data treatment and analysis

A regression equation for the relationship between soil temperature and CO_2 efflux was derived from the data on CO_2 efflux of *Q. glauca* community and *A. koreana* community which was obtained by using the infrared gas analyzer and their soil temperature 10 cm below ground. The data on soil temperature gathered from digital thermometers installed 10 cm below ground in each community was substituted into the regression equation to calculate CO_2 efflux per hour. Furthermore, based on these results, the annual and monthly soil respiration and the amount of carbon emissions from forests were estimated.

Results and Discussion
Monthly variations of soil temperature

The average monthly soil temperature of *Q. glauca* community was 14.3 °C during the study period (Fig. 2). It was 14.1 °C in 2011 and 13.6 °C in 2012. The soil temperature of *Q. glauca* community was the highest in August 2010 reaching 25.4 °C and lowest in January 2011 going down to 2.5 °C. Soil temperature was generally low from January to February, but it was high from July to August. However, it never decreased below 0 °C during the study period. The average monthly soil temperature of *A. koreana* community was 6.8 °C during the study period. It was 6.5 °C in 2011 and 6.2 °C in 2012. The soil temperature of *A. koreana* community was the highest in August 2010 reaching 18.5 °C and lowest in November 2012 going down to –6.5 °C. Soil temperature was generally low from November to April, but it was high from July to September. The temperature difference between the warmest and coldest month was greater in 2012 than 2011.

Soil CO$_2$ efflux analysis

The soil CO_2 efflux of *Q. glauca* community and *A. koreana* community measured by the infrared gas analyzer was formulated into an exponential equation showing positive relationship (Fig. 3). Nonetheless, the exponential CO_2 efflux according to temperature was the same as the existing study (Chen et al. 2013, Darenova et al. 2014). Soil CO_2 efflux was greater in *A. koreana* community than *Q.*

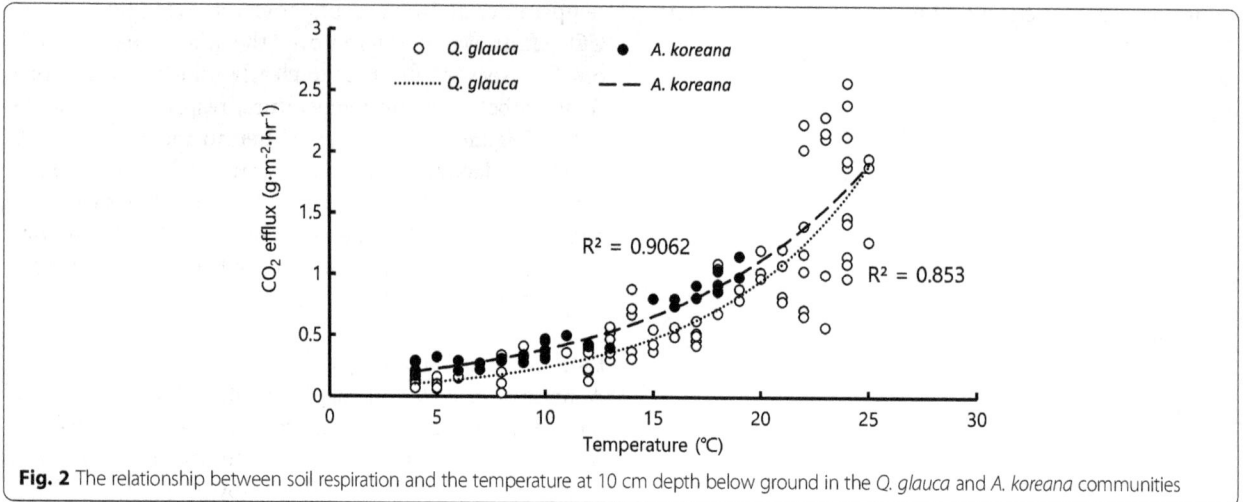

Fig. 2 The relationship between soil respiration and the temperature at 10 cm depth below ground in the *Q. glauca* and *A. koreana* communities

glauca community in all the temperature range. A significant increase in soil CO_2 efflux depending on an increase in the soil temperature in the *Q. glauca* community and *A. koreana* community is related to the biochemical activity of heterotrophs in the soil, and it is known that there is generally an exponential relation between soil CO_2 efflux and soil temperature (Luo and Zhou 2006).

Coefficient of determination (R^2) of the regression equation for soil CO_2 efflux and temperature of *Q. glauca* community and *A. koreana* community was of 0.853 and 0.8419 respectively (Table 2). CO_2 efflux from *Q. glauca* community and *A. koreana* community decreased in winter but increased in summer because soil respiration fluctuates with change in temperature (Fig. 4). Soil respiration of *Q. glauca* community was the lowest over January and February whereas it was the highest in July 2011 and August 2012. On the other hand, soil respiration of *A. koreana* community was the lowest from December 2010 to April 2011, from November 2011 to March 2012, and from November 2012 to December 2012 whereas it was the highest in July 2011 and August 2012.

Q. glauca community and *A. koreana* community had similar lowest monthly CO_2 efflux of about 0.1 g m^{-2} h^{-1}, but it lasted over about 2 months in *Q. glauca* community and about 5~6 months in *A. koreana* community over a year. The highest monthly CO_2 efflux of Q. glauca community was 2.1 g m^{-2} h^{-1} in 2011 and 2.0 g m^{-2} h^{-1} in 2012 whereas *A. koreana* community was 0.8 g m^{-2} h^{-1} in 2011 and 0.9 g m^{-2} h^{-1} 2012. It lasted for a month in both communities in 2011 as well as in 2012. The average monthly CO_2 efflux of *Q. glauca* community was 0.7 g m^{-2} h^{-1}, and *A. koreana* community was 0.4 g m^{-2} h^{-1} during the study period. *Q. glauca* community was about 1.8 times greater than *A. koreana* community. This was caused by the difference in the soil temperature of *Q. glauca* community and *A. koreana* community. CO_2 efflux of *Q. glauca* community over July and August is two times greater than *A. koreana* community due to high soil temperature, but CO_2 efflux of *A. koreana* community is kept at its lowest level from November to March due to low soil temperature. It is presumed that the difference in soil temperature is based on the altitude of *Q. glauca*

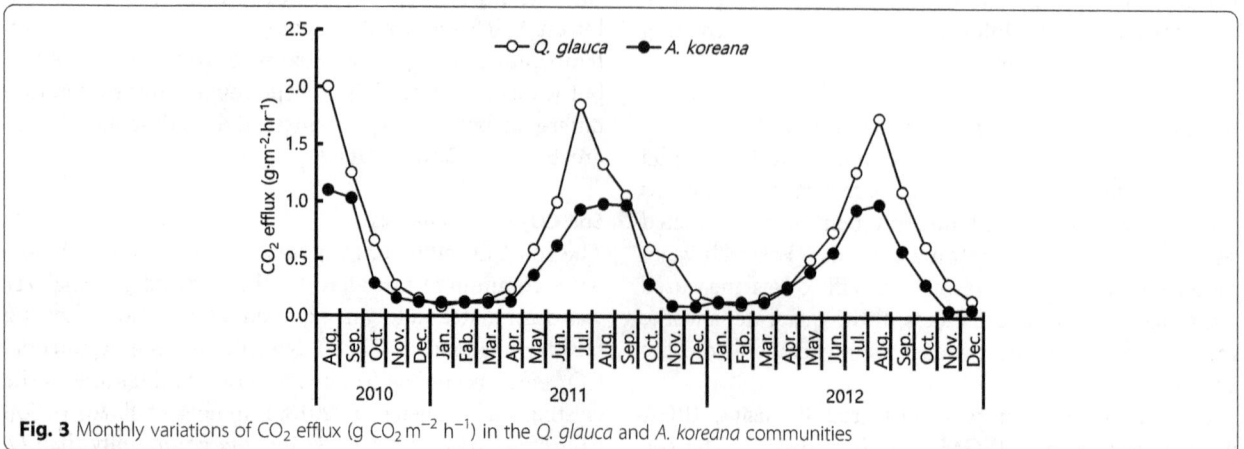

Fig. 3 Monthly variations of CO_2 efflux (g CO_2 m^{-2} h^{-1}) in the *Q. glauca* and *A. koreana* communities

Table 2 Comparison of exponential equations relation to soil respiration and temperature

Community	Equation	a	b	R^2
Q. gluaca	$y = ae^{bx}$	0.0598	0.1381	0.853
A. koreana	$y = ae^{bx}$	0.1201	0.1193	0.8419

a, b are parameters and y, x refers to soil respiration and temperature, respectively

community and *A. koreana* community. This result was the same as that of Kane et al. (2003) which showed a decrease in soil temperature and respiration with increase in altitude even within the same region.

Annual CO_2 efflux of *Q. glauca* community was 56.4 t CO_2 ha^{-1} in 2011, 51.9 t CO_2 ha^{-1} in 2012, and the average was 54.2 t CO_2 ha^{-1} (Table 3). Annual organic carbon efflux was 15.4 t C ha^{-1} in 2011, 14.1 t C ha^{-1} in 2012, and the average was 14.8 t C ha^{-1}. Annual CO_2 efflux of temperate deciduous forest and tropical rainforest is known to range from 4.0 to 10.0 t C ha^{-1} and 8.9 to 15.2 t C ha^{-1} respectively (Raich and Schlensinger 1992, Luo et al. 2006). The average organic carbon efflux of *Q. glauca* community, which is 14.8 t C ha^{-1}, is included in the annual CO_2 efflux range of tropical rainforest. Soil organic carbon efflux of *Q. robur L.* community, a temperate deciduous forest, in Belgium was 6.9 t C ha^{-1} (Yuste et al. 2005), and *Q. mongolica* community, one of main forest

ecosystems in Korea, was 7.7 t C ha^{-1} (Yi et al. 2005) which were greater than the soil organic carbon efflux of *Q. glauca* community.

Annual CO_2 efflux of *A. koreana* community was 35.3 t CO_2 ha^{-1} in 2011, 33.1 t CO_2 ha^{-1} in 2012, and the average was 34.2 t CO_2 ha^{-1} (Table 3). In addition, annual organic carbon efflux was 9.6 t C ha^{-1} in 2011, 9.0 t C ha^{-1} in 2012, and the average was 9.3 t C ha^{-1}. The annual carbon emissions of the *Taxus cuspidate* community in the sub-alpine vegetation area of Mt. Halla in 2012 were 2.9 t CO_2 ha^{-1} (Jang et al. 2017), and they were less than the annual carbon emissions of the *A. koreana* community in Mt. Halla studied during the same period in 2012 (9.0 t CO_2 ha^{-1}).

Average soil CO_2 efflux of *A. koreana* community was greater in this case as the annual soil respiration in alpine zone is generally known to range from 1.5 to 6.0 t C ha^{-1} (Luo et al. 2006). It can be assumed that soil CO_2 efflux of *A. koreana* community is greater than alpine zone because they are distributed in sub-alpine zone. For instance, *A. holophylla* forests distributed in eastern Canada in regions with higher altitude than *A. koreana* community had lower soil CO_2 efflux of 3.5 t C ha^{-1} (Risk et al. 2002).

Relationship between soil respiration and precipitation

Figure 5 shows the regression analysis on monthly CO_2 efflux and precipitation of *Q. glauca* community and *A. koreana* community. The coefficient correlation (R) for monthly CO_2 efflux and precipitation of *Q. glauca* community and *A. koreana* community were 0.64 and 0.67 respectively which showed that precipitation had little effect on monthly CO_2 efflux. Nevertheless, the monthly change in CO_2 efflux in each community displayed similar patterns to the change in monthly precipitation in each study area (Fig. 5). It is known that soil respiration is affected highly by soil temperature and moisture (Davidson et al. 1998, Inclan et al. 2010).

Soil CO_2 efflux can mainly be attributed to respiration of root and heterotroph (Kuzyakov 2006, Saiz et al. 2006). Respiration of heterotrophs is dependent on soil moisture, and they proliferate in proportion to increase in soil water content (Darenova 2014). It has been reported that respiration of heterotrophs increase significantly straight after the rain and gradually subsides afterwards even though the lack of water content accumulated in soil acts as the limiting factor for soil respiration (Liu et al. 2002, Xu et al. 2004). Generally, soil respiration reaches its maximum point when the soil water content is neither too low nor high and optimum water content is known to be at 60% (Suh et al. 2009). In case of pastures on sub-alpine grassland, soil respiration increases as water content increases within the water content range of 10~80% (Moriyama et al. 2013).

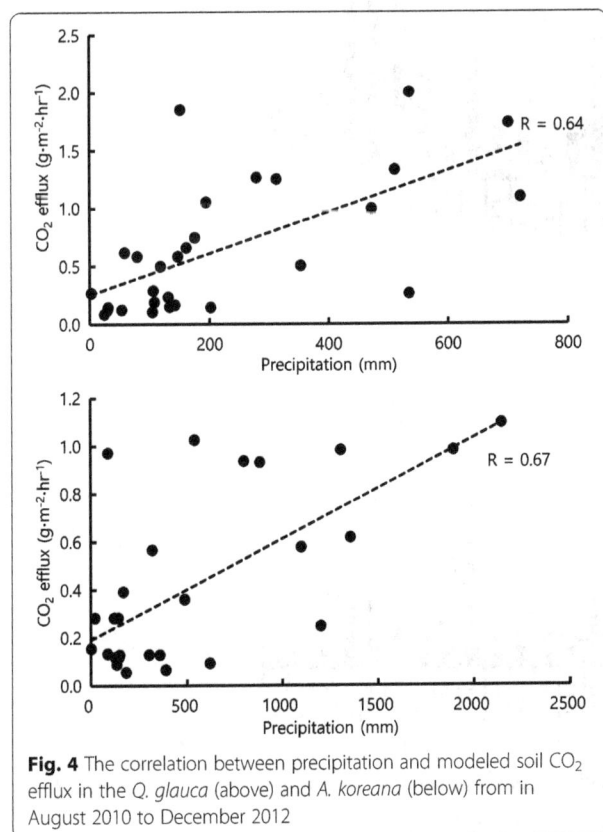

Fig. 4 The correlation between precipitation and modeled soil CO_2 efflux in the *Q. glauca* (above) and *A. koreana* (below) from in August 2010 to December 2012

Table 3 Annual soil respiration (t CO_2 ha^{-1}) and organic carbon (t C ha^{-1}) of the *Q. glauca* and *A. koreana* communities

Community	2010 (Aug.~Dec.)		2011		2012	
	t CO_2 ha^{-1}	t C ha^{-1}	t CO_2 ha^{-1}	t C ha^{-1}	t CO_2 ha^{-1}	t C ha^{-1}
Q. glauca	31.9	8.7	56.4	15.4	51.9	14.1
A. koreana	19.7	5.4	35.3	9.6	33.1	9.0

The seasonal temperature change predominantly affected the monthly changes in soil CO_2 efflux shown in this study considering abovementioned characteristics of soil respiration with respect to soil temperature and water content. But it seems that variation in precipitation, a limiting factor in soil respiration, played a crucial role in soil respiration of *Q. glauca* community and *A. koreana* community.

Conclusions

This study analyzed the yearly and monthly variations of CO_2 efflux in relation to the soil temperature and precipitation in the *Q. glauca* community, a warm-temperate forest and the *A. koreana* community, a sub-

alpine forest (CO_2 is emitted in forests as a result of CO_2 efflux. Although soil respiration is frequently used in research, CO_2 efflux was used in this study). The study results showed a high correlation between CO_2 efflux and either of soil temperature and precipitation. The average soil temperature was 7.5 °C higher in *Q. glauca* community than in *A. koreana* community at the depth of 10 cm below the ground surface. When the CO_2 efflux figures of two forest communities measured during the research were compared, the CO_2 efflux value of *Q. glauca* community was 1.6 times higher than that of *A. koreana* community, and this may be attributed to the temperature difference between the two communities due to the altitude

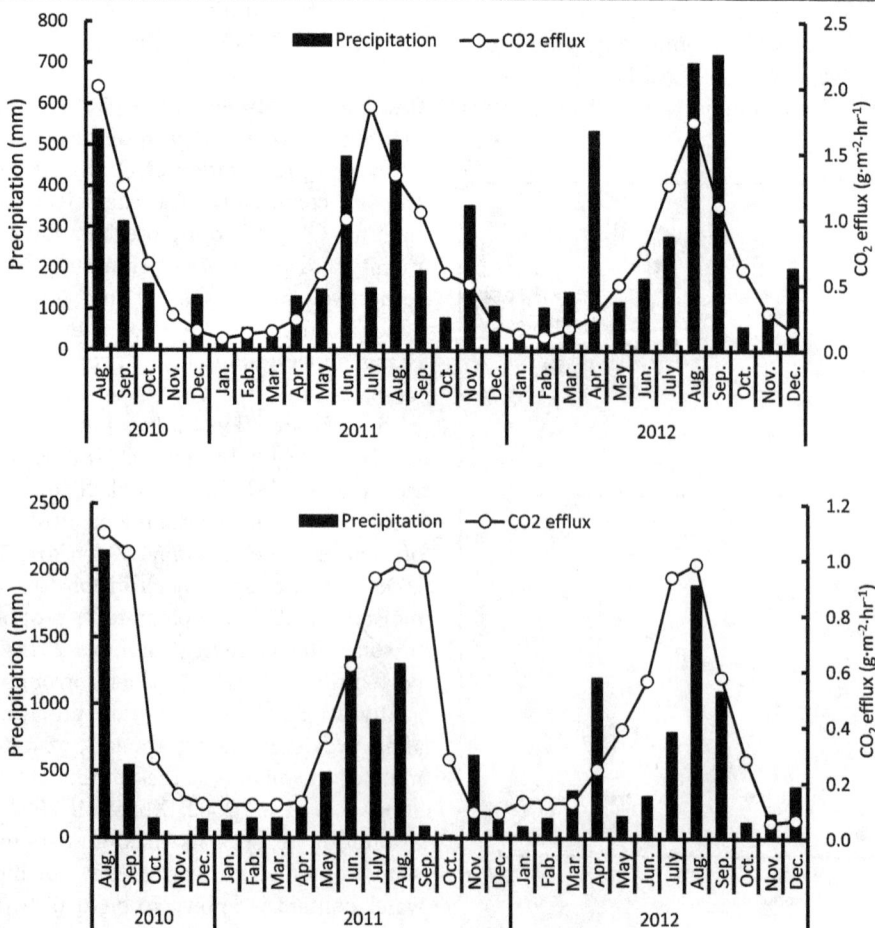

Fig. 5 The monthly variation of precipitation and modeled soil CO_2 efflux in the *Q. glauca* (above) and *A. koreana* (below) from in August 2010 to December 2012

difference. Furthermore, the monthly variation of CO_2 efflux exhibited a pattern similar to the monthly change of precipitation, and this fact may indicate that precipitation has an effect on the soil respiration of each community as a limiting factor. Soil respiration plays a vital role in the global carbon cycle regulation. Specifically, comprehensive and ongoing monitoring research on CO_2 efflux rates of forests and factors affecting them is required in order to estimate the change in the net ecosystem production (NEP) of forests caused by climate change and provide fundamental data for such analyses.

Abbreviations
NEP: Net ecosystem production; NPP: Net primary production

Acknowledgements
This study was supported by the Long-Term Ecological Research Program of the Ministry of the Environment, Republic of Korea.

Funding
This study was conducted with the support of National Institute of Environmental Research, Korea.

Authors' contributions
All authors conducted a survey together during the study period. JHM wrote the manuscript. YYH participated in the design of the study and examined the manuscript. All authors read and approved the final manuscript.

Competing interests
The authors declare that they have no competing interests.

Author details
[1]Division of Ecosystem Services and Research Planning, National Institute of Ecology, Seocheon, South Korea. [2]Department of Biology, Kongju National University, Gongju, South Korea. [3]Division of Education Planning and Management, Nakdonggang National Institute of Biological Resources, Sangju, South Korea.

References
Chen, W., Jia, X., Zha, T., Wu, B., Zhang, Y., Li, C., Wang, X., He, G., Yu, H., & Chen, G. (2013). Soil respiration in a mixed urban forest in china in relation to soil temperature and water content. *European Journal of Soil Biology, 54*, 63–68.

Darenova, E., Pavelka, M., & Acosta, M. (2014). Diurnal deviations in the relationship between CO2 efflux and temperature: a case study. *Catena, 123*, 263–269.

Davidson, E. A., Belk, E., & Boone, R. D. (1998). Soil water content and temperature as independent or confounded factors controlling soil respiration in a temperate mixed hardwood forest. *Global Change Biology, 4*, 217–227.

Inclan, R., Uribe, C., De La Torre, D., Sanchez, D. M., Clavero, M. A., Fernandez, A. M., Morante, R., Cardena, A., Fernandez, M., & Rubio, A. (2010). Carbon dioxide fluxes across the Sierra de Guadarrama, Spain. *European Journal of Forest Research, 129*, 93–100.

Jang, R. H., Jeong, H. M., Lee, E. P., Cho, K. T., & You, Y. H. (2017). Budget and distribution of organic carbon in Taxus cuspidate forest in subalpine zone of Mt. Halla, 41, 4.

Jobbagy, E. G., & Jackson, R. B. (2000). The vertical distribution of soil organic carbon and its relation to climate and vegetation. *Ecological Applications, 10*, 426–436.

Kane, E. S., Pregitzer, K. S., & Burton, A. J. (2003). Soil respiration along environmental gradients in Olympic National Park. *Ecosystems, 6*, 326–335.

Kim, C. S. (2002). *Review on the factors causing change in the subalpine vegetation of Mt. Halla and conservation measures. The proceedings on the conservation and management of subalpine zone in Mt. Halla* (pp. 26–55). Seoul: Institute for Mt. Halla.

Kim, J. U., & Kil, B. S. (1996). Estimation for changes of net primary productivity and potential natural vegetation in the Korean Peninsula by the global warming. *Journal of Ecology and Environment, 19*, 1–7.

Kim, G. B., Lee, K. J., & Hyun, J. O. (1998). Regeneration of seedling under different vegetation types and effects of allelopathy on seedling establishment of A. koreana in the Banyabong Peak, Mt. Jiri. *Journal of Korean Forest Society, 87*, 230–238.

Kim, K. H., Kim, K. Y., Kim, J. K., Sa, D. M., Seo, J. S., Son, B. K., Yang, J. U., Um, K. C., Lee, S. U., Jeong, K. Y., Jeong, J. Y., Jeong, D. Y., Jeong, Y. T., Jeong, J. B., & Hyeon, H. N. (2014). *Soil Science* (p. 471). Seoul: Hyang Mun Press.

Kong, W. S. (1998). The alpine and subalpine geoecology of the Korean Peninsula. *Journal of Ecology and Environment, 21*, 383–387.

Kong WS. (2007). Biogeography of Korean plants. Geobook. Seoul. 335.

Koo, K. A., Park, W. K., & Kong, W. S. (2001). Dendrochronological analysis of A. koreana W. at Mt. Halla, Korea: effects of climate on the Growths. *Korean Journal of Ecology, 24*, 281–288.

Kuzyakov, Y. (2006). Sources of CO2 efflux from soil and review of partitioning methods. *Soil Biology and Biochemistry, 38*, 425–448.

Kwak, J. I., Lee, K. J., Han, B. H., Song, J. H., & Jang, J. S. (2013). A study on the vegetation structure of evergreen broad-leaved forest Dongbaekdongsan (Mt.) in Jeju, Korea. *Korean Journal of Environmental Ecology, 27*, 241–252.

Lee, Y. Y., & Mun, H. T. (2001). A study on the soil respiration in a Quercus acutissima forest. *Korean Journal of Ecology, 24*, 141–147.

Lee, K. J., Won, H. Y., & Mun, H. T. (2012). Contribution of root respiration to soil respiration for Quercus acutissima forest. *Korean Journal of Environmental Ecology, 2012*(26), 780–786.

Liu, X., Wan, S., Su, B., Hui, D., & Luo, Y. (2002). Response of soil CO2 efflux to water manipulation in a tallgrass prairie ecosystem. *Plant and Soil, 240*, 213–223.

Luo, Y., & Zhou, X. (2006). *Soil respiration and the environment* (p. 328). Burlington: Academic.

Moriyama, A., Yonemura, S., Kawashima, S., & Du, M. (2013). Environmental indicators for estimating the potential soil respiration rate in alpine zone. *Ecological Indicator, 32*, 245–252.

Nakane, K. (1995). Soil carbon cycling in a Japanese cedar (Cryptomeria japonica) plantation. *Forest Ecology and Management, 72*, 185–197.

Park, J. C., Yang, K. C., & Jang, D. H. (2010). The movement of evergreen broad-leaved forest zone in the warm temperature region due to climate change in South Korea. *Journal of Climate Research, 5*, 29–41.

Raich, J. W., & Schlesinger, W. H. (1992). The global carbon dioxide flux in soil respiration and its relationship to vegetation and climate. *Tellus, 44B*, 81–99.

Raich, J. W., & Tufekciogl, A. (2000). Vegetation and soil respiration: correlations and controls. *Biogeochemistry, 48*, 71–90.

Risk, D., Kellman, L., & Beltrami, H. (2002). Soil CO2 production and surface flux at four climate observations in eastern Canada. *Global Biogeochemical Cycles, 16*, 1122.

Sabine, C. L., Hemann, M., Artaxo, P., Bakker, D., Chen, C. T. A., Field, C. B., Gruber, N., LeQuere, C., Prinn, R. G., Richey, J. E., Romero-Lankao, P., Sathaye, J., & Valentini, R. (2004). *Current status and past trends of the global carbon cycle. In toward CO2 stabilization: issues, strategies, and consequences* (p. 568). Washington DC: Island Press.

Saiz, G., Byrne, K. A., Butterbach-Bahl, K., Kiese, R., Blujdea, V., & Farrell, E. P. (2006). Stand age related effects on soil respiration in a first rotation Sitka spruce chronosequence in central Ireland. *Global Change Biology, 12*, 1007–1020.

Suh, S., Lee, E., & Lee, J. (2009). Temperature and moisture sensitivities of CO2 efflux from lowland and alpine meadow soils. *Journal of Plant Ecology, 2*, 225–231.

Xu, L., Baldocchi, D. D., & Tang, J. (2004). How soil moisture, rain pulses, and growth alter the response of ecosystem respiration to temperature. *Global Biogeochemical Cycles, 18*, GB4002.

Yi, M. J., Son, Y., Jin, H. O., Park, I. H., Kim, D. Y., Kim, Y. S., & Shin, D. M. (2005). Below-ground carbon allocation of natural Quercus mongolica forests estimated from litterfall and soil respiration measurements. *Korean Journal of Agricultural and Forest Meteorology, 2005*(7), 227–234.

Yuste, J. C., Janssens, I. A., & Ceulemans, R. (2005). Calibration and validation of an empirical approach to model soil CO2 efflux in a deciduous forest. *Biogeochemistry, 2005*(73), 209–230.

18

Spatial distribution patterns of old-growth forest of dioecious tree *Torreya nucifera* in rocky Gotjawal terrain of Jeju Island, South Korea

Sookyung Shin[1], Sang Gil Lee[2] and Hyesoon Kang[1*] (iD)

Abstract

Background: Spatial structure of plants in a population reflects complex interactions of ecological and evolutionary processes. For dioecious plants, differences in reproduction cost between sexes and sizes might affect their spatial distribution. Abiotic heterogeneity may also affect adaptation activities, and result in a unique spatial structure of the population. Thus, we examined sex- and size-related spatial distributions of old-growth forest of dioecious tree *Torreya nucifera* in extremely heterogeneous Gotjawal terrain of Jeju Island, South Korea.

Methods: We generated a database of location, sex, and size (DBH) of *T. nucifera* trees for each quadrat (160 × 300 m) in each of the three sites previously defined (quadrat A, B, C in Site I, II, and III, respectively). *T. nucifera* trees were categorized into eight groups based on sex (males vs. females), size (small vs. large trees), and sex by size (small vs. large males, and small vs. large females) for spatial point pattern analysis. Univariate and bivariate spatial analyses were conducted.

Results: Univariate spatial analysis showed that spatial patterns of *T. nucifera* trees differed among the three quadrats. In quadrat A, individual trees showed random distribution at all scales regardless of sex and size groups. When assessing univariate patterns for sex by size groups in quadrat B, small males and small females were distributed randomly at all scales whereas large males and large females were clumped. All groups in quadrat C were clustered at short distances but the pattern changed as distance was increased. Bivariate spatial analyses testing the association between sex and size groups showed that spatial segregation occurred only in quadrat C. Males and females were spatially independent at all scales. However, after controlling for size, males and females were spatially separated.

Conclusions: Diverse spatial patterns of *T. nucifera* trees across the three sites within the *Torreya* Forest imply that adaptive explanations are not sufficient for understanding spatial structure in this old-growth forest. If so, the role of Gotjawal terrain in terms of creating extremely diverse microhabitats and subsequently stochastic processes of survival and mortality of trees, both of which ultimately determine spatial patterns, needs to be further examined.

Keywords: Dioecy, Gotjawal terrain, Spatial pattern, Spatial segregation of sexes, *Torreya nucifera*

Background

The spatial structure of plants in a population reflects complex interactions of ecological and evolutionary processes (Epperson 2005). Processes that can generate plant spatial patterns include intra- and interspecific interactions (Stoll and Bergius 2005; Nanami et al. 2011),

environmental heterogeneity (Zuo et al. 2008), breeding systems (Bleher et al. 2002), and disturbances (Wolf 2005; Rayburn and Monaco 2011). By analyzing spatial patterns of individuals, it might be possible to identify physical conditions and competitive factors contributing to the spatial structure of a population (Bell et al. 1993; Law et al. 2009; Rayburn et al. 2011). For example, clumped pattern most frequently observed in natural populations has been interpreted as evidence of positive

* Correspondence: hkang@sungshin.ac.kr
[1]Department of Biology, Sungshin University, Seoul 01133, South Korea
Full list of author information is available at the end of the article

interactions among individuals (Callaway 1995), patch distributions of resources (Schenk et al. 2003; Perry et al. 2009), and seed dispersal by gravity (Bleher et al. 2002; Gao et al. 2009).

Previous studies on spatial patterns of plants have focused on hermaphrodite or monoecious species whose parent plants can produce seeds (Nanami et al. 1999; Rayburn et al. 2011; Benot et al. 2013; Cheng et al. 2014). On the other hand, studies on the influences of dioecy consisting of separate male and female plants on spatial structure of tree populations are relatively limited (Gibson and Menges 1994; Hultine et al. 2007; Garbarino et al. 2015). For dioecious plants, spatial distribution of mature males and females is crucial to population survival because they can only reproduce by outcrossing (Bawa 1980; Thomson and Barrett 1981; Osunkoya 1999). Furthermore, females tend to allocate a greater proportion of resources to sexual reproduction than to growth and maintenance compared to males (Lloyd and Webb 1977; Opler and Bawa 1978; Delph 1999; Obeso 2002). Differences in reproductive cost between sexes may result in differential fitness between sexes across environmental gradients. Such differences can subsequently generate spatial segregation of sexes (SSS) (Bierzychudek and Eckhart 1988; Nuñez et al. 2008). In other words, females with resource limitations due to high reproductive costs will occupy more favorable or less stressful conditions regarding elevation (Garbarino et al. 2015), water availability (Ortiz et al. 2002), and/or soil fertility (Lawton and Cothran 2000). For example, female *Acer negundo* (Dawson and Ehleringer 1993), *Juniperus virginiana* (Lawton and Cothran 2000), and *Salix glauca* (Dudley 2006) are found on sites with relatively greater amounts of moisture. However, conflicting results relevant to SSS have recently been documented in several plant species (Ueno et al. 2007, Schmidt 2008, Gao et al. 2009, Forero-Montaña et al. 2010; Garbarino et al. 2015).

Some studies on sex-related size (or age) structures of dioecious species have been published (Goto et al. 2006; Zhang et al. 2010; Gabarino et al. 2015). Reproductive costs are age-specific and reproductive investment is variable during lifespan of plants (Silvertown and Dodd 1999; Montesinos et al. 2006). Nanami et al. (2005) have reported that dioecious individuals tend to shift from clumped distribution to random distribution as tree size increases due to density-dependent mortality caused by intraspecific competition. Therefore, interactions between sex and size in dioecious species should be considered in spatial structure study of dioecious population.

Torreya nucifera (Taxaceae) is a dioecious gymnosperm currently distributed in southern parts of South Korea and Japan. The largest (*n* = 2861) and oldest (mostly 200–400 years old, max. ~ 880 years old) population of *T. nucifera* in the world (*Torreya* Forest hereafter) is located

in Jeju Island, South Korea. In a previous study by Kang and Shin (2012), the *Torreya* population in Jeju Island could be separated into three sites depending on sex ratio and DBH (diameter at breast height). Abiotic heterogeneity influences adaptation activities such as competition, growth, and mortality of individuals, resulting in a unique spatial structure of the population (Scarano 2002; Zuo et al. 2008; Perry et al. 2009). Gotjawal terrain where *Torreya* Forest exists is a unique volcanic area with lava blocks scattered extremely disorderly (Jeon et al. 2012). Thus, Gotjawal terrain with severe topographic heterogeneity generates highly diverse microclimate (Choi and Lee 2015). Furthermore, Gotjawal terrain represents poor soil development and oligotrophic and stressful environment to plants. Therefore, current spatial structure of *Torreya* Forest is likely to reflect the long history of its survival and mortality in such a harsh environment. However, the spatial structure of *Torreya* Forest has not been examined so far.

As plants are sessile, their survival is mostly determined at quite a local scale, even at a scale of a few centimeters, by both biotic and abiotic factors rather than by spatial average of its overall population (Stoll and Prati 2001; Benot et al. 2013). If so, dioecious sexual system of *T. nucifera* trees, size difference among their three sites, and complexity of the terrain should all be considered in the study of spatial structure of *Torreya* Forest in Jeju Island. In this study, the following aspects were examined: (1) the distribution pattern of *T. nucifera* trees according to sex and size at each site, (2) the site differences in those distribution patterns, and (3) the pattern of spatial segregation of sexes and sizes at each site.

Methods
Study species and site
Torreya nucifera (Taxaceae) is a dioecious evergreen gymnosperm that can grow up to 25 m in height. Its pollination occurs in April and green fleshy aril-covered seeds mature in the fall one year after pollination. Various parts of this tree have traditionally been used. For example, its wood has been used for furniture and its seeds have been used for anthelmintic and oriental medicines and oil.

Torreya Forest (Natural Monument No. 374 in South Korea) is located in Jeju Island (33° 29′ N, 126° 48′ E) (Fig. 1a). This forest (44.8 ha in area, 143 m mean a.s.l.) extends 1.4 km in the north-south direction with a width of 0.6 km. It is located between two small volcanoes: Darangshioreum (382.4 m a.s.l.) and Dotoreum (a volcano near the southern end of the population; 284.2 m a.s.l.). In 1999, all *T. nucifera* trees with DBH ≥ 6 cm were tagged and tending was started for 11 plots along trails within the forest (Fig. 1b). *Torreya* Forest terrain is defined as Gujwa-Seongsan Gotjawal, a transition lava zone distributing both pahoehoe and aa lava flows (Jeon et al. 2012). Although

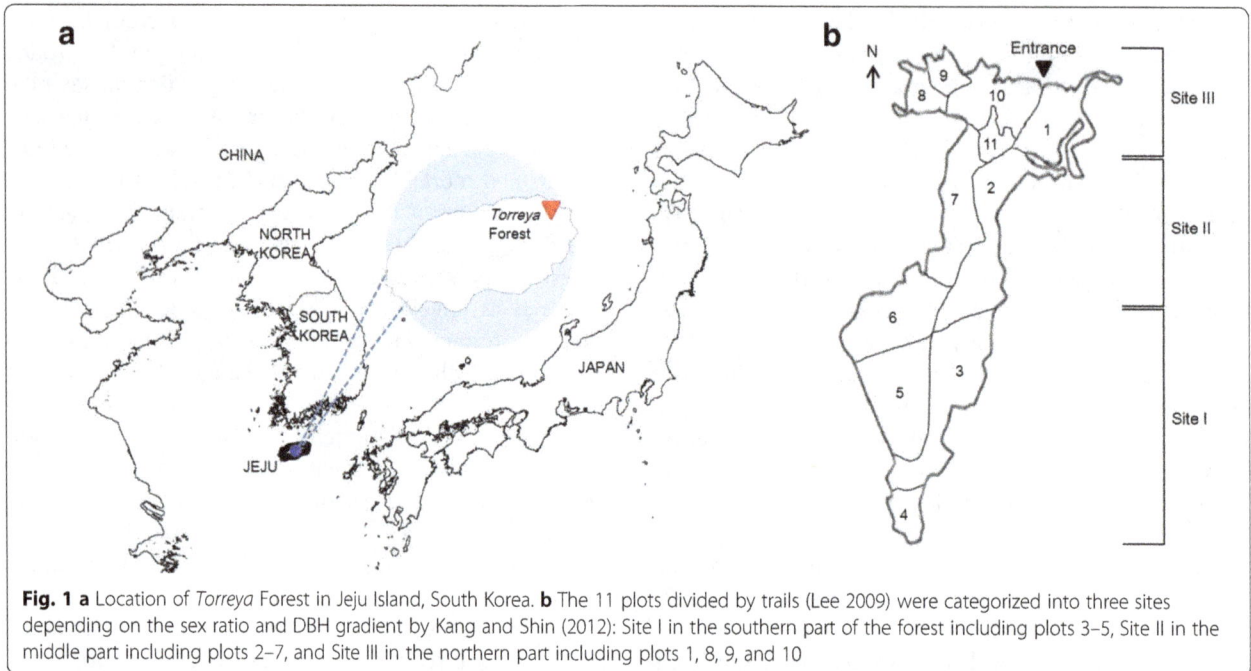

Fig. 1 a Location of *Torreya* Forest in Jeju Island, South Korea. **b** The 11 plots divided by trails (Lee 2009) were categorized into three sites depending on the sex ratio and DBH gradient by Kang and Shin (2012): Site I in the southern part of the forest including plots 3–5, Site II in the middle part including plots 2–7, and Site III in the northern part including plots 1, 8, 9, and 10

T. nucifera is a domonant species in this forest, the Forest consists of diverse evergreen trees with a total of 276 plant taxa, including *Mallotus japonicus*, *Machilus thunbergii*, *Orixa japonica*, and *Polystichum tripteron* [Korea Tree Health Association (KTHA) 1999; Lee 2009; Shin et al. 2010; Choi and Lee 2015]. Figure 2 demonstrates characteristic features of wild *Torreya* Forest. According to 2007–2016 data from Gujwa Regional Meteorological Office close to the *Torreya* Forest, the mean monthly temperature ranges from 5.3 °C in January to 26.7 °C in August (mean annual temperature, 15.7 °C). Its mean annual precipitation is 1774.2 mm with a peak in August (307.8 mm) with mean annual wind speed of 4.0 m/s (Korea Meteorological Administration 2017).

Data collection

The planimetric map was transformed into an image file through high-resolution scanning using a survey map of the location of *T. nucifera* trees (KTHA 1999). Warping was performed by assigning coordinates to each edge. Digitizing process was then used to obtain coordinates for 2861 individuals (1498 males and 1363 females). By inputting attribute values such as sex, DBH, and sites for each individual, a database of location and ecological traits of *T. nucifera* trees was completed. Sex and DBH data were obtained from KTHA (1999) and Lee (2009). *Torreya* Forest exhibited a gradient of sex ratio and DBH from the southern end to the northern end of the forest. In this study, we analyzed spatial patterns of *T. nucifera* trees based on the three sites defined by Kang and Shin (2012). Numbers, sex ratio, and DBH of *T. nucifera* trees in the

three sites are shown in Table 1. ArcGIS Ver. 9.3. (ESRI 2008) was used for all analyses described above.

In *Torreya* Forest, there are somewhat unnatural spaces where *T. nucifera* trees are absent or scant due to the planting of *Pinus thunbergii* trees and various herbaceous plants, paved entrance, and trails. Therefore, a quadrat (160 × 300 m) was established in each of the three sites to represent the typical local condition (Fig. 3a): quadrat A for Site I, quadrat B for Site II, and quadrat C for Site III. The distribution of *T. nucifera* trees according to sex and size (DBH) in each quadrat is shown in Fig. 3b.

Data analyses

T. nucifera trees were categorized into eight groups based on sex (males and females), size (small trees, DBH < 50 cm; and large trees, DBH ≥ 50 cm), and sex by size (small males, large males, small females and large females) for spatial point pattern analysis.

O-ring statistics O(r) was employed to describe the average density of points at a distance of r (Wiegand and Moloney 2004; Law et al. 2009). Value of O(r) was calculated as follows. Around each individual data point, numerous circles with radius r were drawn and the correlation between the average number of individuals within numerous circles and radius r was deduced to determine the O value. O-ring statistics included univariate and bivariate analyses (Zhang et al. 2010). Univariate statistical analysis of $O_{11}(r)$ was used to analyze spatial patterns between individuals in a group while bivariate statistical analysis of $O_{12}(r)$ was used to analyze spatial relationships between two different groups. $O_{11}(r)$ was applied to assess spatial

Table 1 Numbers, sex ratio and DBH of *Torreya nucifera* trees in the three sites of *Torreya* Forest in Jeju Island (adapted from Kang and Shin 2012)

Site	Plot ID	Altitude (m)	Area (ha)	Sex		Sex ratio	Density	Mean DBH (cm)
				Males	Females			
I	3	148	6.4	235	167	0.58 **	62.8	58.9 ± 20.6
	4	154	1.5	16	11	0.59 NS	18.0	40.5 ± 20.6
	5	145	7.1	268	165	0.62 ***	61.0	61.7 ± 19.4
Mean						0.60 ***	57.5	59.7 ± 20.4
II	2	140	6.8	304	293	0.51 NS	87.8	54.2 ± 16.4
	6	144	5.3	194	200	0.49 NS	74.3	56.2 ± 15.9
	7	141	4.7	165	160	0.51 NS	69.2	45.0 ± 16.9
Mean						0.50 NS	78.3	52.5 ± 17.0
III	1	130	4.0	64	47	0.58 NS	27.8	46.2 ± 21.7
	8	129	1.3	23	64	0.26 ***	66.9	57.0 ± 19.0
	9	128	1.3	18	21	0.46 NS	30.0	44.4 ± 17.1
	10	131	3.9	211	235	0.47 NS	114.7	39.4 ± 21.1
Mean						0.46 (*)	65.0	43.0 ± 21.5
Overall mean						0.52 *	67.6	52.4 ± 20.1

Sex ratio = male/(male + female); Density = trees/ha; *DBH* diameter at breast height
(*) $P = 0.0510$; * $P < 0.05$; ** $P < 0.01$; *** $P < 0.0001$; *NS* not significant

patterns of *T. nucifera* trees within two sexes and/or two size groups (e.g., male, female, small, and large) in *Torreya* Forest. $O_{12}(r)$ was used to determine whether there was a spatial relationship between the two sexes and/or the two size groups (e.g., males vs. females and small vs. large trees). To conduct significance test for O(r) value at each distance *r*, null hypothesis was formed using complete spatial randomness (CSR). Then 95% confidence intervals for both univariate and bivariate analyses were computed from 999 Monte Carlo simulations. If O(r) value was above the CSR value, individuals were considered to be clumped. If O(r) value was within 95% confidence intervals, individuals were considered to be randomly distributed. If it was below the CSR value, individuals were considered to be regularly distributed. All calculations and simulations were performed using PROGRAMITA software (Wiegand and Moloney 2004).

Results

Ecological characteristics of the three sites

Ecological characteristics such as sex ratio, size, and density of *T. nucifera* trees in each quadrat are shown in Table 2. *T. nucifera* trees of quadrat A (at Site I) were male-biased (sex ratio = 0.62). Large trees were about four times more abundant than small trees (*n* = 241 vs. 62). The mean DBH was relatively large whereas its average density was the lowest among the three quadrats. On the other hand, *T. nucifera* trees in quadrat B

(at Site II) and C (at Site III) showed a 1:1 sex ratio and somewhat higher density than quadrat A. In addition, in quadrat C, *T. nucifera* trees had the smallest mean DBH, and small trees were more abundant than large trees (*n* = 230 vs. 177).

Univariate spatial analyses

Univariate analyses showed that spatial patterns of *T. nucifera* trees differed between groups (e.g., male vs. female, small vs. large) of each quadrat (Figs. 4, 5 and 6. In quadrat A, all trees and eight groups of *T. nucifera* showed random distribution at all scales (Fig. 4). In quadrat B, all trees demonstrated random distribution at all scales. However, when sex and size were considered, spatial pattern changed. Males were randomly distributed at all scales while females were weakly clumped at 0–17 m (except for 8–10 m). Small trees were randomly distributed at all scales while large trees were clumped (2–23 m) or regularly (54–58 m) distributed at some distances. By assessing univariate patterns of four groups of sex by size, small males and small females were distributed randomly at all scales. However, large males and large females were clumped at 0–16 m scales. Finally, all groups in quadrat C were clumped at short distances. They were distributed randomly or regularly as the distance was increased. When sex and size were combined, males (0–32 m) clumped more strongly than females (0–12 m) in the small tree group whereas females (0–24 m) clumped more strongly than males (0–17 m) in

Fig. 2 Wild *Torreya* Forest in Jeju Island, South Korea. **a** A guideboard of *Torreya* Forest. **b** *T. nucifera* tree reaching 10 m in height. **c** Stand structure of *Torreya* Forest. **d** Rocky Gotjawal terrain of *Torreya* Forest. **e** *T. nucifera* tree rooted on lava blocks

Table 2 Numbers, sex ratio and DBH of *T. nucifera* trees in the three quadrats (160 × 300 m)

		Sex		Sex ratio	Density	Mean DBH (cm)
		Males	Females			
Quadrat A						
DBH	Small	31	31	0.62 ***	63.5	62.5 ± 18.3
	Large	158	85			
Quadrat B						
DBH	Small	72	89	0.53 NS	99.4	54.1 ± 15.7
	Large	181	135			
Quadrat C						
DBH	Small	129	101	0.48 NS	84.8	42.1 ± 21.6
	Large	65	112			

Sex ratio = male/(male + female); Density = trees/ha; DBH, Small = DBH < 50 cm, Large = DBH ≥ 50 cm
*** $P < 0.0001$; *NS* not significant

the large tree group. Thus, spatial patterns of *T. nucifera* trees varied among three quadrats. For example, large females had different distribution types for each quadrat that are random (0–60 m) in quadrat A, clumped (5–16 m) in quadrat B, and regular (48–60 m) in quadrat C (Figs. 4, 5 and 6).

Bivariate spatial analyses

Bivariate analyses for spatial associations between sexes and/or size groups showed that spatial segregation occurred only in quadrat C (Fig. 7). Males and females were spatially independent at all scales. Significant spatial segregation was observed between small and large trees (4–22 m). When combination groups of sex and size were examined, repulsion between groups was notable: small males vs. small females at 50–60 m scales; large males vs. small females at 12–17 m scales; small

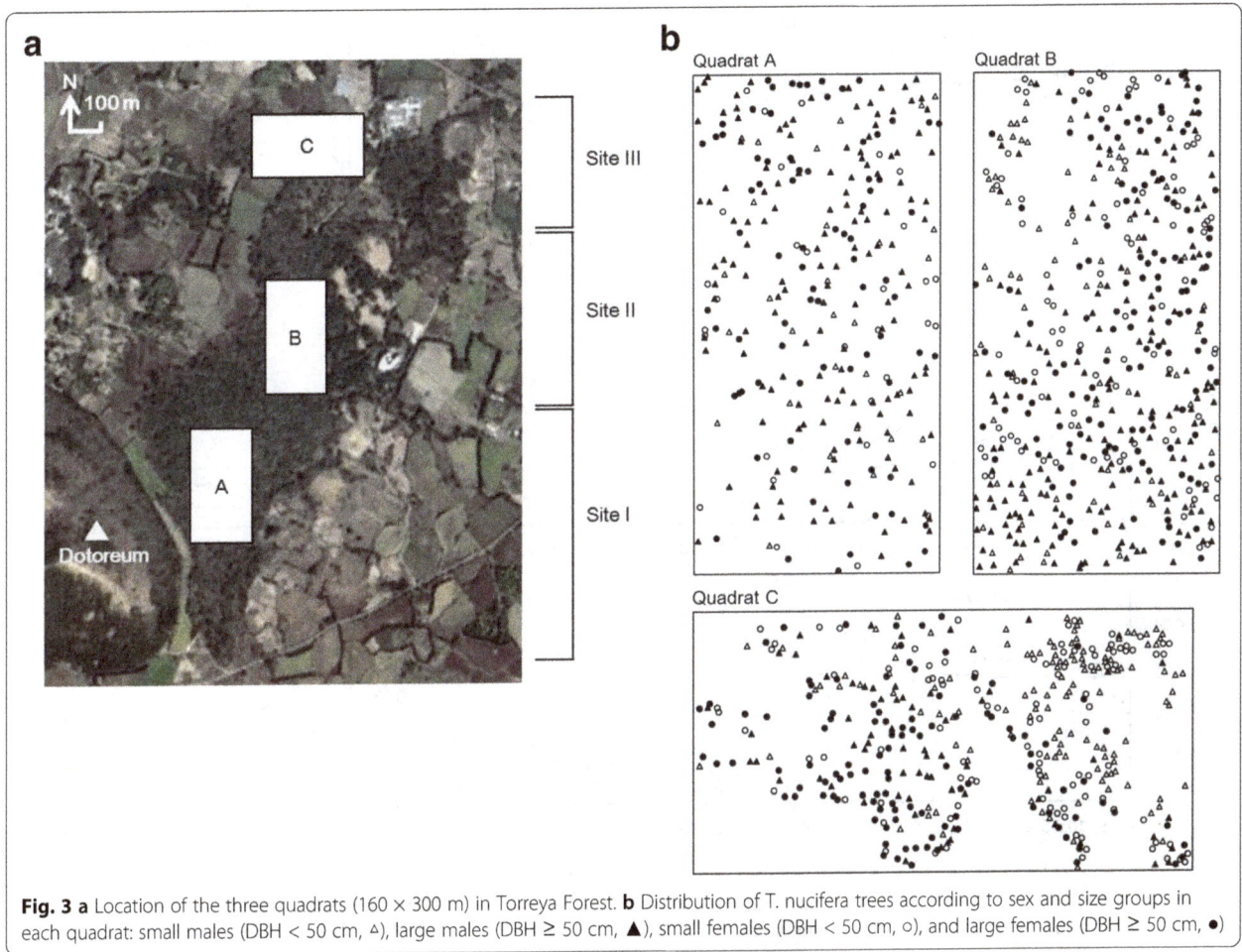

Fig. 3 a Location of the three quadrats (160 × 300 m) in Torreya Forest. **b** Distribution of T. nucifera trees according to sex and size groups in each quadrat: small males (DBH < 50 cm, △), large males (DBH ≥ 50 cm, ▲), small females (DBH < 50 cm, ○), and large females (DBH ≥ 50 cm, ●)

males vs. large females at 0–22 m scales; and large males vs. large females at 37–60 m scales.

Discussion

Spatial patterns of individual *T. nucifera* trees within and among sites

This is the first study to reveal the spatial pattern of old trees that have survived for several hundreds of years in almost natural conditions in Korea. Individual trees of *T. nucifera* in *Torreya* Forest were distributed randomly, clumped, or regularly with each other depending on the sites examined.

Site I at the southern end of *Torreya* Forest, adjacent to a small volcano called Dotoreum (284.2 m a.s.l.), is characterized by relatively large proportion of oldaged trees. Many studies on dioecious species have reported that there is a difference in distribution pattern of individuals according to sex (Zhang et al. 2010; Chen et al. 2014; Garbarino et al. 2015). However, both male and female trees of *T. nucifera* were randomly distributed in quadrat A, not supporting sex-specific segregation between male and female trees. Random distribution of both sexes has also been reported in the old-growth plot

of *Acer barbinerve* (Pan et al. 2010). Several studies in recent years have also emphasized the effects of both biotic and abiotic conditions on plant spatial patterns. For example, old trees tend to be randomly distributed as a result of stochastic mortality and intra- or interspecific competition for resources (Nanami et al. 2005). Dying seedlings under large trees are frequently found in *Torreya* Forest. Even if seedlings were initially clumped around female trees, only small gaps randomly scattered in Gotjawal terrain may allow seedlings to survive and finally can generate a random dispersion pattern.

Site II, corresponding to the middle area of the forest, showed intermediate values in sex ratio and size between Site I and Site III (Kang and Shin 2012). It has been reported that as the size of plant is increased, the clumped pattern is generally weakened (Wang et al. 2010; Cheng et al. 2014). Recently, Garbarino et al. (2015) showed that large males and females of *Taxus baccata*, which is phylogenetically related to *T. nucifera*, were randomly distributed while small trees were mostly clumped. However, in quadrat B, regardless of sexes, small trees of both sexes were randomly distributed while large male and female trees were clumped at 0–16 m scales

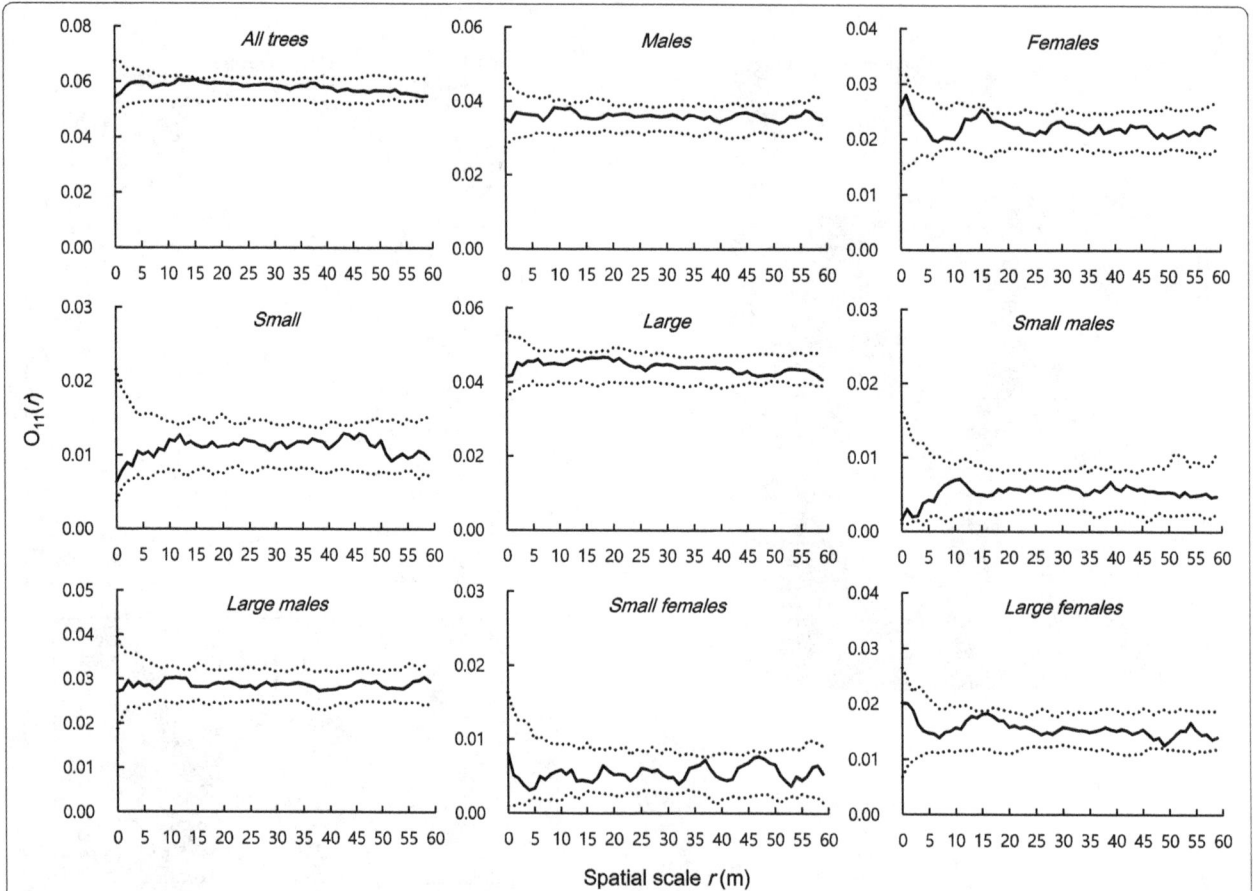

Fig. 4 Spatial patterns of *T. nucifera* trees in univariate analyses in quadrat A. Analyses were conducted for all trees pooled over sex and size groups, separately for each sex and size groups, and then for combination groups of sex and size (e.g., small males, large males, small females, and large females). Solid lines indicate $O_{11}(r)$ value from each analysis; dashed lines indicate 95% confidence envelopes regarding the random spatial structure null hypothesis

(Fig. 5). The clumped patterns of large trees are in contrast to the results as mentioned above. However, several authors suggest that the increase in the clumped pattern with age might be attributable to interspecific competition (Nanami et al. 2011), differential mortality of juveniles (Briggs and Gibson 1992), and clonal structure (Peterson and Squiers 1995). This forest is mixed with diverse broad-leaved plants and *P. thunbergii*. Lee (2005) recognized that interspecific competition greatly reduced the vigor of *T. nucifera* trees. The fact that the vigor of trees, especially female trees, was significantly improved after removal of epiphytes from the canopy of *T. nucifera* trees (Kang and Shin 2012) also supports the importance of interspecific competition. The influence of competition on the clumping of large trees in Site II remains to be tested because we did not examine the density or distance between neighboring trees in this site.

Site III in the northern part of the forest is characterized by relatively small trees and more females than males than in other sites (Kang and Shin 2012).

Individual trees in quadrat C were clustered at short distances and distributed randomly or regularly as distance increased (Fig. 6). Even when combining sex and size, males (0–32 m) clumped more strongly than females (0–12 m) in small trees but females (0–24 m) clumped more strongly than males (0–17 m) in large trees. In other words, consistent spatial patterns across sex and size groups do not exist at all in this site. A simple adaptive explanation for such complex patterns may not be possible at this stage.

In this forest, anthropogenic activities have been controlled for several hundreds of years (Kim 1985, Shin et al. 2010). However, we speculate that Site III might have been sporadically subjected to severe disturbances by human activities such as logging. This is particularly likely because this site (Site III) is located near the entrance to the forest which is very easy to access compared to Site I near a small volcano. The most convincing evidence for artificial planting in this site comes from aerial photos taken from 1967 to 2015 (Fig. 8). In 1967, it appeared that certain locations at Site III were largely free of vegetation.

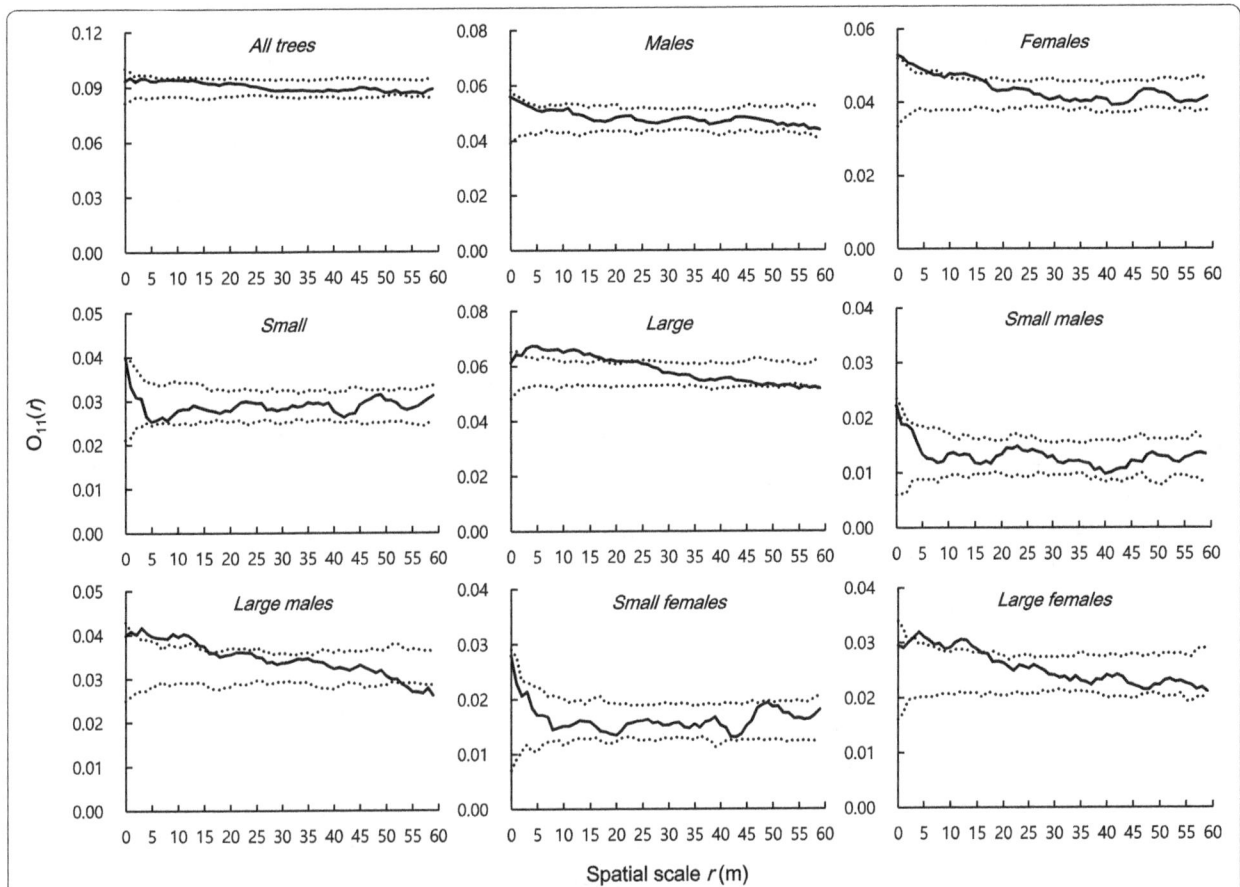

Fig. 5 Spatial patterns of *T. nucifera* trees in univariate analyses in quadrat B. Analyses were conducted for all trees pooled over sex and size groups, separately for each sex and size groups, and then for combination groups of sex and size (e.g., small males, large males, small females, and large females). Solid lines indicate $O_{11}(r)$ value from each analysis; dashed lines indicate 95% confidence envelopes regarding the random spatial structure null hypothesis

Since then, its vegetation coverage has increased significantly until 2015. Such drastic changes in vegetation coverage at those locations are not expected to occur by natural regeneration because it takes about 54 years for the DBH of *T. nucifera* to grow up to 6 cm DBH (Lee 2005). Thus, those photos are highly likely to reflect that artificial planting occurred intensively around 1970s when the Korean government pushed a strong reforestation policy on a nationwide scale. Kang and Shin (2012) provided another piece of evidence for artificial planting in this site. In their study, some plots at Site III are deviated quite a bit from the site's sex ratio and mean size. For example, plot 10 (Fig. 1b) which was almost a vacant area in the aerial photo (Fig. 8a) consists of smaller trees than other plots and its density (114.7 trees/ha) being twice as high as the average population density of 67.6 trees/ha. If *T. nucifera* trees were planted during the 1970s, planting would be concentrated in less rocky localities covered with some soils, perhaps being responsible for clumped spatial patterns at a fine scale as detected in this study. Unfortunately, written documents regarding the reforestation

of *T. nucifera* trees are not available to confirm this inference.

Spatial segregation of sexes in *T. nucifera*

The SSS expected under the hypothesis of high reproductive cost of females has been reported in > 30 species from 24 families of seed plants (Barrett and Hough 2013). In this study, SSS occurred only in Site III, e.g., between small males vs. small females (50–60 m scales), large males vs. small females (12–17 m scales), small males vs. large females (0–22 m scales), and large males vs. large females (37–60 m scales) (Fig. 7). SSS detected only in the site with small trees with the highest density is analogous with the case of *Fraxinus mandshurica*. In dioecious *F. mandshurica*, spatial segregation between sexes was found in the secondary forests, but not in the old-growth (Zhang et al. 2010). They ascribed a lack of SSS in old-growth forest of *F. mandshurica* to sex-specific responses to microenvironments. In this study, Site I trees which were larger than ones in other sites were randomly distributed regardless of sex and size groups.

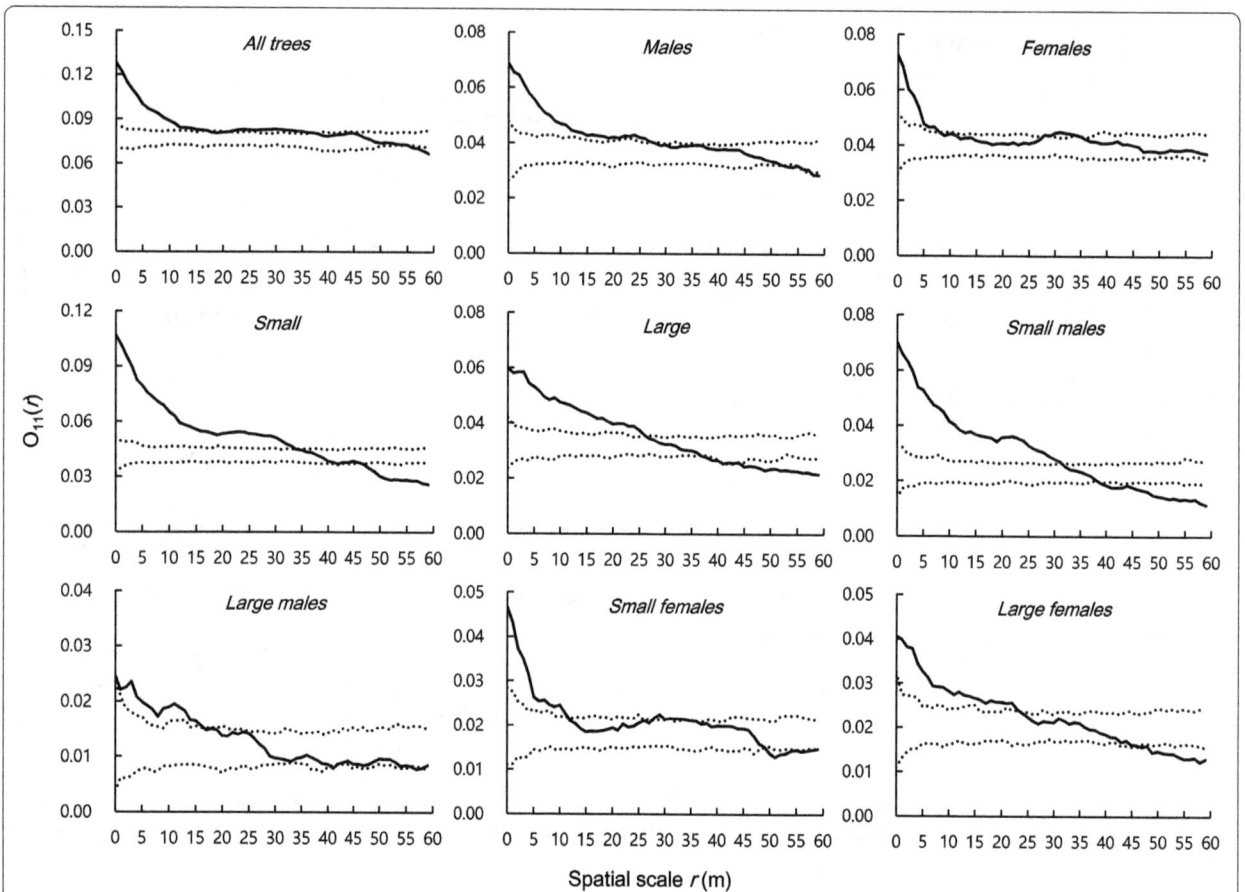

Fig. 6 Spatial patterns of *T. nucifera* trees in univariate analyses in quadrat C. Analyses were conducted for all trees pooled over sex and size groups, separately for each sex and size groups, and then for combination groups of sex and size (e.g., small males, large males, small females, and large females). Solid lines indicate $O_{11}(r)$ value from each analysis; dashed lines indicate 95% confidence envelopes regarding the random spatial structure null hypothesis

Currently, it is unknown whether Site I trees exhibit sex-differential responses to microenvironments.

Influence of the rocky Gotjawal terrain

Diverse spatial patterns of *T. nucifera* trees across the three sites may imply both biotic and abiotic effects in this forest. Most of all, the role of Gotjawal terrain in terms of creating extremely diverse microhabitats, and subsequently, stochastic processes of survival or mortality of trees need to be emphasized. In stressful habitats such as the rocky coasts and alpine forests, spatial structure largely depends on constraining factors such as topography and disturbances (Camarero et al. 2000; Scarano 2002; Humphries et al. 2008). If so, we can draw a couple of inferences regarding the ecological role of Gotjawal terrain. Firstly, it seems that Gotjawal terrain is a major component affecting the distribution pattern of trees. In highly diverse microhabitats at small scales created in Gotjawal terrain, it may be difficult to expect a formation of resource gradient such as moisture or soil nutrients. That means, SSS would be difficult to expect. Furthermore, rocks

and thin soils in Gotjawal terrain may cause stochastic processes over the survival and mortality, and consequently affecting spatial patterns of *T. nucifera* trees. Secondly, *T. nucifera* trees may have to respond to multiple stressors in circumstances with scattered lava blocks and little soil. Thus, diverse spatial patterns of *T. nucifera* trees may be an outcome of biotic as well as abiotic interactions that have occurred for hundreds of years.

Despite that this forest has been known to be a natural forest for such a long time, we found a new evidence for artificial interference in some plots especially near the entrance area. The rocky topography of Gotjawal terrain is certainly a component that has protected the historical and valuable *T. nucifera* trees in Jeju Island. Understanding of the spatial patterns of *T. nucifera* trees will help provide indispensable information on the interactions between trees on a stressful Gotjawal terrain. In order to figure out the biotic and abiotic factors affecting spatial patterns, it is also necessary to understand the tending and planting history in this forest.

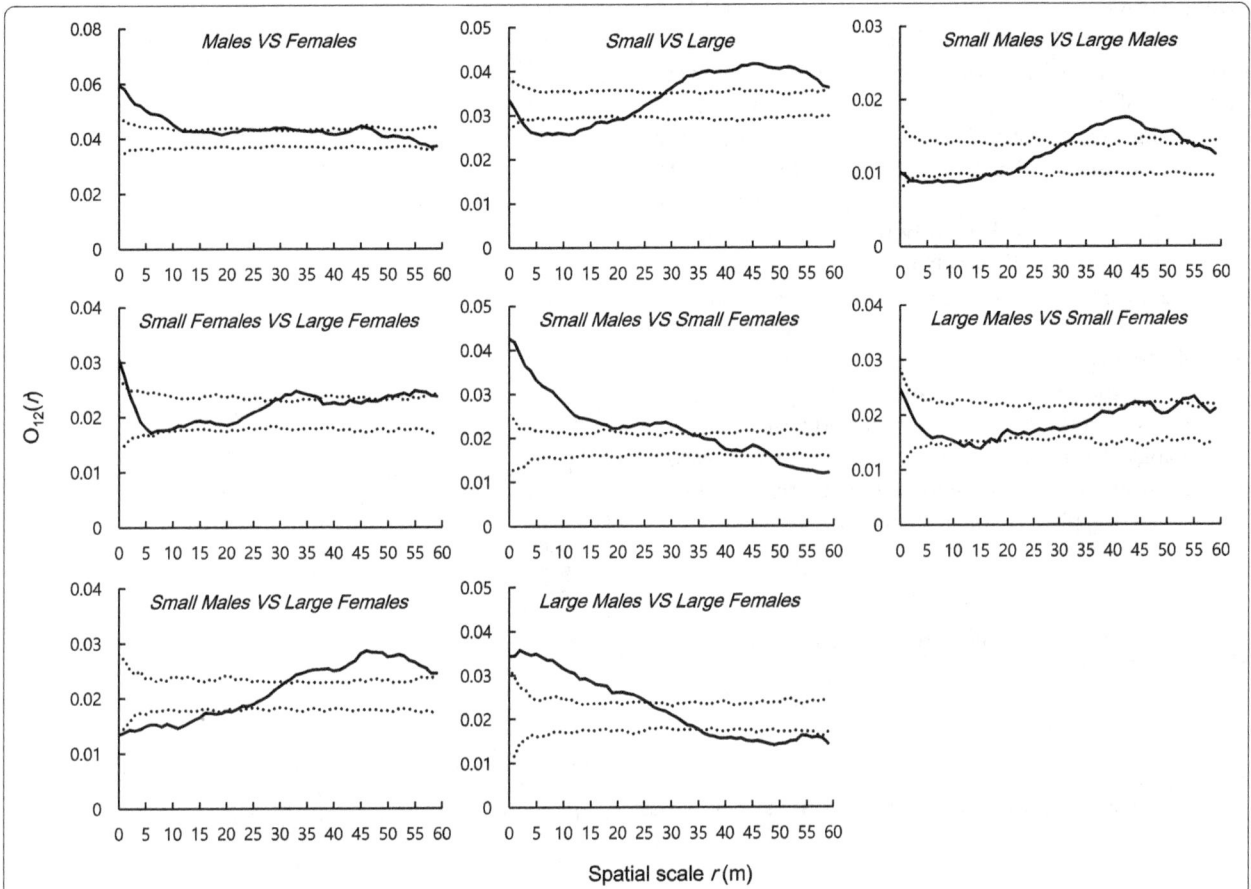

Fig. 7 Spatial patterns of *T. nucifera* trees in bivariate analyses in quadrat C. Analyses were conducted for all trees pooled over sex and size groups, separately for each sex and size groups, and then for combination groups of sex and size (e.g., small males, large males, small females, and large females). Solid lines indicate $O_{12}(r)$ value from each analysis; dashed lines indicate 95% confidence envelopes regarding the random spatial structure null hypothesis

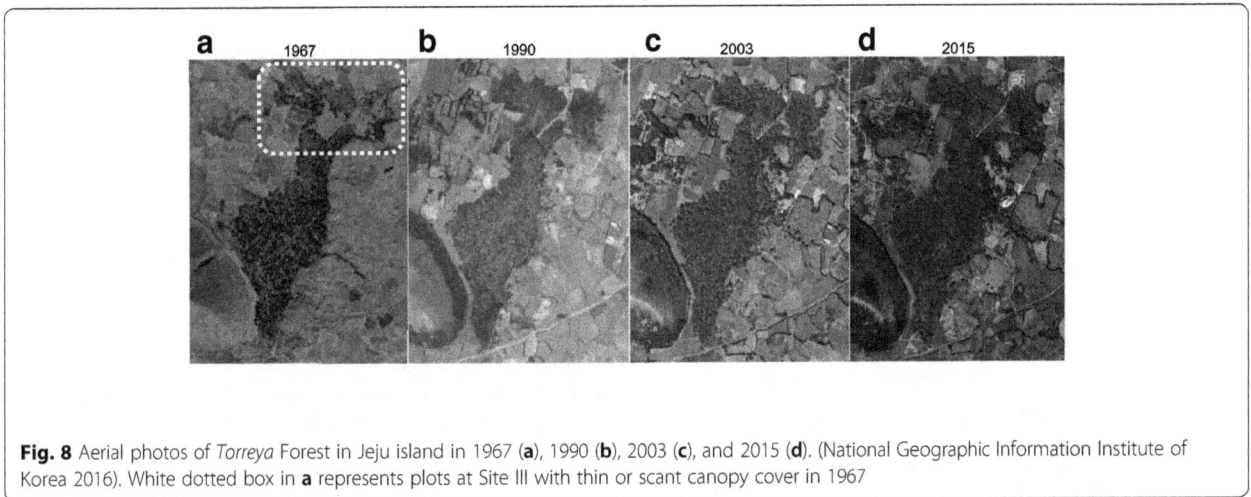

Fig. 8 Aerial photos of *Torreya* Forest in Jeju island in 1967 (**a**), 1990 (**b**), 2003 (**c**), and 2015 (**d**). (National Geographic Information Institute of Korea 2016). White dotted box in **a** represents plots at Site III with thin or scant canopy cover in 1967

Conclusions

Individual trees of dioecious *T. nucifera* were randomly distributed in an old-growth forest with over several hundreds of years of history. However, when the spatial relationship was assessed for sex and size groups, spatial patterns of trees tended to differ among the three sites within the forest, and spatial segregation was notable only in one site of which density was the highest among the three sites and artificial planting occurred several decades ago. Considering the extremely heterogeneous topography and thin soils of Gotjawal terrain, it may not be possible to expect SSS shown in many other dioecious species. We did not directly examine the topographic features and its ecological effects in this study. Incorporation of topological traits in *Torreya* Forest in spatial pattern analyses would be rewarding to identify biotic and abiotic effects affecting plant spatial distribution.

Abbreviations

DBH: Diameter at breast height; KTHA: Korea Tree Health Association; SSS: Spatial segregation of sexes

Acknowledgements

We thank anonymous reviewers for their helpful comments to the drafts of this article.

Funding

This study was supported by the Sungshin University Research Grant to HK (Grant no. 2015-1-11-051/1).

Authors' contributions

All authors conducted a survey together during the study period. SS analyzed the data and drafted the manuscript. SL conducted the fieldwork. HK conceived the study and drafted the manuscript. All authors read and approved the final manuscript.

Competing interests

The authors declare that they have no competing interests.

Author details

¹Department of Biology, Sungshin University, Seoul 01133, South Korea.
²Hankang Tree Hospital, Seongnam 13631, South Korea.

References

Barrett, S. C. H., & Hough, J. (2013). Sexual dimorphism in flowering plants. *Journal of Experimental Botany, 64*, 67–82.

Bawa, K. S. (1980). Evolution of dioecy in flowering plants. *Annual Review of Ecology and Systematics, 11*, 15–39.

Bell, G., Lechowicz, M. J., Appenzeller, A., Chandler, M., DeBlois, E., Jackson, L., Mackenzie, B., Preziosi, R., Schallenberg, M., & Tinker, N. (1993). The spatial structure of the physical environment. *Oecologia, 96*, 114–121.

Benot, M.-L., Bittebiere, A.-K., Ernoult, A., Clément, B., & Mony, C. (2013). Fine-scale spatial patterns in grassland communities depend on species clonal dispersal ability and interactions with neighbours. *Journal of Ecology, 101*, 626–636.

Bierzychudek, P., & Eckhart, V. (1988). Spatial segregation of the sexes of dioecious plants. *The American Naturalist, 132*, 34–43.

Bleher, B., Oberrath, R., & Böhning-Gaese, K. (2002). Seed dispersal, breeding system, tree density and the spatial pattern of trees–a simulation approach. *Basic Applied Ecology, 3*, 115–123.

Briggs, J. M., & Gibson, D. J. (1992). Effect of fire on tree spatial patterns in a tallgrass prairie landscape. *Bulletin of the Torrey Botanical Club, 119*, 300–307.

Callaway, R. M. (1995). Positive interactions among plants. *The Botanical Review, 61*, 306–349.

Camarero, J. J., Gutiérrez, E., & Fortin, M-J. (2000). Spatial pattern of subalpine forest-alpine grassland ecotones in the Spanish Central Pyrenees. *Forest Ecology and Management, 134*, 1–16.

Cheng, X., Han, H., Kang, F., Song, Y., & Liu, K. (2014). Point pattern analysis of different life stage of *Quercus liaotungensis* in Lingkong Mountain, Shanxi Province, China. *Journal of Plant Interactions, 9*, 233–240.

Choi, B-K., & Lee, C-B. (2015). A study on the synecological values of the *Torreya nucifera* Forest (Natural Monument No. 374) at Pyeongdae-ri in Jeju Island. *Journal of the Korean Institute of Traditional Landscape Architecture, 33*, 87–98 (in Korean with English abstract).

Dawson, T. E., & Ehleringer, J. R. (1993). Gender-specific physiology, carbon isotope discrimination, and habitat distribution in boxelder, *Acer negundo. Ecology, 74*, 798–815.

Delph, L. F. (1999). Sexual dimorphism in life history. In M. A. Geber, T. E. Dawson, & L. F. Delph (Eds.), *Gender and sexual dimorphism in flowering plants* (pp. 149–173). Berlin: Springer Berlin Heidelberg.

Dudley, L. S. (2006). Ecological correlates of secondary sexual dimorphism in *Salix glauca* (Salicaceae). *American Journal of Botany, 93*, 1775–1786.

Environmental Systems Research Institute: ESRI. Arc/Info User's Guide. Ver. 9.3. Environmental Systems Research Institute, Redlands, CA, USA; 2008.

Epperson, B. K. (2005). Estimating dispersal from short distance spatial autocorrelation. *Heredity, 95*, 7–15.

Forero-Montaña, J., Zimmerman, J. K., & Thompson, J. (2010). Population structure, growth rates and spatial distribution of two dioecious tree species in a wet forest in Puerto Rico. *Journal of Tropical Ecology, 26*, 433–443.

Garbarino, M., Weisberg, P. J., Bagnara, L., & Urbinati, C. (2015). Sex-related spatial segregation along environmental gradients in the dioecious conifer, *Taxus baccata. Forest Ecology and Management, 358*, 122–129.

Gao, P., Kang, M., Wang, J., Ye, Q., & Huang, H. (2009). Neither biased sex ratio nor spatial segregation of the sexes in the subtropical dioecious tree *Eurycorymbus cavaleriei* (Sapindaceae). *Journal of Integrative Plant Biology, 51*, 604–613.

Gibson, D. J., & Menges, E. S. (1994). Population structure and spatial pattern in the dioecious shrub *Ceratiola ericoides. Journal of Vegetation Science, 5*, 337–346.

Goto, S., Shimatani, K., Yoshimaru, H., & Takahashi, Y. (2006). Fat-tailed gene flow in the dioecious canopy tree species, *Fraxinus mandshurica* var. *japonica* revealed by microsatellites. *Molecular Ecology, 5*, 2985–2996.

Hultine, K. R., Bush, S. E., West, A. G., & Ehleringer, J. R. (2007). Population structure, physiology and ecohydrological impacts of dioecious riparian tree species of western North America. *Oecologia, 54*, 85–93.

Humphries, H. C., Bourgeron, P. S., & Mujica-Crapanzano, L. R. (2008). Tree spatial patterns and environmental relationships in the forest-alpine tundra ecotone at Niwot Ridge, Colorado, USA. *Ecological Research, 23*, 589–605.

Jeon, Y., Ahn, U. S., Ryu, C. G., Kang, S. S., & Song, S. T. (2012). A review of geological characteristics of Gotjawal terrain in Jeju Island: preliminary study. *The Geological Society of Korea, 48*, 425–434 (in Korean with English abstract).

Kang, H., & Shin, S. (2012). Sex ratios and spatial structure of the dioecious tree *Torreya nucifera* in Jeju Island, Korea. *Journal of Ecology and Field Biology, 35*, 111–122.

Kim, Y. S. (1985). Phytogeographic distribution of genus *Torreya* of the world. *Journal of Resource development, 4*, 143–150.

Korea Meteorological Administration. Research for Climate Information. 2017. https://data.kma.go.kr/data/grnd/selectAwsList.do?pgmNo=35. Accessed on Aug. 2017.

Korea Tree Health Association: KTHA. *Torreya nucifera* Forest in Gujwa-eup: conservation and maintenance measures. Bukjeju-gun, Jeju; 1999. (in Korean).

Law, R., Illian, J., Burslem, D. F. R. P., Gratzer, G., Gunatilleke, C. V. S., & Gunatilleke, I. A. U. N. (2009). Ecological information from spatial patterns of plants: insights from point process theory. *Journal of Ecology, 97*, 616–628.

Lawton, R. O., & Cothran, P. (2000). Factors influencing reproductive activity of *Juniperus virginiana* in the Tennessee Valley. *Journal of Torrey Botanical Society, 127*, 271–279.

Lee, S. G. (2005). Genetic conservation strategy in the natural population of *Torreya nucifera* in Kujwa-eup. Cheju: Korea. MS Thesis. Sangji University, Wonju, Korea (in Korean with English abstract).

Lee, S. G. (2009). Studies on the biota, growth characteristics, and vegetational changes in relation to tending care intensity and conservation measures of the *Torreya nucifera* Forest in Gujwa-eup. Jeju: Korea. Ph.D. Dissertation. Sangji University, Wonju, Korea (in Korean with English abstract).

Lloyd, D. G., & Webb, C. J. (1977). Secondary sex characters in plants. *The Botanical Review, 43*, 177–216.

Montesinos, D., Luís, M. D., Verdú, M., Raventós, J., & García-Fayos, P. (2006). When, how and how much: gender-specific resource-use strategies in the dioecious tree *Juniperus thurifera. Annals of Botany, 98*, 885–889.

Nanami, S., Kawaguchi, H., & Yamakura, T. (1999). Dioey-induced spatial patterns of two codominant tree species, *Podocarpus nagi* and *Neolitsea aciculate. Journal of Ecology, 87*, 678–687.

Nanami, S., Kawaguchi, H., & Yamakura, T. (2005). Sex ratio and gender-dependent neighboring effects in *Podocarpus nagi*, a dioecious tree. *Plant Ecology, 177*, 209–222.

Nanami, S., Kawaguchi, H., & Yamakura, T. (2011). Spatial pattern formation and relative importance of intra- and interspecific competition in codominant tree species, *Podocarpus nagi* and *Neolitsea aciculata. Eological Research, 26*, 37–46.

National Geographic Information Institute of Korea. 2016. http://map.ngii.go.kr/ms/map/NlipMap.do. Accessed on Apr 2016.

Nuñez, C. I., Nuñez, M. A., & Kitzberger, T. (2008). Sex-related spatial segregation and growth in a dioecious conifer along environmental gradients in northwestern Patagonia. *Ecoscience, 15*, 73–80.

Obeso, J. R. (2002). The costs of reproduction in plants. *New Phytologist, 155*, 321–348.

Opler, P. A., & Bawa, K. S. (1978). Sex ratios in some tropical forest trees. *Evolution, 32*, 812–521.

Ortiz, P. L., Arista, M., & Talavera, S. (2002). Sex ratio and reproductive effort in the dioecious *Juniperus communis* subsp. *alpina* (Suter) Čelak. (Cupressaceae) along an altitudinal gradient. *Annals of Botany, 89*, 205–211.

Osunkoya, O. O. (1999). Population structure and breeding biology in relation to conservation in the dioecious *Gardenia actinocarpa* (Rubiaceae) - a rare shrub of North Queensland rainforest. *Biological Conservation, 88*, 347–359.

Pan, C., Zhang, C., Zhao, X., Xia, F., Zhou, H., & Wang, Y. (2010). Sex ratio and spatial patterns of males and females of different ages in the dioecious understory tree, *Acer barbinerve*, in a broad-leaved Korean pine forest. *Biodiversity Science, 18*, 292–299 (in Chinese with English abstract).

Perry, G. L. W., Enright, N. J., Miller, B. P., & Lamont, B. B. (2009). Nearest-neighbour interactions in species-rich shrublands: the roles of abundance, spatial patterns and resources. *Oikos, 118*, 161–174.

Peterson, C. J., & Squiers, E. R. (1995). Competition and succession in an aspen-white-pine forest. *Journal of Ecology, 83*, 449–457.

Rayburn, A. P., & Monaco, T. A. (2011). Linking plant spatial patterns and ecological processes in grazed Great Basin plant communities. *Rangeland Ecology & Management, 64*, 276–282.

Rayburn, A. P., Schiffers, K., & Schupp, E. W. (2011). Use of precise spatial data for describing spatial patterns and plant interactions in a diverse Great Basin shrub community. *Plant Ecology, 212*, 585–594.

Scarano, F. R. (2002). Structure, function and floristic relationships of plant communities in stressful habitats marginal to the Brazilian Atlantic rainforest. *Annals of Botany, 90*, 517–524.

Schenk, H. J., Holzapfel, C., Hamilton, J. G., & Mahall, B. E (2003). Spatial ecology of a smalldesert shrub on adjacent geological substrates. *Journal of Ecology, 91*, 383–395.

Schmidt, J. P. (2008). Sex ratio and spatial pattern of males and females in the dioecious sandhill shrub, *Ceratiola ericoides ericoides* (Empetraceae) Michx. *Plant Ecology, 196*, 281–288.

Shin, H., Lee, K., Park, N., & Jung, S. Y. (2010). Vegetation structure of the *Torreya nucifera* stand in Korea. *Journal of Korean Forest Society, 99*, 312–322 (in Korean with English abstract).

Silvertown, J., & Dodd, M. (1999). The demographic cost of reproduction and its consequences in balsam fir (*Abies balsamea*). *The American Naturalist, 154*, 321–332.

Stoll, P., & Bergiou, E. (2005). Pattern and process: competition causes regularspacing of individuals within plant populations. *Journal of Ecology, 93*, 395–403.

Stoll, P., & Prati, D. (2001). Intraspecific aggregation alters competitive interactions in experimental plant communities. *Ecology, 82*, 319–327.

Thomson, J. D., & Barrett, S. C. H. (1981). Selection for outcrossing, sexual selection, and the evolution of dioecy in plants. *The American Naturalist, 118*, 443–449.

Ueno, N., Suyama, Y., & Seiwa, K. (2007). What makes the sex ratio female-biased in the dioecious tree *Salix sachalinensis? Journal of Ecology, 95*, 951–959.

Wang, X., Ye, J., Li, B., Zhang, J., Lin, F., & Hao, Z. (2010). Spatial distributions of species in an old-growth temperate forest, northestern China. *Canada Journal of Forest Research, 40*, 1011–1019.

Wiegand, T., & Moloney, K. A. (2004). Rings, circles, and null-models for point pattern analysis in ecology. *Oikos, 104*, 209–229.

Wolf, A. (2005). Fifty year record of change in tree spatial patterns within a mixed deciduous forest. *Forest Ecology and Management, 215*, 212–223.

Zhang, C., Zhao, X., Gao, L., & von Gadow, K. (2010). Gender-related distributions of *Fraxinus mandshurica* in secondary and old-growth forests. *Acta Oecologica, 36*, 55–62.

Zuo, X., Zhao, H., Zhao, X., Zhang, T., Guo, Y., Wang, S., & Drake, S. (2008). Spatial pattern and heterogeneity of soil properties in sand dunes under grazing and restoration in Horqin Sandy Land, Northern China. *Soil & Tillage Research, 99*, 202–212.

19

Study on the diagnosis of disturbed forest ecosystem in the Republic of Korea: in case of Daegwallyeong and Chupungryeong

Seon-Mi Lee*⬤, Jae-Gyu Cha and Ho-Gyung Moon

Abstract

Background: Baekdudaegan was designated in 2005 as a protected area to prevent destruction and conserve. However, there are many disturbed and destroyed areas. The total disturbed area amounts to 25.9 km^2 (0.94%), including 13.4 km^2 (0.49%) in the core area and 12.5 km^2 (0.45%) in the buffer area. This study aims to classify the vegetation types established in the disturbed areas and diagnose the current conditions for ecological restoration in the forest ecosystem.

Methods: We surveyed the vegetation in the disturbed areas of Daegwallyeong and Chupungryeong and the surrounding natural areas. The survey conducted from July to September 2015 targeted a total of 54 quadrats by Braun-Blanquet method (Daegwallyeong, 22; Chupungryeong, 32). We also investigated the height and coverage of each layer. We classified the vegetation types based on the field data and analyzed the ratio of life form and the exotic plants, species richness, and vegetation index (*Hcl*). The Normalized Difference Vegetation Index (NDVI) was calculated from rapideye satellite imagery in 2014 and 2015.

Results: Vegetation types were classified into 11 groups according to the criteria that included successional sere or plantation at first, followed by developmental stage and origins. As a result of the analysis of the survey data, species richness, vegetation index *(Hcl)*, ratio of tree plants, and the NDVI tended to increase, while the ratio of the exotic plants tended to decrease with the time since disturbance. These indicators had the classified values according to the vegetation types with time since the disturbance.

Conclusions: These indicators can be effectively used to diagnose the conditions of the present vegetation in the disturbed area of the Baekdudaegan area. In addition, the NDVI might be effective for the diagnosis of the disturbed status instead of the human efforts based on the higher spatial resolution of satellite imagery. Appropriate diagnosis of the disturbed forests in the Baekdudaegan area considering the established vegetation types is essential for the elaboration of restoration plans. In addition, restoration target and level should be different according to the disturbed status of restoration site.

Keywords: Baekdudaegan, Ecological restoration, Forest recovery, Vegetation succession, Successional stage, Diagnostic indicator, NDVI

* Correspondence: planteco@nie.re.kr
Division of Ecological Conservation, National Institute of Ecology, 1210
Geumgang-ro, Maseo-myeon, Seocheon-gun, Chungcheongnam-do 33657,
Republic of Korea

Background

According to the Society for Ecological Restoration International Science & Policy Working Group (2004), ecological restoration is practiced to assist the recovery of a degraded, damaged, or destroyed ecosystem, which is the most commonly used definition worldwide. For the successful ecological restoration, survey and analysis of a restoration site are essential. After the obvious diagnosis of the degraded site, it is possible to set up a clear target, goals, and objectives (Rieger et al. 2014).

Baekdudaegan refers to the mountain ranges linked from Mt. Baekdu in the Democratic People's Republic of Korea to Mt. Jiri in the Republic of Korea. This notion was firstly used by Doseon, a monk in the Goryeo dynasty. Afterwards, it was systematized and concretized through several documents, such as *Taekriji* by Jung Hwan Lee, *Seonghosaseol* by Ik Lee, and *Sangyeongpyo* by Gyeong Jun Shin in the Choseon dynasty (Lee and Kwon 2002).

The Baekdudaegan protected area was designated and notified by the Korea Forest Service in 2005 as one of the core ecological axes in the Republic of Korea. The purposes of designation are to prevent undiscerning destruction of the environment and conserve the natural environment (National Law Information Center 2015). It includes five provinces, six cities, and five counties. The total area increased from 263,427 ha in 2005 to 275,646 ha in 2016 (Korea Forest Service 2016). Baekdudaegan is one of the three main axes of Korean Peninsula along with the Demilitarized Zone (DMZ) and the coastal belt (AKECU 2010). It plays the important roles as biodiversity hotspots and ecological corridors of wildlife species (Korea Forest Service 2008).

In spite of the status of the protected area, there are many disturbed areas. The total damaged area amounts to 25.9 km^2 (0.94%), including 13.4 km^2 (0.49%) in the core area and 12.5 km^2 (0.45%) in the buffer area (National Institute of Ecology 2015). Several studies addressed the damaged status (Cho 2012; Choi et al. 2014; Kwon and Lee 2003; Kwon et al. 2004; Lee et al. 2007a; Oh and Lee 2003) and restoration (Ahn et al. 2009; Daegu Gyeongbuk Development Institute 2008; Kim et al. 2008). These studies were mostly subjected to survey on the trial deterioration and studies on the diagnosis of the vegetation after disturbance or degradation are rare.

Several studies on the diagnosis using several indicators have been conducted in South Korea. For instance, Myung et al. (2002) diagnosed the stream vegetation through the diversity of plant community, exotic plant species, actual vegetation map, and vegetation profiles. Similarly, Kim (2009) developed the index consisting of vegetation diversity (the number of plant communities), ratio of exotic and annual plant species, vegetation profiles, and species richness (the number of plant species).

Furthermore, Nam (2015) studied species diversity, ratio of exotic plant and annual plant species, and vegetation. These studies were usually performed in the streams or rivers, but rarely addressed a forest ecosystem. Meanwhile, the first attempt to suggest a restoration practice according to the restoration level based on the damage degree from air pollution in South Korea was made by Kim et al. (2015) who selected vegetation types and damage degree as indicators.

This study aims to classify the vegetation types established in the disturbed areas and diagnose the current conditions for making use of ecological restoration after disturbances in the forest ecosystem.

Methods

Study area

From the time-series analysis by using the landuse maps in 2003 and 2013, we selected two study areas that were the most disturbed and destroyed (National Institute of Ecology 2015) (Fig. 1). One is Daegwallyeong in Pyeongchang county, Gangwon province (latitude N 37° 42′~N 37° 46′ and longitude E 128° 40′~E 128° 44′), and the other is Chupungryeong in Gimcheon city, Gyeongsangbuk province (latitude N 36° 11′~N 36° 12′ and longitude E 127° 59′~E 128° 03′). The mean annual precipitation and temperature during 30 years (1981~2010) were 1898.0 mm and 6.6 °C at Daegwallyeong Weather Station and 1187.1 mm and 11.7 °C at Chupungryeong Weather Station, respectively (Korea Meteorological Administration 2015). Daegwallyeong and Chupungryeong belong to the middle province as a floristic region and the cool temperate broad-leaved forest as a plant formation (Lee and Yim 1978).

Disturbance characteristics

Restoration attempts to return an ecosystem to the original state (Society for Ecological Restoration International Science & Policy Working Group 2004) and the power of an ecosystem to do so is called resilience (Connell and Slatyer 1977; Greipsson 2011). Vegetation starts to recover naturally when the anthropogenic disturbances ceased or are neglected (Egler 1952). The main disturbance types were pastures in Daegwallyeong and fields in Chupungryeong (National Institute of Ecology 2015). In Daegwallyeong, Samyang pasture was developed from 1972 around 850~1400 m above the sea level (Noh et al. 2013). In addition, the wind power generators have been recently set up after the intact forests in this area were destroyed. Meanwhile, the main disturbance type was the fields in Chupungryeong (National Institute of Ecology 2015). Mostly, the fields that were far from the cities or difficult to access showed the tendency to be abandoned. Reduced agricultural activities resulted in forest recovery.

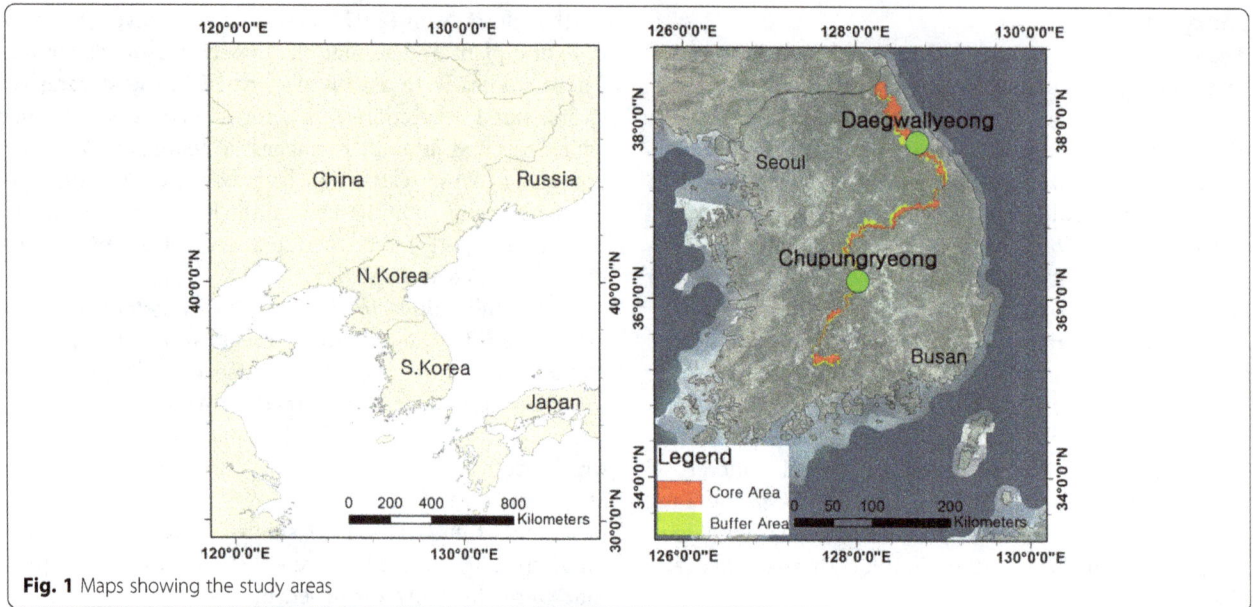

Fig. 1 Maps showing the study areas

In this study, we surveyed the vegetation in the following five disturbance types that are abandoned or neglected: quarry, fields, paddy fields, and pastures, and plantations. We did not include the sites which are now in use after disturbances. In addition, the reason why we included the plantations is the positive effects of their ecological restoration reported in previous studies (Kim et al. 2013a; Lee et al. 2004a; Shin 2005).

Vegetation survey

We investigated the vegetation survey using the Braun-Blanquet method (Braun-Blanquet 1964) in the disturbed area and surrounding natural area as a reference. The vegetation survey conducted from July to September 2015 targeted a total of 54 quadrats, including 22 quadrats in Daegwallyeong and 32 quadrats in Chupungryeong. The quadrant size was 2 m × 2 m dominated by herbs, 5 m × 5 m dominated by shrubs, and 10 m × 10 m or 15 m × 15 m dominated by trees. We also investigated the height and coverage of herb, shrub, subtree, and tree layers.

All plant species that appeared in each plot were identified by Lee (2014) and exotic plant species were identified by Park (2009) and Lee et al. (2011). Unidentified plants in the fields were collected and identified in the laboratory.

Data analysis

We divided the vegetation structure into the herb, shrub, subtree, and tree layers, which means the state of vegetation development. Then, the ratio of herbs, shrubs, and subtrees or trees was analyzed among the vegetation types. In addition, based on the surveyed data of the height and coverage of each layer, we calculated the

vegetation index (Hcl) (formula 1) (Lee et al. 2004b). Additionally, we divided it by 100.

$$Hc/ = \sum (H/ * Vcl)(Hl : \text{height of layer(m)}) \quad (1)$$

The differences of vegetation index among vegetation types were tested with one-way ANOVA (SPSS, version 20.0).

Satellite analysis

To analyze the vegetation recovery according to vegetation types, Normalized Difference Vegetation Index (NDVI) was calculated from rapideye satellite imagery of Daegwallyeong in 2015 and Chupungryeong in 2014. NIR is the wavelength value of the near-infrared ray and RED is the wavelength value of the red visible ray.

$$\text{NDVI} = (\text{NIR} - \text{RED})/(\text{NIR} + \text{RED})$$

The value of NDVI ranged from −1 to 1, theoretically. The areas not covered with vegetation have the value close to 0. Meanwhile, the areas that are well developed with vegetation have the value close to 1 (Norman et al. 2014). Relative values to the highest value of the NDVI were compared.

Results

We categorized the vegetation types into 11 groups based on the field data. We made a division into three levels. Level 1 was mainly divided by successional sere including xerarch succession and hydrarch succession, plantation, and natural area as a reference. Level 2 was classified by developmental stages including herb, shrub, subtree, or tree according to the status of vegetation development. Level 3 was divided by origins of species such as exotic or indigenous characteristics (Table 1). Each type was expressed by remarks of Table 1 hereafter.

Rank-abundance curve according to the vegetation types is shown in Fig. 2. The number of species of the disturbed areas tended to increase with the lapse of time since abandonment or restoration practice such as plantation. The number of species in natural areas ranged from 44 to 84. In addition, the number of species in P2 and H3 was almost similar to the corresponding number in the natural areas. The number of species in X types (X1, *X2*, and X3) was lower than that of H types (H1, H2, and H3).

Vegetation index *(Hcl)*, which was calculated by the mean coverage and height of tree, subtree, shrub, and herb layers, was high in the natural areas, and the values of P2 and H3 were similar to the corresponding values in the natural areas. However, the mean of P2 and H3 was still far from that of the natural area (see Fig. 4). A division of stratification through the value of mean and the distribution range of *Hcl* were clearly observed.

In the result of analysis on the growth form of plant species, percentages of tree species increased and those of herb species decreased with time since abandonment or restoration practice such as plantation. The percentage of tree species in X3 and P2 tended to be similar to that of N types (N1, N2, and N3) than to other types. However, H types (H1, H2, and H3) did not obviously show that tendency (Fig. 3). By contrast, the ratio of exotic plants was high in X1 and P1 and decreased with time since abandonment or plantation. On the other hand, ratio of exotic plant species emerged in H3 (Fig. 5).

NDVI of each type, extracted from rapideye satellite imagery, is shown in Fig. 6. As the result of comparison,

it also increased with time since abandonment or plantation. The value of P2 and H3 types tended to be similar to N types (N1, N2, and N3). However, standard deviation showed a high value. As a result of one-way ANOVA test, there are significant differences at the 0.001 level among vegetation types (see Table 2).

Discussion

An ecological restoration project involves several developmental processes, such as planning, design, implementation, and aftercare. Each step includes major actions, such as an initiate project, site analysis, SWOT-C, refined goals and objectives, project scope, concept design, project plans, installation, monitoring, remediation, project close-out, and projects added to the ongoing program (Rieger et al. 2014). It is a cost-effective way based on the site analysis and diagnosis evaluation to improve restoration effects (Lee et al. 2005; Lee et al. 2007b).

Vegetation can be easily surveyed to provide valuable information about disturbances and environmental processes (Amoros et al. 2000). In the present study, we classified the vegetation types for making use of restoration practice in the forest ecosystem. We analyzed five indicators related to vegetation development to diagnose the disturbed forest ecosystem in Daegwallyeong and Chupungryeong, namely, species richness; vegetation index *(Hcl)*; the ratio of the exotic species and that of the herb, shrub, and tree species; and NDVI. These indicators were used in numerous studies related to forest recovery after disturbance. Specifically, species richness (namely, the number of species) was used by Kim et al. (2013a, b), Lee et al. (2002), Lee (2006), and Lindborg and Eriksson (2004); vegetation index was employed by Lee et al. (2004b) and Lee (2006); the ratio of exotic species was used by Kim et al. (2013b); the ratio of annual herbs or tree species was employed by Burt and Rice (2009), Grime (2001), and Lee (2006). Finally, Cai et al. (2015), Leon et al. (2012), and van Leeuwen (2008) used NDVI. Sizeable differences in the most indicators were observed between X1 and N types (N1, N2, and N3). In addition, our results confirmed that most indicators of H3 and P2 have similar conditions to those of N types as references (see Figs. 2, 3, 4, 5 and 6). However, the ratio of the exotic plants in H3 was exceptional. Generally, exotic species disappear when trees make a crown and provide a shade (Korea National Park Research Institute 2014). Although *Salix koreensis*, which is the dominant plant in H3, developed to the tree layer and made a shade, the ratio of exotic species emerged, while H1 and H2 did not (see Fig. 5). Exotic species that invaded in H3 were just two: *Erigeron annuas* and *Bidens frondosa*. It was assumed that *E. annuas* was invaded temporarily from the surrounding areas according to the decrease of

Table 1 Categorization of the vegetation types into 11 groups based on the field data

Level 1	Level 2	Level 3	Remark	No. of plots
Xerarch succession	Herb	Exotic	X1	14
		Indigenous	X2	6
	Shrub	Indigenous	X3	4
Hydrarch succession	Herb	Indigenous	H1	3
	Shrub	Indigenous	H2	3
	Subtree/Tree	Indigenous	H3	3
Plantation	Shrub	Exotic/indigenous	P1	3
	Subtree/Tree	Exotic/indigenous	P2	4
Natural area	Tree	Indigenous (Quercusmongolica)	N1	9
		Indigenous (Q. serrata)	N2	3
		Indigenous (Q. acutissima)	N3	3

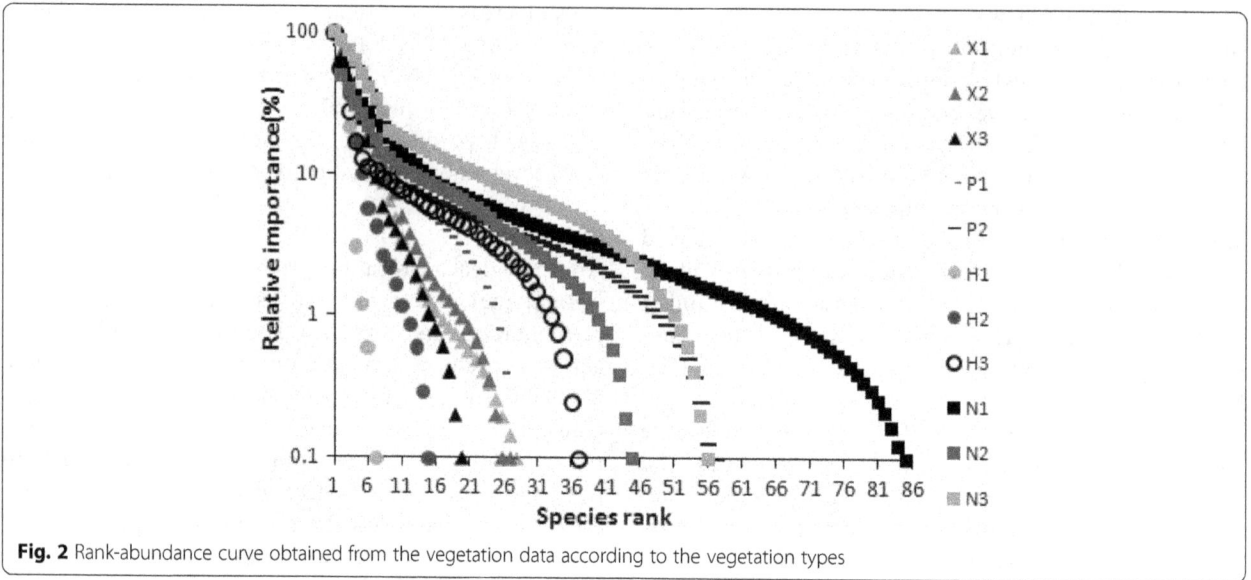

Fig. 2 Rank-abundance curve obtained from the vegetation data according to the vegetation types

soil water contents since the time after abandonment. In addition, we assumed that *B. frondosa* was a remnant from the early stage of abandoned paddy fields. Many previous studies reported that *B. frondosa* showed up or was dominant in paddy fields in South Korea (Lee 1997; Kim et al. 2011). However, *B. frondosa* has heliophilous seeds (Shin et al. 1999). They might disappear when the vegetation development go on and make a shade more densely. Meanwhile, the ratio of exotic species in N1, N2, and N3 types showed up slightly. Exotic plants that were entered into the intact natural forests were mainly *Robinia pseudoacacia* and *Larix kaempferi* which are the major afforestation tree species in South Korea. We found only very few of them in the intact natural forests. It is supposed that these exotic plants invaded from the surrounding plantations.

Meanwhile, the vegetation index (*Hcl*) was developed to estimate the vegetation developmental processes in the area severely disturbed by fire. It well explained the development of vegetation stratification. It was proved that the vegetation index showed a high correlation with the time since fire disturbance (Lee et al. 2004b). Processes and speed of the vegetation development after disturbance might vary according to the disturbance types and location conditions. However, most disturbed areas are under the vegetation development immediately after the disturbances ceased (Egler 1952). Therefore, the vegetation index is an effective indicator for confirming the vegetation developmental processes, as it is calculated by the height and coverage of each layer. We confirmed that the vegetation index showed the low values in the early developmental stage with herb layers

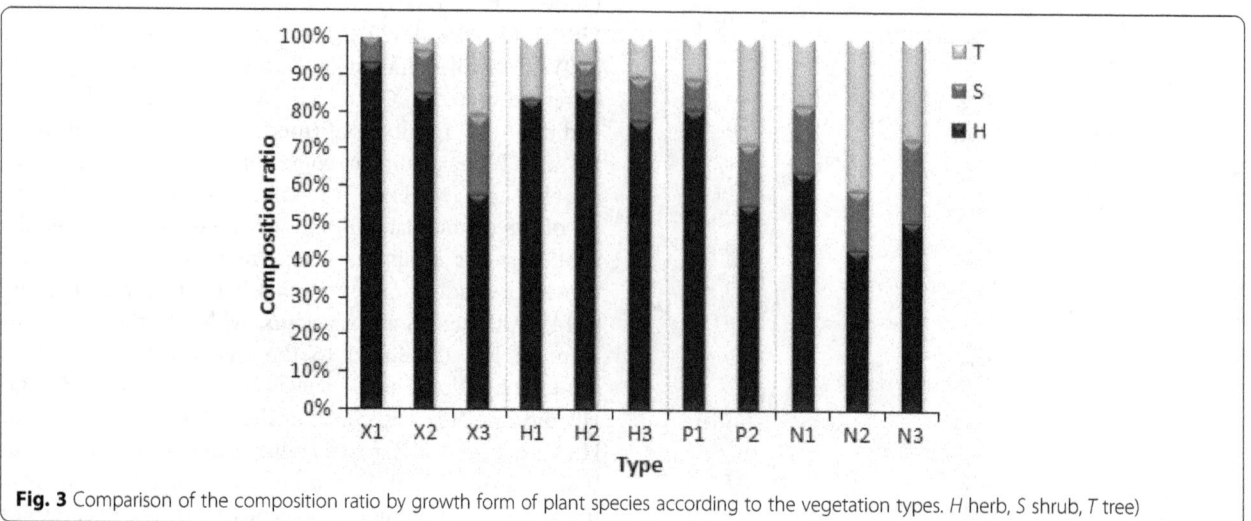

Fig. 3 Comparison of the composition ratio by growth form of plant species according to the vegetation types. *H* herb, *S* shrub, *T* tree)

Table 2 Results of one-way ANOVA for vegetation index calculated from the vegetation coverage and height

Vegetation types	N	Mean	Std	F	p
X1	14	1.1	0.6	37.102	0.000*
X2	6	1.3	0.3		
X3	4	0.8	0.1		
H1	3	1.0	0.1		
H2	3	3.3	1.5		
H3	3	11.7	1.0		
P1	3	2.6	0.2		
P2	4	11.7	7.3		
N1	9	14.3	3.1		
N2	3	14.2	1.0		
N3	3	18.0	1.8		

*$p < 0.001$

(X1, X2, and H1) ranging from 0.5 to 2.1. That of the later developmental stage with subtree or tree layers (H3 and P2) showed higher values ranging from 5.6 to 22.3. These results are similar trend with those reported by Lee et al. (2004b). Therefore, studies on the quantification of the vegetation index and analyzing its correlation with the time since disturbance are needed in the future.

The NDVI has been shown to be the most effective index to quickly and simply identify the vegetated area and its condition (Agone and Bhamare 2012). It has been widely used to monitor vegetation (Benedetti and Rossini 1993; Kinyanjui 2011; Meneses-Tovar 2011; Wessels 2005) and to find the historical information of vegetation cover changes (Peters et al. 2002). Very low values of NDVI (0.1 and below) were observed in the barren areas, such as rocks, sand, or snow, while moderate values (0.2 to 0.3) were observed in shrub and grassland areas. In addition, high values (0.6 to 0.8) correspond to temperate and tropical rain forests (Weier and Herring 2000). Therefore, there is a correlation between vegetation types and various sites (Meneses-Tovar 2011). In fact, field survey is the most accurate way to identify the disturbed status. However, it is difficult to cover broad areas, because this approach is time-consuming, expensive, and requires more people who can conduct the field survey (Alshaikh 2015). In an attempt to make use of the NDVI for the diagnosis, in the present study, we tried to diagnose the disturbed status by comparing the values of NDVI calculated from rapideye satellite imagery. Our results confirmed the effect of NDVI. Specifically, H3, P2, N1, N2, and N3 which are types dominated by the tree or subtree layer had higher values than other types. However, the difference between herb dominated types (X1, X2, and H1) and shrub dominated types (X3, H2, and P1) was not significant (see Fig. 6). It can be inferred that the damaged area sampled from the field was narrow and the spatial resolution of rapideye satellite imagery is low to detect spatial variability (Wessels 2005). Therefore, in further research, a higher spatial resolution of satellite imagery is necessary and the number of samples should be sufficient. In addition, the NDVI from the low resolution of satellite imagery is suitable for a broadly degraded area.

The indicators suggested in the present study are broadly applicable for other disturbed areas, including many disturbance types in a forest ecosystem. However,

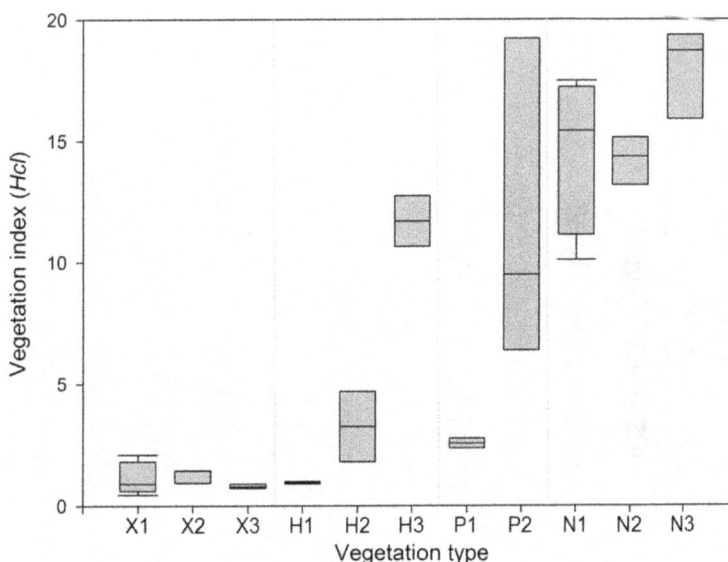

Fig. 4 Comparisons of vegetation index (Hcl) according to the vegetation types

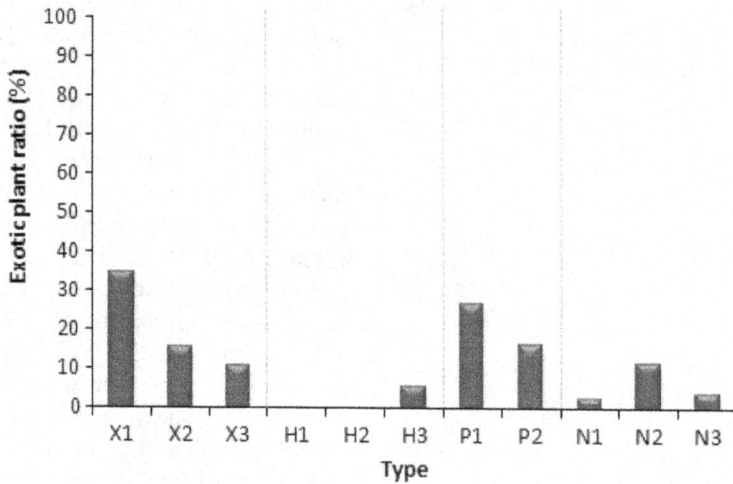

Fig. 5 The ratio of exotic plants to the total number of species of the each vegetation type

specific indicators for the each disturbance type should be chosen. In addition, a more precise survey should be practiced for a restoration project of a specific area. For a successful restoration, a restoration plan based on the accurate survey and diagnosis of the degraded area in the Baekdudaegan has to be proposed. In addition, restoration target and level should be different according to the disturbed status of restoration site.

Conclusions

An ecological restoration project involves several developmental processes, such as planning, design, implementation, and aftercare. Diagnosis evaluation of degraded forest considering the established vegetation types is essential for the elaboration of restoration plans. In the present study, we classified the vegetation

types according to the developmental stages, species composition, and characteristics of dominant species. Species richness, vegetation index *(Hcl)*, the ratio of tree plants, and NDVI tended to increase, but the ratio of exotic plants tended to decrease with the time since abandonment. In addition, we confirmed a correlation between vegetation types and the values of the NDVI. Therefore, we suggested the five indicators for diagnosis. These indicators are applicable for broadly degraded areas, including many disturbance types in a forest ecosystem. However, specific indicators for each disturbance type should be chosen. In addition, a more precise survey should be practiced for a restoration project of a specific area. A restoration plan based on the correct diagnosis of the degraded area in the Baekdudaegan area has to be proposed for successful restoration.

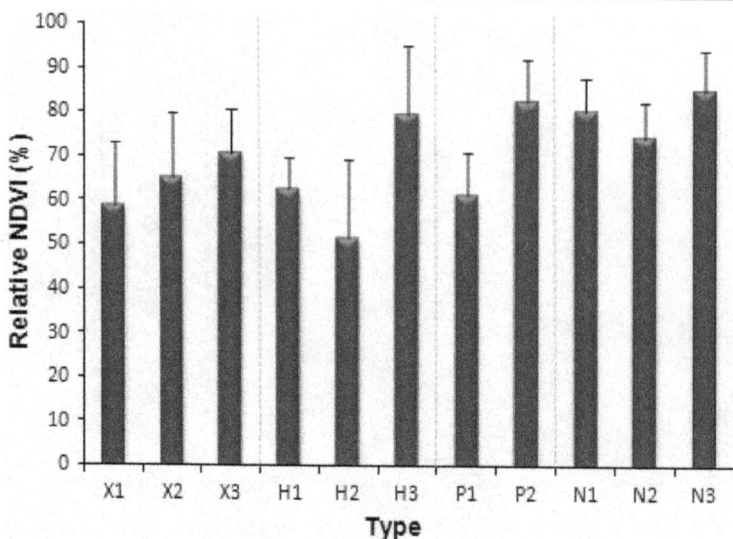

Fig. 6 Relative NDVIs according to the vegetation types from the rapideye satellite imagery

Acknowledgements
We sincerely appreciated Hyeon Ho and Myeong from the Korea National Park Research Institute for reviewing willingly this manuscript.

Funding
This study was supported by the National Institute of Ecology titled "Conservation and Restoration Based Research on the Core Ecological Axis of the Korean Peninsula" in 2015.

Authors' contributions
SML carried out the field survey and data analysis and drafted the manuscript. JGC analyzed the NDVI from the rapideye satellite imagery. HGM supported the field survey and entered the data into a computer. All authors read and approved the final manuscript.

Competing interests
The authors declare that they have no competing interests.

References
Agone, V., & Bhamare, S. M. (2012). Change detection of vegetation cover using remote sensing and GIS. *Journal of Research and Development, 2*, 1–12.

Ahn, T. M., Kim, I. H., Lee, J. Y., Kim, C. K., Chae, H. S., Lee, Y., Min, S. Y., & Kim, M. W. (2009). Development of participatory ecological restoration system through integrative categorization of disturbed areas in BaigDooDaeGahn. *The Korea Society for Environmental Restoration and Revegetation Technology, 12*, 11–22 (in Korean with English abstract).

AKECU (ASEAN-Korea Environmental Cooperation Unit). (2010). *Biodiversity and natural resources conservation in protected areas of Korea and the Philippines* (p. 31). Seoul: GeoBook.

Alshaikh, A. Y. (2015). Space applications for drought assessment in Wadi-Dama (West Tabouk), KSA. *Egyptial Journal of Remote Sensing & Space Sciences, 18*, 43–53.

Amoros, C., Bornette, G., & Henry, C. P. (2000). A vegetation-based method for ecological diagnosis of riverine wetlands. *Environmental Management, 25*, 211–227.

Benedetti, R., & Rossini, P. (1993). On the use of NDVI profiles as a tool for agricultural statistics: the case study of wheat yield estimate and forecast in Emilia Romagna. *Remote Sensing of Environment, 45*, 311–326.

Braun-Blanquet, J. (1964). *Pflanzensoziologie* (p. 865). Wien: Grundzuge der Vegetationstunde.

Burt, J. W., & Rice, K. J. (2009). Not all ski slopes are created equal: disturbance intensity affects ecosystem properties. *Ecological Applications, 19*, 2242–2253.

Cai, H., Yang, X., & Xu, X. (2015). Human-induced grassland degradation/restoration in the central Tibetan Plateau: the effects of ecological protection and restoration projects. *Ecological Engineering, 83*, 112–119.

Cho, W. (2012). Deterioration status of closed-trail of national parks on the Baekdudaegan mountains, South Korea. *Korean Journal of Environment and Ecology, 26*, 827–834 (in Korean with English abstract).

Choi, S. M., Kweon, H. K., Lee, J. W., Choi, Y. H., & Choi, T. J. (2014). A study on deterioration of ridge trail in Jeongmaek. *Korean Journal of Environment and Ecology, 28*, 450–456 (in Korean with English abstract).

Connell, J. M., & Slatyer, R. O. (1977). Mechanisms of succession in natural communities and their role in community stability and organization. *American Naturalist, 111*, 1119–1144.

Daegu Gyeongbuk Development Institute. (2008). *Research on the ecological restoration plan of Ihwaryeongin the Baekdudaegan* (p. 37) (in Korean).

Egler, F. E. (1952). Vegetation science concepts I. Initial floristic composition, a factor in old-field vegetation development with 2 figs. *Vegetatio, 4*, 412–417.

Greipsson, S. (2011). *Restoration ecology* (pp. 104–105). Canada: Jones & Bartlett Learning, LLC.

Grime, J. P. (2001). *Plant strategies and vegetation processes* (2nd ed.). Chichester: Wiley.

Kim, H. S. (2009). *Diagnostic evaluation on the Korea rivers for restoration of vegetation* (Master's degree of Seoul Women's University, p. 61) (in Korean with English abstract).

Kim, N. C., Nam, U. J., & Shin, K. J. (2008). A study on the slope ecological restoration models for the Baekdu-mountain range. *The Korea Society for Environmental Restoration and Revegetation Technology, 11*, 72–84 (in Korean with English abstract).

Kim, S. K., Kim, S. Y., Won, J. G., Shin, J. H., & Kwon, O. D. (2011). Prediction of rice yield loss and economic threshold level by densities of *Sagittaria trifolia* and *Bidens frondosa* in direct-seeding flooded rice. *Korean Journal of Weed Science, 31*, 340–347 (in Korean with English abstract).

Kim, S. M., An, J. H., Lim, Y. K., Pee, J. H., Kim, G. S., Lee, H. Y., Cho, Y. C., Bae, K. H., & Lee, C. S. (2013a). Ecological changes of the *Larix kaempferi* plantations and the restoration effects confirmed from the results. *Korean Journal of Ecology and Environment, 46*, 241–250 (in Korean with English abstract).

Kim, S. M., An, J. H., Lim, Y. K., Pee, J. H., Kim, G. S., Lee, H. Y., Cho, Y. C., Bae, K. H., & Lee, C. S. (2013b). The effects of ecological restoration confirmed in the *Pinus koraiensis* plantation. *Journal of Agriculture & Life Science, 47*, 19–28 (in Korean with English abstract).

Kim, G. S., Pee, J. H., An, J. H., Lim, C. H., & Lee, C. S. (2015). Selection of air pollution tolerant plants through the 20-years-log transplanting experiment in the Yeocheon industrial area, southern Korea. *Animal Cells and Systems, 19*, 208–215.

Kinyanjui, M. J. (2011). NDVI-based vegetation monitoring in Mau forest complex, Kenya. *African Journal of Ecology, 49*, 165–174.

Korea Forest Service. (2008). *Forest policy white book* (p. 334) (in Korean).

Korea Forest Service. (2016). *Notification of area extension of Beakdudaegan protected area* (in Korean).

Korea Meteorological Administration. (2015). http://www.kma.go.kr (in Korean) (26 Dec 2015 accessed).

Korea National Park Research Institute. (2014). *Study on the management plan of specific exotic species-exotic plants* (p. 142) (in Korean).

Kwon, T. H., & Lee, J. W. (2003). Trail deterioration on the ridge of the Baekdudaegan—a case of the trail between Manbokdae and Bokseongijae. *Korean Journal of Environment and Ecology, 16*, 465–474 (in Korean with English abstract).

Kwon, T. H., Lee, J. W., & Kim, D. W. (2004). Trial deterioration and managerial strategy on the ridge of the Baekdudaegan: a case of the trial between Namdeogyusan and Sosagogae. *Korean Journal of Environment and Ecology, 18*, 175–183 (in Korean with English abstract).

Lee, H. S. (1997). *Studies on germination characteristics of Bidens tripartite and Bidens frondosa* (Master's thesis of Kyunghee University, p. 25) (in Korean with English summary).

Lee, K. S. (2006). Changes of species diversity and development of vegetation structure during abandoned field succession after shifting cultivation in Korea. *Journal of Ecology and Field Biology, 29*, 227–235 (in Korean with English abstract).

Lee, T. B. (2014). *Illustrated flora of Korea* (p. 1828). Seoul: Hyangmunsa (in Korean).

Lee, J. W., & Kwon, T. H. (2002). Present states and management proposal of the Baekdudaegan-area in Doraegijae~Pijae. *Korean Journal of Environmental Ecology, 15*, 420–430 (in Korean with English abstract).

Lee, W. T., & Yim, Y. J. (1978). Studies on the distribution of vascular plants in the Korea peninsula. *Korean Journal of Plant Taxonomy, 8*(Supplement), 1–33 (in Korean with English abstract).

Lee, C. S., You, Y. H., & Robinson, G. R. (2002). Secondary succession and natural habitat restoration in abandoned rice fields of central Korea. *Restoration Ecology, 10*, 306–314.

Lee, C. S., Cho, H. J., & Yi, H. B. (2004). Stand dynamics of introduced black locust (*Robinia pseudoacacia* L.) plantation under different disturbance regimes in Korea. *Forest Ecology and Management, 189*, 281–293.

Lee, K. S., Choung, Y. S., Kim, S. C., Shin, S. S., Ro, C. H., & Park, S. D. (2004). Development of vegetation structure after forest fire in the east coastal region, Korea. *Korean Journal of Ecology, 27*, 99–106 (in Korean with English abstract).

Lee, C. S., Cho, Y. C., Shin, H. C., Moon, J. S., Lee, B. C., Bae, Y. S., Byun, H. G., & Yi, H. (2005). Ecological response of streams in Korea under different management regimes. *Water Engineering Research, 6*, 131–147.

Lee, C. S., Cho, Y. C., Shin, H. C., Lee, S. M., & Cho, H. J. (2007). Effect of partial habitat restoration by a method suitable for riverine environments in Korea. *Journal of Ecology and Field Biology, 30*, 171–177.

Lee, D. K., Song, W. K., Jeon, S. W., Sung, H. C., & Son, D. Y. (2007). Deforestation patterns analysis of the Baekdudaegan Mountain Range. *The Korea Society for Environmental Restoration and Revegetation Technology, 10*, 41–53 (in Korean with English abstract).

Lee, Y. M., Park, S. H., Jung, S. Y., Oh, S. H., & Yang, J. C. (2011). Study on the current status of naturalized plants in South Korea. *Korean Journal of Plant Taxonomy, 41*, 87–101. in Korean with English abstract.

Leon, J. R. R., van Leeuwen, W. J. D., & Casady, G. M. (2012). Using MODIS-NDVI for the modeling of post-wildfire vegetation response as a function of environmental conditions and pre-fire restoration treatments. *Remote Sensing, 4*, 598–621.

Lindborg, R., & Eriksson, O. (2004). Effects of restoration on plant species richness and composition in Scandinavian semi-natural grasslands. *Restoration Ecology, 12*, 318–326.

Meneses-Tovar, C. L. (2011). NDVI as indicators of degradation. *Unasylva, 238*, 39–46.

Myung, H., Kwon, S. Z., & Kim, C. H. (2002). Diagnosis of vegetation for the ecological rehabilitation of streams—the case of the Namhanriver. *Journal of Korean Institute of Landscape Architecture, 30*, 98–106 (in Korean with English abstract).

Nam, G. B. (2015). *Diagnostic evaluation about Han river basin and restoration planning based on its evaluation* (Master's degree of Seoul Women's University, p. 83) (in Korean with English abstract).

National Institute of Ecology. (2015). *Conservation and restoration based research on the core ecological axis of the Korean peninsula* (in Korean with English abstract).

National Law Information Center. (2015). http://www.law.go.kr/lsSc.do?menuId= 0&p1=&subMenu=1&nwYn=1&section=&tabNo=&query=%EB%B0%B1% EB%91%90%EB%8C%80%EA%B0%84%20%EB%B3%B4%ED%98%B8%EC%97% 90%20%EA%B4%80%ED%95%9C%20%EB%B2%95%EB%A5%A0#undefined (in Korean) (26 Dec 2015 accessed).

Noh, T. H., Han, B. H., Kim, J. Y., Lee, M. Y., & Yoo, K. J. (2013). Actual vegetation and structure of plant community in Daegwallyeong ranch, Gangwon-do (Province). *Korean Journal of Environment and Ecology, 27*, 579–591 (in Korean with English abstract).

Norman, L., Villarreal, M., Pulliam, H. R., Minckley, R., Gass, L., Tolle, C., & Coe, M. (2014). Remote sensing analysis of riparian vegetation response to desert marsh restoration in the Mexican Highlands. *Ecological Engineering, 70*, 241–254.

Oh, K. K., & Lee, J. E. (2003). Literatures review for the flora, vegetation and environmental management in the Baekdudaegan—from Cheonwangbong in the Jirisan to Hyangjeokbong in the Deokyusan. *Korean Journal of Environment and Ecology, 16*, 475–486 (in Korean with English abstract).

Park, S. H. (2009). *Colored illustration of naturalized plants of Korea* (p. 602). Seoul: Ilchokak (in Korean).

Peters, A. J., Walter-Shea, E. A., Ji, L., Viña, A., Hayes, M., & Svoboda, M. D. (2002). Drought monitoring with NDVI-based standardized vegetation index. *Photogrammetric Engineering & Remote Sensing, 68*, 71–75.

Rieger, J., Stanley, J., & Traynor, R. (2014). *Project planning and management for ecological restoration* (pp. 7–11). Washiington: Island Press.

Shin, H. C. (2005). *The effects of ecological restoration confirmed in the Pitch pine (Pinus rigida) plantation* (Master's degree of the Seoul Women's University, Seoul, Korea, p. 42) (in Korean with English abstract).

Shin, H. J., Shin, J. S., Kim, J. H., Kim, H. Y., Lee, I. J., Shin, D. H., & Kim, K. U. (1999). Study on seed germination of *Bidens tripartite* L. and *Bidens frondosa* L. *Agricultural Research Bulletin of Kyungpook National University, 17*, 53–57 (in Korean with English abstract).

Society for Ecological Restoration International Science & Policy Working Group. (2004). *The SER international primer on ecological restoration*. Tucson: Society for Ecological Restoration International. www.ser.org.

Van Leeuwen, W. J. D. (2008). Monitoring the effects of forest restoration treatments on post-fire vegetation recovery with MODIS multitemporal data. *Sensors, 8*, 2017–2042.

Weier, W., & Herring, D. (2000). *Measuring vegetation (NDVI&EVI)*. http:// earthobservatory.nasa.gov/Features/MeasuringVegetation/.

Wessels, K. J. (2005). *Monitoring land degradation in southern Africa by assessing changes in primary productivity* (Ph.D. Dissertation of the University of Maryland, p. 146).

20

Establishment strategy of a rare wetland species *Sparganium erectum* in Korea

Seo Hyeon Kim[1], Jong Min Nam[1] and Jae Geun Kim[1,2]*

Abstract

Background: To reveal establishment strategy of *Sparganium erectum*, we tried to find realized niche of adults through field survey and effects of water level on the establishment process through mesocosm experiments.

Results: In the field survey, the height and coverage of community living in deeper water were greater than those of community living in shallow water. There was no statistically significant difference ($p > 0.05$) in the means of water and soil properties between the two communities. In mesocosm experiments, we found no correlation between water levels and germination rates, but *S. erectum* seedlings have characteristics of post germination seedling buoyancy when *S. erectum* seeds germinated in inundation conditions. Shoot height, total leaf length, and survival rates of sinking seedlings in shallow water levels at −5, 0, and 5 cm were higher than those in deeper water levels at 10 and 20 cm. Floating seedlings established in water levels of 3 and 6 cm only. The seedlings could live up to 6 weeks in floating state but died if they were unable to establish.

Conclusions: The water level around adult *S. erectum* communities in the field were different from the water level at which *S. erectum* seedlings can survive in the mesocosm experiments. The findings provided not only understanding of *S. erectum* habitat characteristics but also evidence to connect historical links between the early seedlings stage and adult habitat conditions. We suggested the logical establishment strategy of *S. erectum* based on the data.

Keywords: Adult niche, Habitat environment, Macrophyte, Regeneration niche, Seedling establishment

Background

Most macrophytes are capable of both sexual and asexual reproduction (Yang and Kim 2016). Sexual reproduction is important for the colonization and maintenance of macrophyte populations. Vegetative reproduction is also important because populations are usually maintained by this method rather than by seed production. The environmental ranges of living adults and establishing seedlings have been emphasized because the range determines species distribution and abundance in vegetation (Grubb 1977; Grime 2006; Jeong and Kim 2017). Grubb (1977) stated that the environmental gradient affecting seed dispersal, germination, and seedling survival is the regeneration niche, while the environmental gradient affecting adult survival is the habitat niche. These two niches are not mutually exclusive. Thus, habitat and regeneration niches must be analyzed simultaneously to understand the relationships between plants and their habitat environments (Collins and Good 1987, Yang and Kim 2017).

Various environmental factors including water depth, salinity, light, temperature, and nutrient concentration influence different plant stages (Clarke and Allaway 1993; Coops and van der Velde 1995; Kim et al. 2013). Above all, water depth is the most important factor affecting seedling establishment and growth for submerged and amphibious plants (Seebloom et al. 1998; Kwon et al. 2007). Water depth can create favorable or unfavorable conditions for the germination and establishment of various species (Eriksson 1989; Kim et al. 2013). Each species has diverse properties that enable them to establish, survive, and colonize according to the water depth (Grace 1987). Thus water depth is important for seedling establishment and growth to sustain the species populations.

Sparganium erectum L., a perennial macrophyte, is widely distributed in Europe but is designated as a vulnerable species in South Korea and an endangered

* Correspondence: jaegkim@snu.ac.kr
[1]Department of Biology Education, Seoul National University, Seoul 08826, South Korea
[2]Center for Education Research, Seoul National University, Seoul 08826, South Korea

species in Japan (Cook 1962; National Institute of Biological Resources 2012; National Museum of Nature and Science 2017). *S. erectum* is mainly distributed in banks of river and canals and forms a continuous belt (DeKlerk et al. 1997; Whitton et al. 1998; Takahashi et al. 2000). This plant prefers slow water flow and fine sediment in riverine areas. Also, *S. erectum* grows from on wet mud to in water to a depth of 100 cm (Cook 1962; Asaeda et al. 2010; Kaneko and Jinguji 2012). Persisting necrotic leaves and stems of *S. erectum* during winter can accumulate fine sediments that influence both the physical environment of habitats and the retention of seeds (Pollen-Bankhead et al. 2011; O'Hare et al. 2012). These could be exploited to create mesohabitats for other plants and animals and they contribute to physical and biological habitat diversification in rivers (Friedman et al. 1996; Abbe and Montgomery 2003). Thus, *S. erectum* is worthy as an ecosystem engineer in riverine areas (Gurnell et al. 2006; Gurnell 2007; Asaeda et al. 2010; Liffen et al. 2011; O'Hare et al. 2012).

Habitat niche of species living in shallow water such as *Persicaria thunburgii* and *Cicuta virosa* are similar for adult and seedling (Kim et al. 2013; Shin and Kim 2013; Shin et al. 2013). However, many species such as *Typha* and *Phramites* spp. that live in deep water adopt different strategies over their lifetime (Shipley et al. 1989; Kwon et al. 2006; Hong et al. 2012). Adults of these species can live in deep water, but seedlings can only endure shallow water. The water depth for *S. erectum* adults shares similar habitat environments as these emergent plants. Therefore, it was predicted to differ from the water depth for seedlings of the plant. *S. erectum* mainly reproduces using rhizomes regardless of water level, but most seeds can be only produced at stable water levels (Cook 1962). Seeds of *S. erectum* usually fall into the water and disperse by hydrochory (Pollux et al. 2009). *Sparganium* spp. produce two types of seeds: short floating seeds that sink within 4 weeks (approximately 71% of all seeds), and long floating seeds that float for at least 6 months (approximately 28% of all seeds) (Pollux et al. 2009). Thus, *S. erectum* germination and seedling growth require an underwater or

saturated germination environment (Cook 1962). *Sparganium* spp. are also considered to exhibit different growth forms depending mainly on water depth and is especially reported to change their growth form from submerged to emergent during the early seedling phase (Kankaala et al. 2000; Riis et al. 2000; Asaeda et al. 2010; Kim and Kim 2015). However, the effects of water depth on seed germination and seedling establishment as well as growth are unknown.

In this study, we surveyed the water level, accompanying species, and water and soil environments in habitats. We also evaluated the effect of water level on the early life stages through mesocosm experiments. To know the germination condition, seed germination experiments were done with different seed storage conditions, temperatures, photoperiods, and water levels. We also investigated how the water level influences growth responses (growth, mortality, and stand development) of two seedling types (planted versus floating) when water level was the only environmental factor controlled.

Methods
Field study
Vegetation survey

We searched *S. erectum* habitats through literature and field surveys in South Korea and selected the largest habitat of *S. erectum* in South Korea at the Gosancheon stream of the Mankyeong river in Wanju (N 35° 56′ 36.8″ E 127° 10′ 18.9″, altitude 42 m). We could not find any more habitats where *S. erectum* was dominant species.

We investigated three habitats of *S. erectum* for environment evaluations. *S. erectum* communities grouped into two based on coverage of *S. erectum*: communities A and B. The coverage of *S. erectum* was over 50% in community A and mingled with other species, mainly with *Leersia japonica*, *Typha orientalis*, and *Persicaria thunbergii*. Community A had the lowest water depth and there was no inflow of river water except during floods. Community A was not directly connected to the main stream of the Mankyeong river and had no direct exposure to waves (Fig. 1). Community B was monospecific

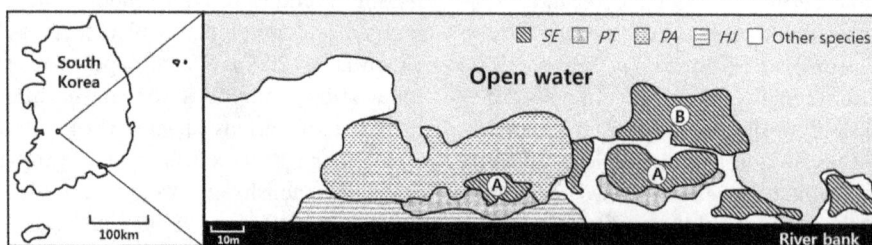

Fig. 1 Habitat of *S. erectum* at Gosancheon in Mankyeong River, South Korea. The figure on the *right* represents a magnification of the figure on the *left*. *A* and *B* show *S. erectum* communities. Community A lives in shallow water and community B lives in deep water. Community A was dominated by *S. erectum* and other species were *Leersia japonica*, *Typha orientalis*, and *Persicaria thunbergii*. Community B was a pure stand of *S. erectum*. SE: *Sparganium erectum* community, PT: *Persicaria thunbergii* community, PA: *Phragmites australis* community, HJ: *Humulus japonicas* community

stands of *S. erectum* and formed a belt and rarely had accompanying species. Community B extended to the main stream but the velocity of the moving water was so slow that it rarely affected the population.

Plant sociological analysis including coverage, density, and height was performed at 30 quadrats of 1 m × 1 m (10 quadrats in community A and 20 quadrats in B) based on a modification of the Braun-Blanquet method (Mueller-Dombois and Ellenberg 2003; Kim et al. 2004). We measured water depth at each quadrat using a 1-m stick ruler.

Analyses of water and soil properties

We collected eight and three water samples for communities A and B, respectively, and sampled twice in July and September before and after flooding. pH was measured with a pH meter (model AP 63; Fisher, USA) and electron conductivity (EC) was measured with a conductivity meter (Corning Checkmate model 311; Corning, USA) in the field. NO_3-N, NH_4-N, and PO_4-P were analyzed by the hydrazine method (Kamphake et al. 1967), indo-phenol method (Murphy and Riley 1962), and ascorbic acid reduction method (Solorzano 1969), respectively. K^+, Ca^{2+}, Na^+, and Mg^{2+} were measured using an atomic absorption spectrometer (Model AA240FS; Varian, USA).

We collected eight and three soil samples for communities A and B, respectively, when the plants reached maximum growth. We sampled submerged bed sediments until 3 cm below the surface of the sediments. Soil organic matter contents were analyzed by the loss on ignition method (Boyle 2004). NO_3-N and PO_4-P analyses were performed by the same method of water analysis (Kamphake et al. 1967, Solorzano 1969).

Statistical analysis

To compare habitat environmental characteristics, water and soil properties were analyzed and statistical analysis was performed with *t*-test at the 5% significance level, using SPSS ver. 20.0 software (SPSS, Inc., Chicago, IL, USA).

Mesocosm experiments
Germination experiment

Seeds were collected on August 2 in 2012. After collection, the seeds were cleaned with distilled water. Two third of them were stored in wet condition (a sealed plastic bag filled with wet cotton) and one third were dried in room temperature for 1 week and stored in dry condition. In each condition, seeds were divided into three groups and stored in three temperatures (−20 °C, 4 °C, 25 °C). Germination rate was calculated after 30-day incubation and germination criterion was the exposure of longer cotyledon than 0.1 cm.

To know the effect of storage period on germination, germination tests were done in growth chamber with 35 °C/light for 12 h and 25 °C/dark for 12 h by sowing three replicates of 20 seeds on Petri dishes containing three pieces of filter paper (Baskin and Baskin 1998). Average light intensity in the chamber was 127($\pm$16.5) μmol/m^2.

To know the effect of temperature and photoperiod on germination, germination tests were conducted under alternating temperature regime at 12-h intervals (35/25, 30/20, 25/15, 20/10, 15/5 °C), at different photoperiods (12 h light/12 h dark and 24 h dark). Each condition has 10 seeds stored in 4 °C/wet condition and repeated three times.

To know the effect of the water level, we selected the seeds that sank, thus germination occurred in inundation (water level, 20 cm) or waterlogged conditions (water level, 0 cm). This experiment was conducted in a greenhouse in June 2013, using a stainless steel tank (150 cm × 80 cm × 50 cm). We compared germination rates and germination characteristics between the two conditions. For each water condition, 10 pots were prepared. Each pot (Φ 15 cm; height 9 cm) was surrounded by plastic film to prevent moving out of seedling after germination and contained 20 seeds. Germination responses were recorded after 2 weeks. A seed was considered to have germinated if any part of the leaf had emerged from the seed coat. We calculated germination rate and seedling (after germination) buoyancy rate in the inundation condition.

Seedling establishments

Seeds used for seedling establishment experiments were collected on September 5, 2013, in the Mankyeong river, South Korea (N35° 56′ 36.8″ E 127° 10′ 18.9″, 42 m). They were stored in a sealed plastic bag filled with wet cotton at 4 °C in the dark until February 2014, representing 5 months of wet and cold stratification for breaking seed dormancy. The seedling establishment experiments were divided into two types according to seedling types (floating seedlings and sinking seedlings). Seeds germinated and grew in the greenhouse until they became seedlings of 4 ± 1 cm length. In both experiments, the water levels were based on a preliminary experiment and field survey conducted in 2012. The water levels in these treatments were maintained by weekly additions of tap water. We monitored growth responses such as the shoot height, total leaf length, survival rate, and total dry mass. We measured the early response in the first 2 months every 2~3 days, then measured once every week or two for the next 2 months. Algae is lethal to *S. erectum* seedlings, therefore, any algal growth was carefully removed with a plankton net. Plants were harvested in October when the seeds were mature and the total plant dry mass was measured. To compare plant growth differences among the treatments at the end of the life cycle, the harvested parts were classified into shoots and roots then dried at 60 °C for over 48 h. Afterward, the plants' total dry weights were measured.

Sinking seedling experiments After germination, only the seedlings that sank were used for the sinking seedling experiments. Sinking seedlings were planted in plastic pots (Φ 15 cm; height 9 cm) and the evaluations were adjusted to account for 1 week of transplant stress. The experimental water level gradient included treatments in which the water levels were −5, 0, 5, 10, and 20 cm, with 10 replicates of each water level for a total of 50 pots. The water level gradients were made by plastic stairs in a stainless steel tank (150 cm × 80 cm × 50 cm) and ten pots were located at each water level gradient. 1350 g of soil comprised of 1:8 mixtures of nursery soil (NH_4-N 350 mg/kg, PO_4-P 400 mg/kg; Pungnong, Korea) and sand was added to each pot.

Floating seedling establishments After germination, only the seedlings that floated were used for the floating seedling experiments. Floating seedlings were placed in plastic pots (Φ 15 cm; height 9 cm) surrounded by plastic film. Soil comprised of 1:8 mixtures of nursery soil (NH_4-N 350 mg/kg, PO_4-P 400 mg/kg; Pungnong, Korea) and sand was added up to 3 cm thick in each pot. The experimental water level gradient included treatments in which the water levels were 3, 6, 9, and 12 cm, with 18 replicates for each water level, and a total of 72 pots. We placed six pots in each plastic container, with holes drilled in the containers to maintain the water level. The seedling establishment standard stipulated that seedlings remain stationary despite water flow. A seedling was considered dead if no part of the leaf was green.

Statistical analysis Statistical analyses were performed with t-test at the 5% significance level in the germination experiments. We performed one-way ANOVA at the 5% level based on Duncan's test in the seedling experiments. We used SPSS ver. 20.0 software for all statistical analyses (SPSS, Inc.; Chicago, IL, USA).

Results
Field study
S. erectum germinated in early May. Flowers were produced in June and July and the plants bore fruit in August and September. The average height was highest during the flowering periods and gradually declined over the months the seeds ripened in most of the quadrats (Fig. 2a). Coverage of community A was higher than that of community B in the early growing season, but coverage of community B increased and surpassed community A after July (Fig. 2b). Coverage of community B in October remained higher than that of community A because *S. erectum* shoots reemerge throughout the year except

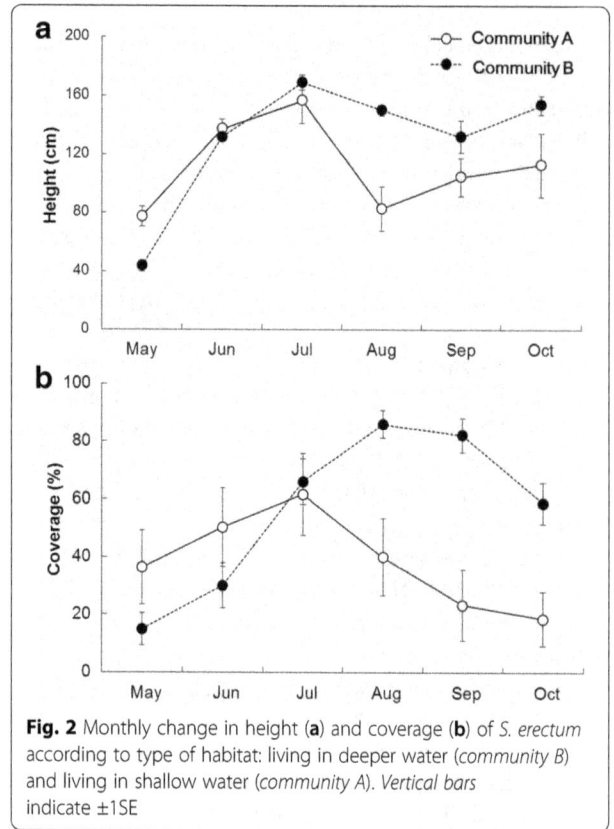

Fig. 2 Monthly change in height (**a**) and coverage (**b**) of *S. erectum* according to type of habitat: living in deeper water (*community B*) and living in shallow water (*community A*). *Vertical bars* indicate ±1SE

in winter and can survive under water. Height and coverage of community B were also higher than those of community A after August, beyond the maximum growth months.

The water depth of community B was deeper than that of community A year round (Fig. 3). Water depths in community A and B in August increased simultaneously compared to other months because of a lot of rain during the investigation period.

Mean water and soil environmental properties at *S. erectum* habitats are shown in Tables 1 and 2. There were no significant differences between the two communities in all investigated soil environmental factors at $p \leq 0.05$. Water

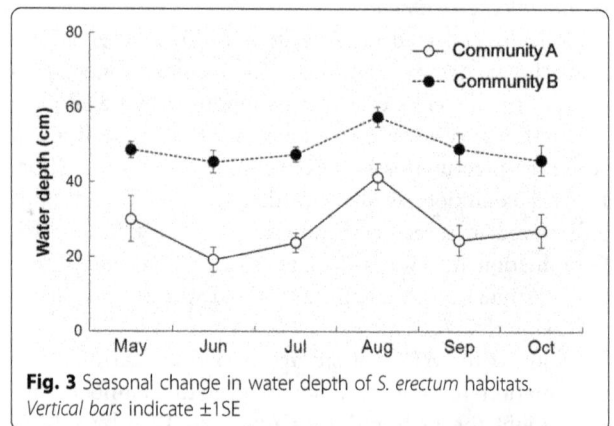

Fig. 3 Seasonal change in water depth of *S. erectum* habitats. *Vertical bars* indicate ±1SE

Table 1 Soil environmental properties at *S. erectum* habitats in June 2013 (A, $n = 8$; B, $n = 3$)

Factor	Community A	Community B	p value	t value
Soil texture	Sandy loam	Sandy loam	–	–
Loss-On-Ignition (%)	0.2 (0.2)	0.1 (0.1)	0.180	1.452
NO_3-N (mg/kg)	4.6 (1.7)	5.2 (1.1)	0.584	−0.569
PO_4-P (mg/kg)	28.5 (16.0)	24.0 (13.3)	0.677	0.430

Values are means (S.D.)

environmental factors in the two communities showed no significant differences at $p \leq 0.05$, with the exception of pH and EC before the flooding in July. The pH value of community B, in the deeper water depth areas, was higher, whereas the EC of community B was lower than that of community A. pH, EC, and cation contents in community B were significantly different in September after flooding. EC and cation contents in community A were higher than those of community B due to inflow of muddy water.

Mesocosm experiments
Germination experiments
When the *S. erectum* seeds were collected, they floated on the water surface. However, most of the seeds (98.1%) sank after exposure to cold and wet conditions for 3 months. Germination rate was 0% in all seeds stored in dry condition. Among seed stored in wet condition, only seeds stored in 4 °C were germinated and others were not. Germination rate of seeds stored for 3 months was 100 and 44.4% for 1 month stored and 82.3% for 2 months stored (Fig. 4).

Seeds were germinated only in three conditions (30/20/12, 25/15/12, 20/10/12) of temperature and photoperiod (Fig. 5). The highest germination rate was in the condition of 30/20(°C)/12. There was no germination under high temperature of 20 °C or 24 h dark condition.

Two weeks after sowing, the cumulative germination rate and post germination seedling buoyancy at each water level were recorded (Table 3). Cumulative germination rates between inundation and waterlogged treatments were not different ($p = 0.207$). 38.4% of *S. erectum* seedlings have characteristics of post germination seedling buoyancy when germinated in inundation conditions.

Effects of water level on seedling establishment

Sinking seedling experiment Clear differences in the effects of water levels on seedling growth were observed (Fig. 6 and Table 4). Leaf extension was very rapid in submerged conditions compared with other conditions during the first 3 weeks. After 3 weeks, early height growth rate at 0 cm water level was higher than at other water levels and seedlings at 0 cm water level reached maximum shoot height first. The same tendency was seen in seedlings at 5 and −5 cm. The early height growth was delayed in water level groups over 10 cm until the seedlings emerged from below the water surface. The deeper the water level, the later the seedlings broke the water surface (Table 4). However, the shoot height at each water level eventually reached a maximum.

Total leaf length representing growth quantity differed at each water level (Fig. 6b). Total leaf length at 5 cm water level regimes was higher than at other water regimes. After seedlings emerged from below the water surface, they reached their maximum heights and tillering began (Table 4). Tillering began at water levels over 10 cm for seedlings that had yet to reach their maximum height. Tillering tended to occur later in inundation conditions. After tillering, the total leaf length increased considerably at each water level.

Survival rate was 100% for the first 10 weeks, then fell to 90% at the water level of 10 cm. The survival rates at

Table 2 Water environmental properties in *S. erectum* habitats in July and September 2013 (A, $n = 8$; B, $n = 3$)

Region	July				September			
	Community A	Community B	p value	t value	Community A	Community B	p value	t value
pH	6.15 (0.11)	6.54 (0.10)	***	−5.402	6.30 (0.23)	6.82 (0.15)	**	−3.625
EC (µs/cm)	192.9 (25.7)	101.6 (10.9)	***	5.806	269.3 (117.5)	86.0 (1.6)	*	2.611
Turbidity (NTU)	2.2 (1.5)	1.0 (0.3)	n.s	1.364	22.1 (53.3)	1.2 (0.2)	n.s	0.672
NO_3-N (mg/L)	6.2 (3.4)	1.8 (0.08)	n.s	2.155	4.2 (4.0)	1.4 (0.6)	n.s	1.918
NH_4-N (mg/L)	0.1 (0.1)	0.1 (0.0)	n.s	0.502	0.5 (0.9)	0.0 (0.0)	n.s	0.906
PO_4-P (mg/L)	0.0 (0.0)	0.0 (0.0)	n.s	0.358	0.0 (0.0)	0.0 (0.0)	n.s	1.528
K^+ (mg/L)	16.6 (9.7)	8.2 (1.2)	n.s	1.439	26.3 (8.4)	8.1 (2.7)	**	3.587
Na^+ (mg/L)	8.8 (4.2)	5.4 (0.9)	n.s	1.371	10.9 (1.9)	4.7 (1.6)	***	4.933
Ca^{2+} (mg/L)	19.5 (8.1)	13.4 (1.4)	n.s	1.245	35.8 (12.9)	15.5 (3.1)	*	2.628
Mg^{2+} (mg/L)	4.5 (2.2)	1.6 (0.5)	n.s	2.126	3.7 (1.8)	0.0 (0.0)	***	5.651

Values are means (S.D.)
***, $p < 0.001$; **, $0.001 < p < 0.01$; *, $0.01 < p < 0.05$; n.s = no significance, $0.05 < p$

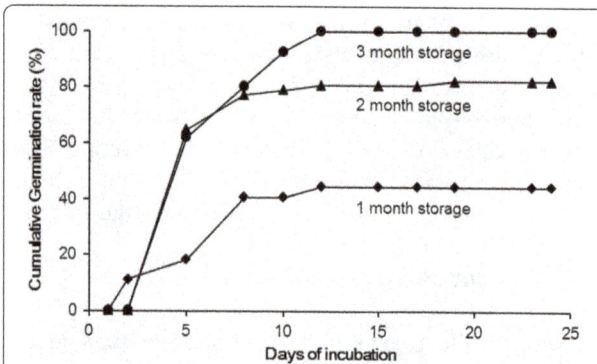

Fig. 4 Cumulative germination rate of seeds stored in 4 °C/wet condition according to storage period

−5 and 0 cm fell to 50% within 3 weeks thereafter and remained at that level. Survival rate at 20 cm remained at 100% for 5 weeks before dropping to 10% by 12 weeks. Seedling survival rates in submergence treatments (water level 5, 10, and 20 cm) were higher than those in non-submergence treatments. The long submerged leaves were very fragile and thin and could not survive when exposed to air conditions. When leaves were exposed to air conditions as floating leaves spread on the water surface, the seedlings began to produce erect and thick leaves and their growth rates increased.

The total biomasses of the planted seedlings differed among water levels (Fig. 7). Although the total dry masses at −5, 0, and 5 cm showed no statistical differences, the total dry mass at the 0 cm water level (19.73 ± 5.40 g) was higher than at others. The total dry mass at submerged conditions with 10 and 20 cm water level were significantly lower than the others ($p < 0.05$). Overall, total growth was higher at water levels near the water surface compared with deeper water levels.

Floating seedling experiment After seedlings rooted in substrate, the total leaf length increased (Fig. 8a). However,

there was a difference in total leaf length at 3 and 6 cm water levels. Survival rates decreased at all water levels within 2 weeks of seedling establishment (Fig. 8b). If seedlings could not rooted in substrate, they all died. After 6 weeks, the respective survival rates were 39 and 17% at water levels of 3 and 6 cm. All seedlings unable to establish at these water levels died. All seedlings at water levels of 9 and 12 cm died. The maximum water level for the survival of floating seedlings was approximately 6 cm.

Total dry masses at 9 and 12 cm water level regimes were not measured because all seedlings died. There was no statistically significant difference between the total dry mass associated with 3 and 6 cm water levels ($p > 0.05$) (Fig. 9).

Discussion

Multiple plant species have the potential to coexist at a given location along environmental gradients because each species differs in their adult niche (Whittaker 1960; Shreve 1922; Byun et al. 2008; Hong and Kim, 2014). Therefore, environmental factors have an impact on the distribution of species in wetlands. However, abiotic factors alone cannot fully dictate species distributions. Many species have limited movement between locations and prior occupation affects species distributions and community structure (Grace 1987; Cornell and Lawton 1992; Tilman 1997). Thus, identifying the abiotic conditions in areas where the species live and simultaneously identifying environmental conditions at suitable habitats available for establishment are important as these reflect the current environmental conditions as well as historical recruitment events (Seabloom et al. 2001).

This study showed the range of the soil and water properties of *S. erectum* habitat. Except for water depth, the chemical characteristics in *S. erectum* habitats were similar to those of common coexisting species, *Typha*, *Phragmites*, *Persicaria*, and *Humulus* spp. in Korean wetlands (Kang and Joo 1999; Kwon et al. 2006; Lee et

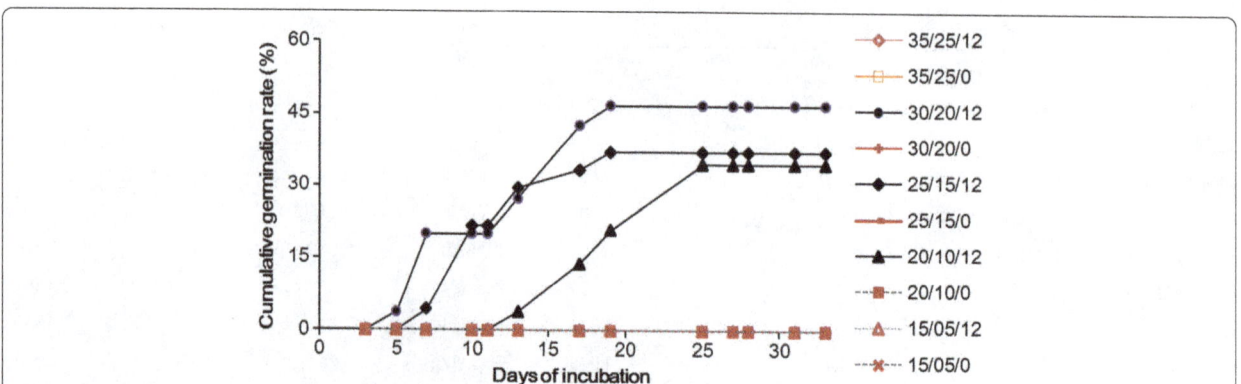

Fig. 5 Cumulative germination rate of seeds stored in 4 °C/wet condition according to alternating temperature regime and photoperiod. Condition indicates high temperature/low temperature (°C)/photoperiod (h)

Table 3 Effects of water conditions on the germination (%) and post germination seedling buoyancy (%) (0 cm, $n = 10$; 20 cm, $n = 10$)

Water level	Waterlogged condition (0 cm)	Inundation condition (20 cm)	p value	t value
Cumulative germination rate (%)	65 ± 5.2	71 ± 4.3	n.s	0.885
Post germination seedling buoyancy (%)	–	38.4 ± 4.9	–	–

Values are means ± S.D.
n.s = no significance, 0.05 < p

Table 4 The dates that seedlings overcame the water surface, reached maximum height, and started tillering in the planted seedling experiment

Water level	Date that seedlings overcome water surface	Date that seedlings reach maximum height	Date that seedlings start tillering
20 cm	53.0	73.0	85.0
10 cm	42.9 ± 1.5	80.0 ± 6.2[a]	88.5 ± 3.5[a]
5 cm	25.0 ± 2.3	70.4 ± 5.2[b]	58.1 ± 4.4[b]
0 cm	–	56.2 ± 2.0[c]	39.0 ± 1.0[c]
−5 cm	–	61.0 ± 0.0[bc]	47.8 ± 1.1[bc]

Values are means ± S.D. Different letters in the table indicate significant differences at the 5% level based on Duncan's test among groups of means. 20 cm water level was excluded from statistical analysis because there was only one sample from that water level

al. 2007; Kim and Kim 2009; Kim et al. 2012). This is in agreement with previous reports that the distribution of adult wetland plants is dominated by a single environmental gradient of water depth (Spence 1982). *S. erectum* was mainly distributed at deeper water depth than the other species. Although water level depends on precipitation, *S. erectum* was not distributed at and had low coverage at shallow water depths (−10~10 cm) where species abundance was high as many species live and are distributed at −20~30 cm water depths (Coops et al. 1996; Kang and Joo 1999; Kim et al. 2013; Kwon et al. 2006, Jeon et al. 2013). This suggests that shallow water depths were quickly occupied by early starting annual plant species and adult *S. erectum* lagged far behind the annual plant species.

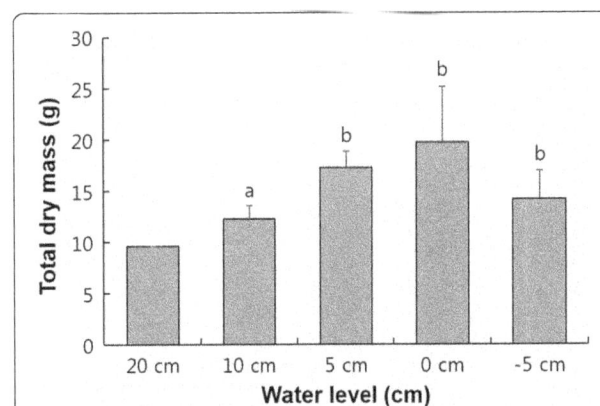

Fig. 6 Growth responses of shoot height (**a**), total leaf length (**b**), and survival rate (**c**) of planted seedlings by water level. *Vertical bars* indicate ± SE

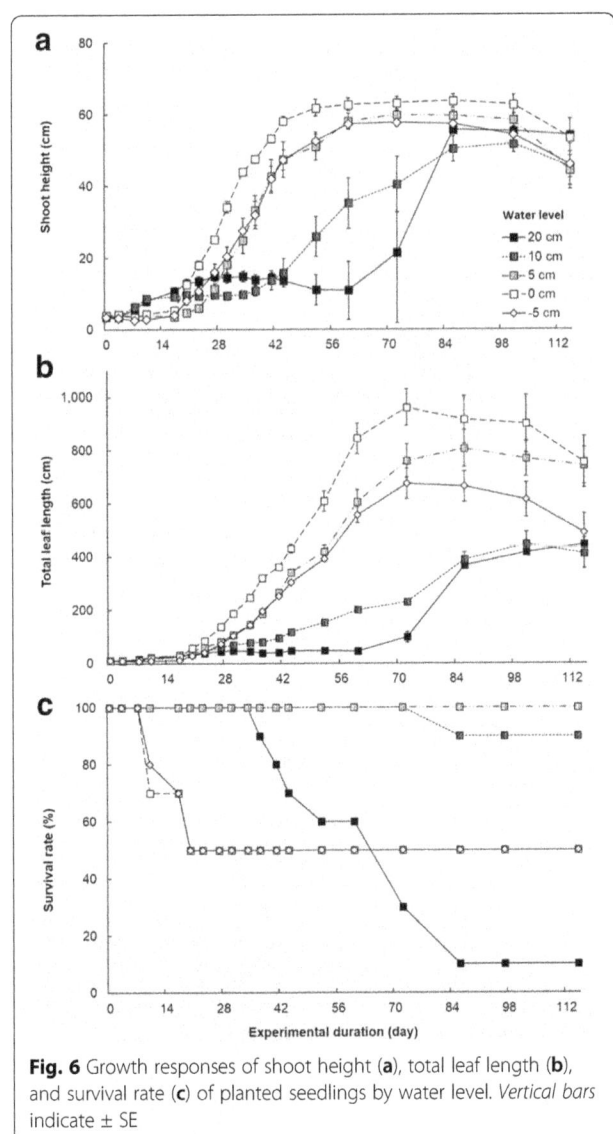

Fig. 7 Mean total biomass for planted seedlings of *S. erectum* harvested at the end of the experiment. *Vertical bars* indicate ±SE. *Different letters* indicate significant differences at the 5% level based on Duncan's test among the groups of means. Total dry mass at the 20 cm water level was excluded from statistical analysis because there was only one sample

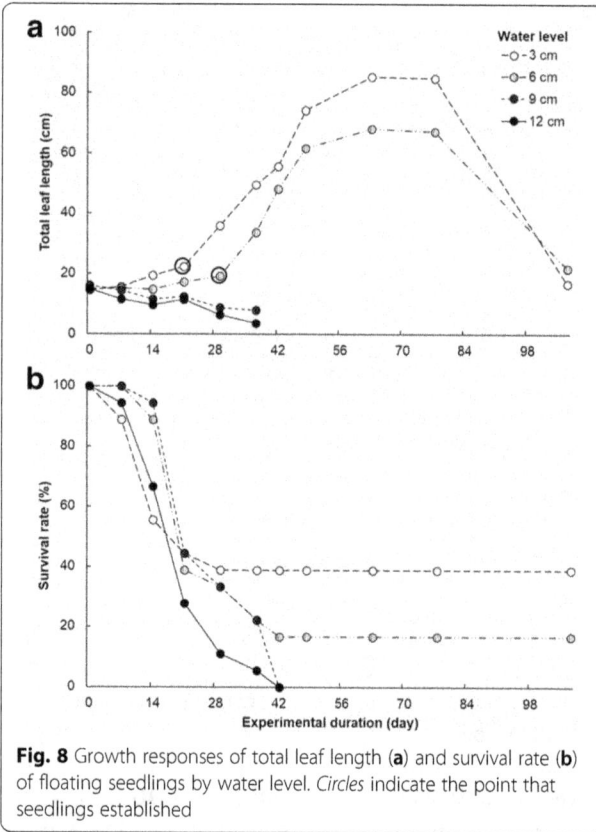

Fig. 8 Growth responses of total leaf length (**a**) and survival rate (**b**) of floating seedlings by water level. *Circles* indicate the point that seedlings established

We observed that plant height and coverage of *S. erectum* community in shallow water decreased more rapidly in deep water with time. Adults of *S. erectum* are mainly distributed in deep waters and the seeds fall in deep waters after ripening. Therefore, seeds are dispersed by hydrochory. Our results indicated that germination potential of seeds was very low at dry condition. Also, most seeds sink into the bottom of water after floating for 3 months. However, seeds buried in bottom soil could not germinate because of darkness. Once the seeds

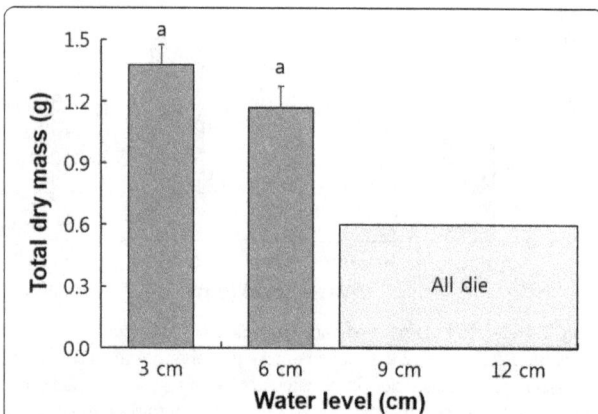

Fig. 9 Mean total biomass for floating seedlings of *S. erectum* harvested at the end of the experiment. *Vertical bars* indicate ±SE

started to germinate in inundation conditions, a number of the seedlings began to float. Thus, the floating seedlings move through the water flow while sinking seedlings establish where they fall. Our finding that seedlings float after germination is not surprising. Previous studies have shown that amphibious plant seedlings like *Triglochin procerum*, *Philydrum lanuginosum*, and *Helmholtzia glaberrima* float after germinating underwater (Nicol and Ganf 2000; Prentis et al. 2006). Some amphibious species have flexible seed germination strategies for two reasons. First, seeds can germinate under both flooded and non-flooded conditions like the family *Pontederiaceae* (Pons 1982). Second, when water depths fluctuate according to seasonal variations and different periods of time, germination is not seriously affected (Prentis et al. 2006).

After germination, floating seedlings survived for 1 or 1.5 months with prolonged floatation. Stable water depths increased the overall submergence of the seedlings, reducing CO_2 and O_2 availability and providing stress in the form of phytotoxicity and hypoxia. This resulted in seedlings with depleted carbohydrates, which prevents rhizome growth and leads to reduced shoot growth (Rea 1996). Therefore, stable water depths also obstruct regeneration from seed (Sand-Jensen et al. 1992; Clevering et al. 1995; Rea 1996). *S. erectum* seedlings could not establish below 6 cm water levels and most of them were able to establish from 3 to 6 cm. The rate of seedling establishment was 20~40% and the seedlings grew after successful establishment. The amount of post germination seedling buoyancy indicates that a high percentage of seedlings can establish if the water level stabilizes. These seedlings are often at the water surface in flooded conditions, which may be advantageous to reduce inundation stress (van der Valk 1981). Also, they have the ability to move to more favorable conditions for establishment (Nicol and Ganf 2000).

After germination, sinking seedlings have modified morphological traits such as leaf elongation and increased shoot: root ratios in response to the total submergence period (Cooling et al. 2001). This can increase the percentage that reaches the water surface. If seedlings cannot reach the water surface, plant growth decreases (Haslam 1970; Waters and Shay 1992; Coops et al. 1996). But once seedlings grew taller than the water surface, establishment, survival rates, and total leaf length were higher than for floating seedlings. With sinking seedlings, the seeds move to hydrochory, fall in the water column and germinate where they sink. The survival rate and growth rate of *S. erectum* is higher than other plants such as *Typha* and *Phramites* spp. Survival rates at water logged conditions were lower than at other water levels, but the total dry mass was higher. Seedlings of major accompanying species of *S. erectum*

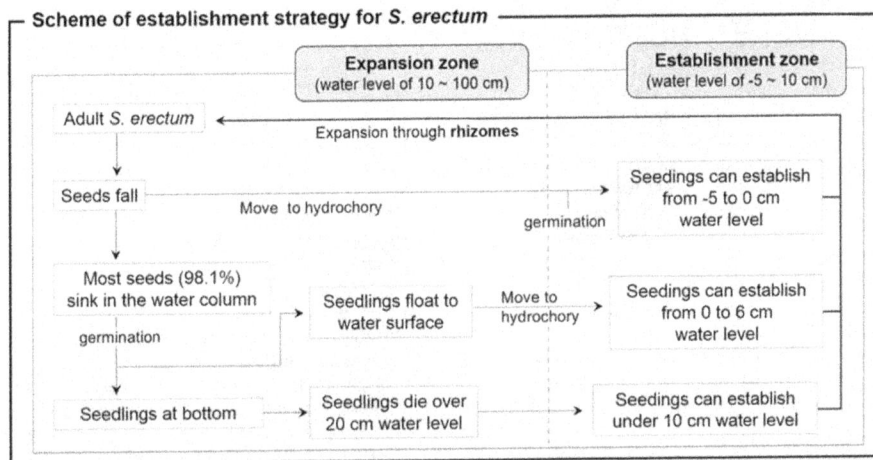

Fig. 10 Scheme of establishment strategy of *S. erectum*

such as *P. thunbergii*, *P. australis*, and *H. japonicus* grow well in water logged conditions but seedling survival rates and growth rates are reduced at water levels more than twice their heights (Mauchamp et al. 2001; Kim et al. 2013; Choo et al. 2015). It seems that *S. erectum* is more competitive than its major accompanying species, particularly in inundation conditions as opposed to waterlogged condition.

Conclusions

The water level around adult *S. erectum* populations in the field were different from the level in which the seedlings survived in the mesocosm experiments. Adult *S. erectum* live at deeper water depths than other species that share the same habitat niche, while seedlings of *S. erectum* can establish only at shallow water levels regardless of the seedling type (Fig. 10). This means that the survival strategies differ at each life stage and advance the process of establishment and growth. After floating on the water surface, then reaching and establishing on favorable conditions such as exposed mudflats and sediment topographies, the seedlings cannot survive in competition with other species such as annual plant species at shallow water levels where those species are greater. Consequently, *S. erectum* living in shallow water eventually die out and only *S. erectum* that can regenerate through rhizomes and live at deeper water depths survive. These results not only explain *S. erectum* habitat characteristics but also provide evidence to connect life-historical links between the early seedling stage and adult habitat conditions. When *S. erectum* is planted at shallow water depths in wetlands, the wetlands cannot maintain the landscape and there is a possibility the species will die out at shallow water levels. Finally, this study contributes practical information for the life cycle of *S. erectum* in wetland.

Abbreviation
EC: Electron conductivity

Acknowledgements
The authors would like to thank Eunguk Kim for assistance in mesocosm experiments.

Funding
This study was funded by the Korea Ministry of Environment (MOE) as "public technology program based on Environmental Policy" (2016000210003) and by the Basic Science Research Program through the National Research Foundation of Korea (NRF-2015R1D1A1A01057373).

Authors' contributions
KSH participated in the design of the study, carried out field study, mesocosm experiments, and data analyses, and wrote the manuscript draft. KJG conceived the study, participated in the design of the study, edited manuscript draft, and secured funding. NJM participated in field work. All authors read and approved the final manuscript.

Competing interests
The authors declare that they have no competing interests.

References
Abbe, T. B., & Montgomery, D. R. (2003). Patterns and processes of wood debris accumulation in the Queets river basin, Washington. *Geomorphology*, *51*(1), 81–107.
Asaeda, T., Rajapakse, L., & Kanoh, M. (2010). Fine sediment retention as affected by annual shoot collapse: *Sparganium erectum* as an ecosystem engineer in a Lowland Stream. *River Research and Applications*, *26*(9), 1153–1169.

Baskin, C. C., & Baskin, J. M. (1998). *Seeds: Ecology, Biogeography, and Evolution of Dormancy and Germination*. New York: Academic Press.

Boyle, J. (2004). A comparison of two methods for estimating the organic matter content of sediments. *Journal of Paleolimnology, 31*, 125–127.

Byun, C., Kwon, G. J., Lee, D., Wojdak, J. M., & Kim, J. G. (2008). Ecological assessment of plant succession and water quality in abandoned rice fields. *Journal of Ecology and Field Biology, 31*(3), 213–223.

Choo, Y. H., Nam, J. M., Kim, J. H., & Kim, J. G. (2015). Advantages of amphicarpy of *Persicaria thunbergii* in the early life history. *Aquatic Botany, 121*, 33–38.

Clarke, P. J., & Allaway, W. G. (1993). The regeneration niche of the grey mangrove (*Avicennia marina*): effects of salinity, light and sediment factors on establishment, growth and survival in the field. *Oecologia, 93*, 548–556.

Clevering, O. A., Van Vierssen, W., & Blom, C. W. P. M. (1995). Growth, photosynthesis and carbohydrate utilization in submerged *Scirpus maritimus* L. during spring growth. *New Phytology, 130*(1), 105–116.

Collins, S. L., & Good, R. E. (1987). The seedling regeneration niche: habitat structure of tree seedlings in an oak-pine forest. *Oikos, 48*(1), 89–98.

Cook, C. D. K. (1962). *Sparganium erectum* L. (S-Ramosum Hudson, Nom Illeg). *Journal of Ecology, 50*(1), 247–255.

Cooling, M. P., Ganf, G. G., & Walker, K. F. (2001). Leaf recruitment and elongation: an adaptive response to flooding in *Villarsia reniformis*. *Aquatic Botany, 70*(4), 281–294.

Coops, H., van den Brink, F. W. B., & van der Velde, G. (1996). Growth and morphological responses of four helophyte species in an experimental water-depth gradient. *Aquatic Botany, 54*(1), 11–24.

Coops, H., & van der Velde, G. (1995). Seed dispersal, germination and seedling growth of six helophyte species in relation to water-level zonation. *Freshwater Biology, 34*, 13–20. doi:10.1111/j.1365-2427.1995.tb00418.x.

Cornell, H. V., & Lawton, J. H. (1992). Species interactions, local and regional processes, and limits to the richness of ecological communities: a theoretical perspective. *Journal of Animal Ecology, 61*, 1–12.

DeKlerk, P., Janssen, C. R., & Joosten, J. H. J. (1997). Patterns and processes in natural wetland vegetation in the Dutch fluvial area: A palaeoecological study. *Acta Botanica Neerlandica, 46*(2), 147–159.

Eriksson, O. (1989). Seedling dynamics and life histories in clonal plants. *Oikos, 55*, 231–238.

Friedman, J. M., Osterkamp, W., & Lewis Jr., W. M. (1996). The role of vegetation and bed-level fluctuations in the process of channel narrowing. *Geomorphology, 14*(4), 341–351.

Grace, J. B. (1987). The impact of preemption on the zonation of two *Typha* species along lakeshores. *Ecological Monographs, 57*(4), 283–303.

Grime, P. J. (2006). *Plant Strategies, Vegetation Processes, and Ecosystem Properties* (2nd ed.). New York: Wiley.

Grubb, P. J. (1977). The maintenance of species-richness in plant communities: the importance of the regeneration niche. *Biological Reviews, 52*(1), 107–145.

Gurnell, A. M. (2007). Analogies between mineral sediment and vegetative particle dynamics in fluvial systems. *Geomorphology, 89*(1), 9–22.

Gurnell, A. M., Van Oosterhout, M., De Vlieger, B., & Goodson, J. (2006). Reach-scale interactions between aquatic plants and physical habitat: River Frome, Dorset. *River Research and Applications, 22*(6), 667–680.

Haslam, S. M. (1970). The performance of *Phragmites communis* Trin. in Relation to Water-supply. *Annals of Botany, 34*, 867–877.

Hong, M. G., & Kim, J. G. (2014). Role and effects of winter buds and rhizome morphology on the survival and growth of common reed (Phragmites australis). *Paddy and Water Environment, 12*(Suppl 1), S203–S209. doi:10.1007/s10333-014-0445-z.

Hong, M. G., Nam, J. M., & Kim, J. G. (2012). Occupational strategy of runner reed (*Phragmites japonica* Steud.): change of growth patterns with developmental aging. *Aquatic Botany, 97*(1), 30–34.

Jeon, S. H., Kim, H., Nam, J. M., & Kim, J. G. (2013). Habitat characteristics of sweet flag (*Acorus calamus*) and their relationships with sweet flag biomass. *Landscape and Ecological Engineering, 9*(1), 67–75.

Jeong, T. S., & Kim, J. G. (2017). *Parnassia palustris* population differences in three Korean habitat types. *Landscape and Ecological Engineering, 13*(1), 93–105. doi:10.1007/s11355-016-0305-7.

Kamphake, L. J., Hannah, S. A., & Cohen, J. M. (1967). Automated analysis for nitrate by hydrazine reduction. *Water Research, 1*, 205–216.

Kaneko, K., & Jinguji, H. (2012). Effects of environmental factors on *Sparganium emersum* and *Sparganium erectum* colonization in two drainage ditches with different maintenance. *Agricultural Sciences, 3*, 538–544.

Kang, H. C., & Joo, Y. K. (1999). The structural characteristics in natural wetlands and fitted depth zones of *Phramites japonica* (in Korean). *Journal of the Korean Institute of Traditional Landscape Architecture, 17*(4), 191–200.

Kankaala, P., Ojala, A., Tulonen, T., Haapamäki, J., & Arvola, L. (2000). Response of littoral vegetation on climate warming in the boreal zone; an experimental simulation. *Aquatic Ecology, 34*(4), 433–444.

Kim, D. H., Choi, H., & Kim, J. G. (2012). Occupational strategy of *Persicaria thunbergii* in riparian area: rapid recovery after harsh flooding disturbance. *Journal of Plant Biology, 55*(3), 226–232.

Kim, S., & Kim, J. G. (2009). *Humulus japonicus* accelerates the decomposition of *Miscanthus sacchariflorus* and *Phragmites australis* in a floodplain. *Journal of Plant Biology, 52*(5), 466–474.

Kim, S. H., & Kim, J. G. (2015). Analysis of environmental characteristics for habitat conservation and restoration of near threatned *Sparganium japonicum*. *Journal of the Korean Society of Environmental Restoration Technology, 18*, 37–51.

Kim, D. H., Kim, H. T., & Kim, J. G. (2013). Effects of water depth and soil type on the survival and growth of *Persicaria thunbergii* during early growth stages. *Ecological Engineering, 61*, 90–93.

Kim, J. G., Park, J. H., Choi, B. J., Sim, J. H., Kwon, G. J., Lee, B. A., Lee, Y. W., & Ju, E. J. (2004). *Method in Ecology*. Seoul: Bomoondang (in Korean).

Kwon, G. J., Lee, B. A., Nam, J. M., & Kim, J. G. (2006). The optimal environmental ranges for wetland plants: 1. *Zizania Latifolia* and *Typha angustigolia* (in Korean). *Journal of the Korean Society of Environmental Restoration Technology, 9*, 72–88.

Kwon, G. J., Lee, B. A., Nam, J. M., & Kim, J. G. (2007). The relationship of vegetation to environmental factors in Wangsuk stream and Gwarim reservoir in Korea: II. Soil environments. *Ecological Research, 22*, 75–86.

Lee, B. A., Kwon, G. J., & Kim, J. G. (2007). The optimal environmental ranges for wetland plants:II. *Scirpus tabernaemontani* and *Typha latifolia*. *Journal of Ecology and Field Biology, 30*(2), 151–159.

Liffen, T., Gurnell, A. M., O'Hare, M. T., Pollen-Bankhead, N., & Simon, A. (2011). Biomechanical properties of the emergent aquatic macrophyte *Sparganium erectum*: Implications for fine sediment retention in low energy rivers. *Ecological Engineering, 37*(11), 1925–1931.

Mauchamp, A., Blanch, S., & Grillas, P. (2001). Effects of submergence on the growth of *Phragmites australis* seedlings. *Aquatic Botany, 69*(2–4), 147–164.

Mueller-Dombois, D., & Ellenberg, H. (2003). *Aims and Methods of Vegetation of Ecology*. New York: Blackburn Press.

Murphy, J., & Riley, J. P. (1962). A modified single solution method for the determination of phosphate in natural waters. *Analytica Chemica Acta, 27*, 31–36.

National Institute of Biological Resources. (2012). Red Data Book of Endangered Vascular Plants in Korea. Ministry of Environment (in Korean).

National Museum of Nature and Science. (2017). Global red list of Japanese threatened plants. https://www.kahaku.go.jp/english/research/db/botany/redlist/list/list_05_254_1.html. Accessed 29 July 2017.

Nicol, J. M., & Ganf, G. G. (2000). Water regimes, seedling recruitment and establishment in three wetland plant species. *Marine and Freshwater Research, 51*(4), 305–309.

O'Hare, J. M., O'Hare, M. T., Gurnell, A. M., Scarlett, P. M., Liffen, T., & McDonald, C. (2012). Influence of an ecosystem engineer, the emergent macrophyte *Sparganium erectum*, on seed trapping in lowland rivers and consequences for landform colonisation. *Freshwater Biology, 57*(1), 104–115.

Pollen-Bankhead, N., Thomas, R. E., Gurnell, A. M., Liffen, T., Simon, A., & O'Hare, M. T. (2011). Quantifying the potential for flow to remove the emergent aquatic macrophyte *Sparganium erectum* from the margins of low-energy rivers. *Ecological Engineering, 37*(11), 1779–1788.

Pollux, B. J. A., Verbruggen, E., Van Groenendael, J. M., & Ouborg, N. J. (2009). Intraspecific variation of seed floating ability in *Sparganium emersum* suggests a bimodal dispersal strategy. *Aquatic Botany, 90*(2), 199–203.

Pons, T. L. (1982). Factors affecting weed seed germination and seedling growth in lowland rice in Indonesia. *Weed Research, 22*, 155–161.

Prentis, P. J., Meyers, N. M., & Mather, P. B. (2006). Significance of post-germination buoyancy in *Helmholtzia glaberrima* and *Philydrum lanuginosum* (Philydraceae). *Australian Journal of Botany, 54*(1), 11–16.

Rea, N. (1996). Water depths and *Phragmites*: decline from lack of regeneration or dieback from shoot death. *Folia Geobotanica, 31*(1), 85–90.

Riis, T., Sand-Jensen, K., & Vestergaard, O. (2000). Plant communities in lowland Danish streams: species composition and environmental factors. *Aquatic Botany, 66*(4), 255–272.

Sand-Jensen, K., Pedersen, M. F., & Nielsen, S. L. (1992). Photosynthetic use of inorganic carbon among primary and secondary water plants in streams. *Freshwater Biology, 27*(2), 283–293.

Seabloom, E. W., Moloney, K. A., & van der Valk, A. G. (2001). Constraints on the establishment of plants along a fluctuating water-depth gradient. *Ecology, 82*(8), 2216–2232.

Seabloom, E. W., van der Valk, A. G., & Moloney, K. A. (1998). The role of water depth and soil temperature in determining initial composition of prairie wetland coenoclines. *Plant Ecology, 138*(2), 203–216.

Shin, C. J., & Kim, J. G. (2013). Ecotypic differentiation in seed and seedling morphology and physiology among *Cicuta virosa* populations. *Aquatic Botany, 111*, 74–80.

Shin, C. J., Nam, J. M., & Kim, J. G. (2013). Comparison of environmental characteristics at *Cicuta virosa* habitats, an endangered species in South Korea. *Journal of Ecology and Environment, 36*, 19–29.

Shipley, B., Keddy, P., Moore, D., & Lemky, K. (1989). Regeneration and establishment strategies of emergent macrophytes. *Journal of Ecology, 77*, 1093–1110.

Shreve, F. (1922). Conditions indirectly affecting vertical distribution on dessert Mountains. *Ecology, 3*(4), 269–274.

Solorzano, L. (1969). Determination of ammonia in natural waters by the phenolhypochlorite method. *Limnology and Oceanography, 14*, 799–801.

Spence, D. H. N. (1982). The zonation of plants in freshwater lakes. In A. Macfadyen & E. D. Ford (Eds.), *Advances in Ecological Research* (pp. 37–125). New York: Academic.

Takahashi, H., Sato, T., & Volotovsky, K. A. (2000). A quantitative comparison of distribution patterns in four common sparganium species in Yakutia, Eastern Siberia. *Acta Phytotaxonomica et Geobotanica, 51*(2), 155–168.

Tilman, D. (1997). Community invasibility, recruitment limitation, and grassland biodiversity. *Ecology, 78*(1), 81–92.

van der Valk, A. (1981). Succession in wetlands: a Gleasonian approach. *Ecology, 62*(3), 688–696.

Waters, I., & Shay, J. M. (1992). Effect of water depth on population parameters of a *Typha glauca* stand. *Canadian Journal of Botany, 70*, 349–351.

Whittaker, R. H. (1960). Vegetation of the Siskiyou mountains, Oregon and California. *Ecological Monographs, 30*(3), 279–338.

Whitton, B. A., Boulton, P. N. G., Clegg, E. M., Gemmell, J. J., Graham, G. G., Gustar, R., & Moorhouse, T. P. (1998). Long-term changes in macrophytes of British rivers: 1. River wear. *The Science of the Total Environment, 210*(1–6), 411–426.

Yang, Y. Y., & Kim, J. G. (2016). The optimal balance between sexual and asexual reproduction in variable environments: A systematic review. *Journal of Ecology and Environment, 40*, 12. doi:10.1186/s41610-016-0013-0.

Yang, Y. Y., & Kim, J. G. (2017). The life history strategy of *Penthorum chinense*: implication for the restoration of early successional species. *Flora, 233*, 109–117. doi:10.1016/j.flora.2017.05.017.

21

Relationship between the sexual and the vegetative organs in a *Polygonatum humile* (Liliaceae) population in a temperate forest gap

Byeong-Mee Min

Abstract

Background: The aim of this study was to clarify the relationship between the sexual reproduction and the resource allocation in a natural *Polygonatum humile* population grown in a temperate mixed forest gap. For this aim, the plant size, the node which flower was formed, the fruiting rate, and the dry weight of each organ were monitored from June 2014 to August 2015.

Results: Firstly, in 3–13-leaf plants, plants with leaves ≤ 8 did not have flowers and in plants with over 9 leaves the flowering rate increased with the number of leaves. Among plants with the same number of leaves, the total leaf area and dry weight of flowering plants were larger than those of non-flowering plants. The minimum leaf area and dry weight of flowering plants were 100 cm^2 and 200 mg, respectively. Secondary, the flowers were formed at the 3rd~8th nodes, and the flowering rate was highest at the 5th node. Thirdly, cumulative values of leaf properties from the last leaf (the top leaf on a stem) to the same leaf rank were greater in a plant with a reproductive organ than in a plant without a reproductive organ. Fourthly, fruit set was 6.1% and faithful fruit was 2.6% of total flowers. Biomasses of new rhizomes produced per milligram dry weight of leaf were 0.397 ± 190 mg in plants that set fruit and 0.520 ± 0.263 mg in plants that did not, and the difference between the 2 plant groups was significant at the 0.1% level.

Conclusions: *P. humile* showed that the 1st flower formed on the 3rd node from the shoot's base. And *P. humile* showed the minimum plant size needed in fruiting, and fruiting restricted the growth of new rhizomes. However, the fruiting rate was very low. Thus, it was thought that the low fruiting rate caused more energy to invest in the rhizomes, leading to a longer rhizome. A longer rhizome was thought to be more advantageous than a short one to avoid the shading.

Keywords: Flowering rate, Fruiting, Leaf area, Leaf dry weight, *Polygonatum humile*, Rhizome

Background

In temperate deciduous forests, understory herbaceous plants grow under insufficient irradiance after canopy closure (Emborg 1998, Augusto et al. 2003). These plants use diverse strategies to adapt to the low-light environment. Spring ephemerals complete life cycles before canopy closure, and shade-tolerant plants slowly grow (Schemske et al. 1978, Houle 2002, Legner et al.

2013). In any case, photosynthetic substances produced during the growing season are limited, and understory plants use the pertinent energy strategy to ensure their survival and reproduction. In particular, a young perennial herbaceous plant invests energy in growth or survival rather than in reproduction until its size is to a certain extent (Silvertown 1982, Hartnett 1990). These plant species reproduce asexually and sexually. Asexual reproduction is closely related to growth, and generally, reproductive potential is evaluated by the sexual one. Reproductive potential of perennial herbaceous species

Correspondence: bmeemin@hanmail.net
Department of Science Education, Dankook University, Yongin 16890, South Korea

depends on plant size rather than on its age (Klinkhamer et al. 1992). When clonal plants reach the reproductive stage, sexual reproduction and vegetative propagation may compete for resources in the plant (Cook 1983). In perennial herbaceous plants, resource allocation to asexual and sexual reproduction is an indicator of adaptive plasticity and a strategy for survival, and this topic has been addressed many times in population ecology (Jerling 1988, de Kroon and Schieving 1991). Moreover, information on reproduction strategies is essential to predict for population dynamics in the future (Silvertown 1982). In other words, in clonal perennial herbaceous plants, energy budget strategies for asexual and sexual reproduction determine the entire life cycle, and the plant's response resulting from these strategies is the basis for interpreting environmental adaptation or genetic output (Stephenson 1981). Microenvironmental adaptation to insufficient sunlight is transiently reflected by morphological and physiological diversity. However, the energy budget in a plant cannot be precisely divided into that for asexual and sexual reproduction because of methodological and physiological difficulties (Klinkhamer et al. 1990). Putting aside respiration during seed production, total net production by a plant does not remain at its disposal, since photosynthetates are lost because of organ death, herbivore predation, and physical damage. Thus, to estimate the energy used for sexual reproduction, an indirect method has been used: sexual organs are removed and weighed, and biomasses are compared for flowering and non-flowering plants (Klinkhamer et al. 1990, Ehrlén and van Groendael 2001). This method has been tested in diverse settings by many researchers (Samon and Werk 1986, Reekie and Bazzaz 1987, Verburg et al. 1996). They used the term "reproductive efforts" with respect to the plant's reproductive investment. Reproductive efforts are the rate of the overall resources (biomass or nutrients) used for reproductive organs or seeds.

Polygonatum humile (Liliaceae) is generally distributed in grasslands or at the periphery of a shrub in Korea (Choung 1991, Jang 2002) or in grasslands and sand dunes in Japan (Hasegawa and Kudo 2005). Thus, this species needs full sunlight. And *P. humile* is a clonal perennial herb, which reproduces asexually by rhizomes and sexually by seeds (Choung 1991, Hasegawa and Kudo 2005). Unlike other species of genus *Polygonatum*, the shoots of *P. humile* are erect and its rhizomes are long (Jang 2002, Hasegawa and Kudo 2005). The total length of the rhizome which grows for about 3.3 years and is connected in 1 line is 47.6 ± 25.6 cm a shoot, and rhizomes grow at a rate of 15.5 ± 4.4 cm a year (Hasegawa and Kudo 2005). Each plant has 2 rhizomes (Choung 1991, Jang et al. 1998) or 1~4 rhizomes (Jang 2002). The

flower germinates from the axilla of the leaf (node). The number of flowers is 19.0 (Jang et al. 1998) or 2.7 ± 1.1 (Hasegawa and Kudo 2005) per stem and 2~6 (Jang et al. 1998) or 1 (Jang 2002) or 1~2 (Lee 2003) per node. The reported number of seeds per fruit is 6.1 ± 2.71 (Hasegawa and Kudo 2005) or many (Jang 2002). Thus, the number of flowers produced by *P. humile* varied according to the reports. Besides the studies mentioned above, there are few ecological studies on *P. humile*. Moreover, there are much fewer pharmaceutical and agricultural studies for *P. humile* than for other *Polygonatum* species (Seo et al. 2011).

The aim of this study was to clarify the relationship between the sexual reproduction and the resource allocation in a natural *P. humile* population grown in a temperate mixed forest gap. For this aim, the plant size, the node which flower was formed, the fruiting rate, and the dry weight of each organ were monitored from June 2014 to August 2015. The properties of plant sizes were the leaf area, the dry weight of leaf, root and rhizome, and the length of stem and rhizome.

Methods

The study area was located at Yeongheung-ri, Yeongwal-eup, Yeongwal-gun, Gangwon province (37° 11′ 35.7″ N, 128° 28′ 04.0″ E). The study site was 385 m a.s.l. (slope, 5°; direction, 175°). Annual mean air temperature and precipitation in an average year were 10.8 °C and 1224.1 mm, respectively, and those in 2015 were 12.5 °C and 676 mm, respectively, on Yeongwal Meteorological Station. *P. humile* grew in a gap created to move a grave to another place (radius, 10 m) and was momentarily shaded by trees in the south. *P. humile*'s patch was about a 5 m × 6 m circle. Except for *P. humile*, the herb layer was 5% in coverage and composed of *Aster scaber, Carex lanceolata, Festuca ovina*, and *Zoysia japonica*. The surrounding forest had 4 layers of vegetation. The tree layer was composed of *Pinus densiflora, Quercus mongolica*, and *Quercus dentata* and had 90% coverage. There were *P. densiflora, Quercus serrata, Prunus sargentii*, and *Euonymus sieboldiana* and 40% coverage in the subtree layer. There were *Exochorda serratifolia, Pourthiaea villosa*, and *Rhododendron mucronulatum* in the shrub layer and *C. lanceolata, Leibnitzia anandria*, and *A. scaber* in the herb one. The coverage was 10% in the former and 1% in the latter. The litter layer was 10-cm deep and a layer of soil was 10-cm deep. However, the soil of the study area where the *P. humile* population was distributed had been disturbed by the work in relocating the grave but had settled by the time of the study. A field survey was carried out for 2 purposes. One was to check the changes in the sexual organ. A quadrat of 2 × 2 m was chosen on April 25, 2015. Each flowering plant of *P. humile* was numbered by using a plastic rod, and the phenological stages of the sexual organ were checked every week till June 27, 2015. The phenological stages were based on Hasegawa and Kudo (2005). At the end of August 2015, all numbered plants were dug out. The other was to conform the difference between flowering plants and non-flowering ones, so that all plants of *P. humile* were divided into 2 groups and the properties of the plant organ analyzed. In the lab, the plants were divided into the root and shoot systems. The roots were separated according to the age of the rhizome, and the rhizomes were numbered R_0 (this year's rhizome), R_1 (the last year's rhizome), and R_2 (2-years-ago rhizome). The leaf ranks on a stem were numbered from the lowest (the leaf closest to the soil surface) or the 1st leaf on a stem (L_1) to the highest (the last leaf on a stem). The position of the leaf was coincided with a node, so that the leaf rank on a stem was the same as the node one. And flowers germinated in the axilla of leaf (node). Thus, L_1 was the same as the 1st node. The number of leaves on a stem was from 3 to 13. However, 13-leaf plants (13-L) were not sufficiently sampled and were excluded from the analysis. Length, width, and/or area of the rhizomes, leaves, and stems were measured, followed by oven-drying at 85 °C for 48 h and weighing on a Hansung Analytical & Precision Balance, Model MARK M (1-mg unit), and Model YMJ-C Digital Leaf Area Meter (0.0001-cm^2 unit). The total leaf area, total leaf dry weight, and specific leaf area of a plant were divided into the 10 classes. That was, the total leaf area, the total leaf dry weight, and the specific leaf area were divided by 20 cm^2 (range of 59–260 cm^2), 100 mg (range of 99–1252 mg), and 30 cm^2 g^{-1} (range of 203–511 cm^2 g^{-1}), respectively. The cumulative values of leaf properties of a stem were calculated for the following reasons and methods. In this study, I made 2 assumptions for the resource allocation during the main growth season. Firstly, the photosynthetic substance produced in the leaf moved from the last leaf to the new rhizome (R_0). Secondary, the resource provided to the reproductive organ (flower or fruit) was synthesized in posterior leaves (the last leaf rank direction) rather than in the anterior ones (L_1 direction), based on the reproductive organ. Thus, it was thought that leaves in the posterior nodes based on the reproductive organ had a profound effect on seed production. Cumulative values of leaf properties were summed for the

flowering plants from the last leaf to the closest leaf rank having a reproductive organ (L_R). Types of L_R were L_3, L_4, L_5, L_6, L_7, and L_8, but L_3 was excluded in the analysis. To compare, the cumulative values of leaf properties for the non-flowering plants were summed from the last leaf to the leaf which its rank was the same as L_R but did not have a reproductive organ. This time, the specific leaf area (SLA) was not cumulative but mean values. Significance of the differences was verified by Student's t test, and correlation coefficients (CCs) between the 2 factors were calculated by using Pearson's equation ($y = ax + b$).

Results and discussion
Relationship between the sexual organ and the plant size
The flower bud of *P. humile* appeared at the same time as the leafing on April 20, 2015, and the last petal was attached until May 30, 2015. And the fruit was 1st observed on May 21, 2015, and remained on the node at the end time of the field survey (September 10, 2015). Two thirds of *P. humile*'s shoots have sprouted, and small flower buds at the nodes were observed by April 25, 2015. The 1st fruit was firstly observed on May 20, 2015, and all corollas fell down by May 27, 2015 (Table 1). Thus, corollas attached at the nodes were observed during 37 days in this *P. humile* population. After June 2, 2015, the peduncle (after flower fall) or fruit was observed. However, because the 3 stages (flower budding, flowering, and flower withering) could not be easily distinguished, it was determined whether a white corolla was attached at the node or not. In Japan, *P. humile*'s shoots sprouted in mid-May, and it flowers from June 5 to June 22 at high latitude (42° 39′ N) (Hasegawa and Kudo 2005). Thus, the flowering time of *P. humile* is very diverse on the habitat. The fruit set was 6.1% of the total flowers, and 2.6% of those were over 110 mg (the fruit where the diameter was over 8 mm, faithful fruit) and the remainder (the fruit where the diameter was about below 5 mm, 3.5%) below 64 mg (Table 1). The criterion of whether the fruit was faithful followed Lee (1996). By Lee (1996), *P. humile*'s fruit is 8–9 mm in diameter. However, the fruit set was 49–61% (Hasegawa and Kudo 2005) or 75.9% (Jang et al. 1998). Thus, the fruit set of *P. humile* was much lower in this study than in the other 2 studies, likely because the plants

Table 1 Fruit set in the *P. humile* population 2014 and 2015 (size of fruit)

Date	No. of flower	Size of fruit (rate)		Total
		Over 110 mg (faithful)	Under 64 mg (unfaithful)	
Aug. 22, 2014	148	5 (3.5%)	7 (4.7%)	12 (8.1%)
Aug. 26, 2015	196	4 (2.5%)	5 (2.6%)	9 (4.1%)
Total (rate)	344	9 (2.6%)	12 (3.5%)	21 (6.1%)

examined in this study grew in a forest gap, where sunlight was insufficient, whereas plants examined in the other 2 studies grew in full sunlight. Although sexual reproduction could be limited, the study site was over 5 cm in the litter layer and soil water was not in shortage by shading. Thus, because of insufficient sunlight, *P. humile* growing in a forest gap might prefer asexual reproduction rather than the sexual one. The number and the largest length of rhizome per shoot were 1~3 and 25.5 cm, respectively, as the previous report.

The minimum number of leaves needed for flowering (at least 1 flower per stem) was 9 (Fig. 1). The flowering rate increased with the number of leaves, and that of 13-leaf plants was 100%. There were no flowering plants with a total leaf area per plant below 100 cm^2. The flowering rate increased with leaf area, and plants with leaf area over 240 cm^2 had 100% flowering rate. The flowering rate increased with leaf dry weight and was 5% at 200 mg and 100% at >800 mg. However, the flowering rate showed 2 peaks with an increase of SLA and was almost not affected by SLA values. As a result, the flowering rate of *P. humile* increased with leaf size. For the same number of leaves, each organ of the flowering plants was larger than that of the non-flowering plants, but SLA was not. Moreover, of the 11 properties, the differences between the 2 groups were significant at the 0.1~5% level in 6 properties in 9-leaf plants, in 10 in 10-leaf plants, and in 6 in 11-leaf plants (Fig. 2). Leaf area

and dry weight significantly differed between the 2 groups at the 0.1~5% level regardless of the number of leaves. However, SLA and rhizome size significantly differed between the 2 groups at the 0.1~5% level for 10-leaf plants only. The differences in stem length and dry weight between the 2 groups were significant at the 0.1~1% level in the 9- and 10-leaf plants. According to these results, the *P. humile* population showed the following properties. Generally, a minimum plant size is required before sexual reproduction occurs (Hirose and Kachi 1982, Schmid and Weiner 1993). Plant flowering in the field was positively correlated with shoot size (Verburg et al. 1996). However, the relationship between plant size and vegetative reproductive effort has received little attention and data are scarce, and methodological and physiological difficulties remained (as mentioned above). First, *P. humile* required a minimum number of leaves for sexual reproduction. And the number of leaves might be an indicator of plant size. However, the number of leaves was not an appropriate indicator because the flowering rate was not 100% even in 12-leaf plants which almost reached maximum size. Moreover, if the number of leaves and flower buds had been determined before the shoot sprouting, the flowering plants might have already invested more energy from the rhizome into shoot growth than the non-flowering plants regardless of the fruit set. In other words, plants that formed

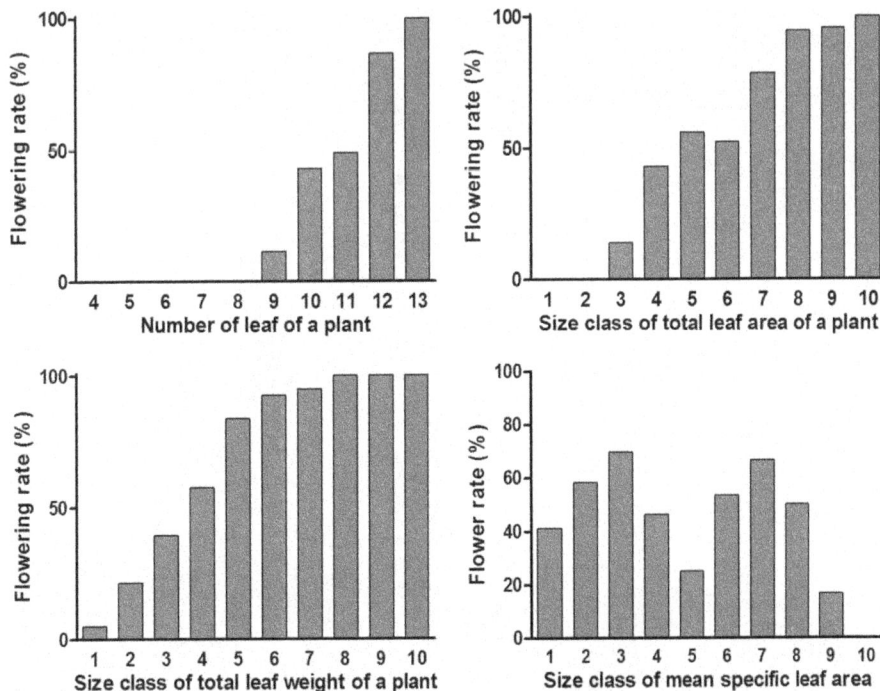

Fig. 1 Flowering rate along the number of leaves (*n* = 412) and the size classes of the total leaf area, the total leaf dry weight, and the mean SLA (*n* = 246) of a plant

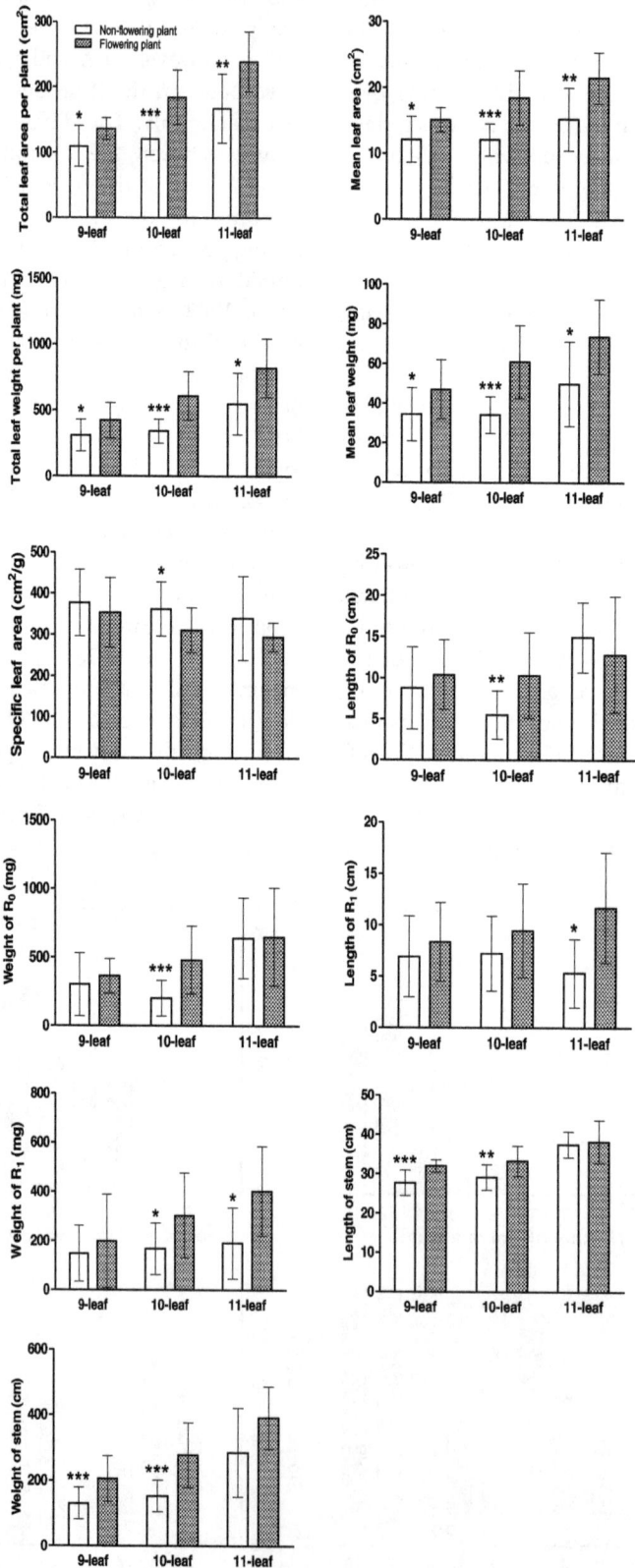

Fig. 2 Comparison of organ sizes between the flowering and the non-flowering plants which had the same number of leaves on June 28, 2014. *, **, and *** indicate the significant level at 5, 1, and 0.1%, respectively

flower buds probably invested energy into other organs rather than only in the increase in the number of leaves. Second, stem length in plants that had ≥ 11 leaves was maximal, whether the plants flowered or not. Unlikely other species in genus *Polygonum*, *P. humile* shoot erects. Third, the flowering rate increased with the total leaf area but was 100% at the largest class only. Thus, stem length and total leaf area were not good indicators of flowering plants. Fourth, the increase of total leaf dry weight was proportioned to the flowering rate. Thus, total leaf dry weight was thought to be the most proper indicator of the relationship between the plant size and flowering.

Leaf ranks (or nodes) with flower

Each leaf rank (node) had a flower or not. And the flower was from the 3rd node (or L_3) to the 6th node (or L_6) in 9-leaf plants and from L_3 to L_8 in 10~12-leaf plants. Thus, there was no flower at 2 nodes from the bottom and at 2~4 nodes from the top. The flowering rate was highest at the L_5 in 9~12-leaf plants and decreased toward the bottom (in the direction of the L_1) and to the top (in the direction of the last leaf) (Fig. 3). The resource provided to reproductive organ (flower or fruit) was synthesized in posterior leaves (the last leaf direction) rather than in the anterior ones (the L_1 direction), based on a reproductive organ. Cumulative values of leaf properties were summed for the flowering plants from the last leaf to the closest leaf rank having a reproductive organ (L_R). To compare, cumulative values of leaf properties for the non-flowering plants were summed from the last leaf to the leaf where its rank was the same as L_R but did not have a reproductive organ. At each leaf rank, the cumulative values of leaf area and leaf dry weight were larger in the flowering plants (L_R plants) than in the non-flowering plants (non-L_R plants) (Fig. 4), and the differences between the 2 groups were almost significant at the 5% level. In particular, the cumulative values of leaf properties from the L_{12} to the L_8 in 12-L were larger in plants with L_R at the L_8 than in plants without a reproductive organ to the L_8. On the fact that the flowers were mainly in the middle of the stem, it was thought that flowers secured 2~4 leaves supplying photosynthetic substances for fruit setting, whereas the 2 or 3 leaves at the bottom might have provided energy to the root system (rhizome and root) regardless of the fruit set. However, if photosynthetic substances are transported both up and down, sexual reproduction and rhizome growth might compete for resources. Generally, in clonal plants, sexual reproduction and vegetative propagation may compete for resources within the plant (Cook 1983), and *P. humile* is a clonal plant. Energy used for flower bud germination can be

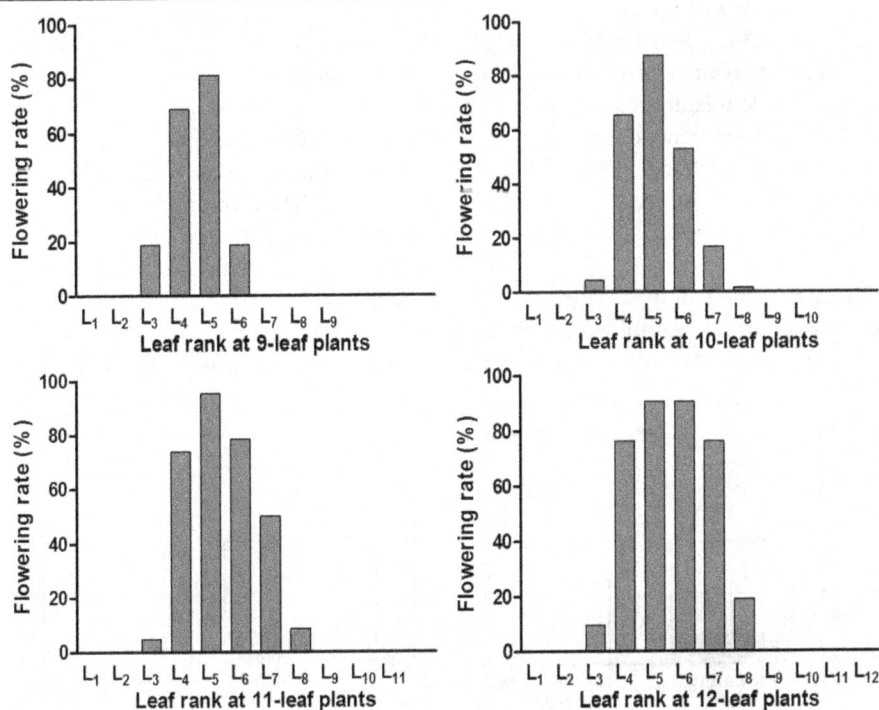

Fig. 3 Of the flowering plants (plants which had 1 or more flowers), flowering rate along the leaf rank (or node) in each number of leaves (9-L, 10-L, 11-L, and 12-L) of a plant

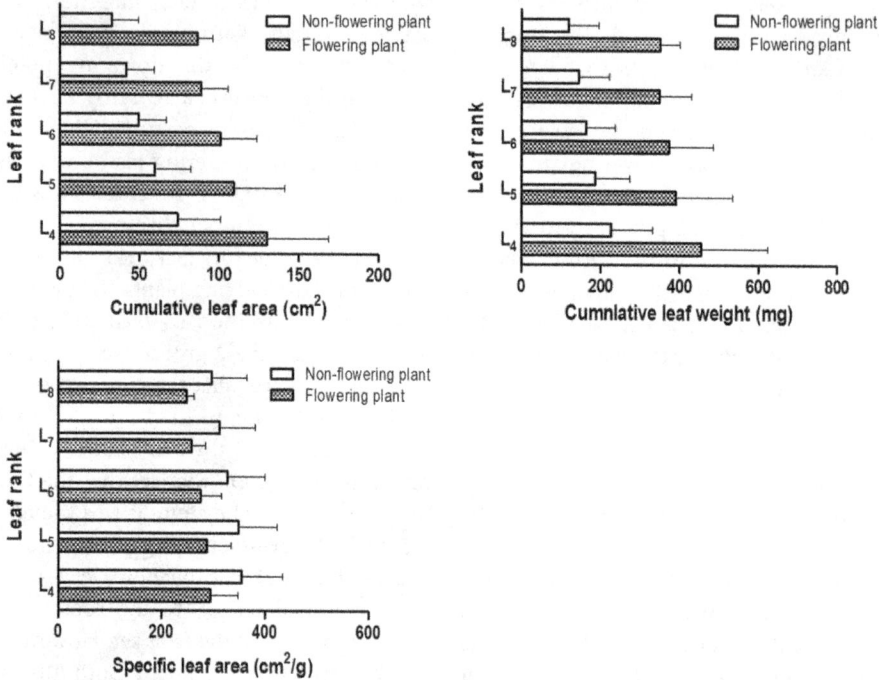

Fig. 4 Mean SLA, cumulative values of the leaf area, and the leaf dry weight in flowering and non-flowering plants. Flowering plants: plants which had a reproductive organ at each leaf rank (LR); non-flowering plants: plants which had no reproductive organ from the last leaf to the leaf rank equal to LR. LR was the closest leaf rank to the last one and had a reproductive organ

brought from the last year's rhizome, but energy for the fruit set must be transported from the nearest leaves (Stephenson 1981). Jang (2002) reported that *P. humile* produced 2~4 rhizomes. Production of ramets decreases the risk of genet mortality. Sexual reproduction varied negatively with both growth and vegetative propagation (Worley and Harder 1996). Shade-tolerant temperate forest herbs devote energy to sexual reproduction to a lesser extent than grassland herbs (Bierzychudek 1982). Thus, *P. humile* in a forest gap might keep at least 2~3 leaves for asexual reproduction by rhizome growth.

Resource budget of new rhizome and fruit
The dry weight of new rhizome per milligram of leaf was 0.397 ± 0.190 mg in plants that set fruit and

0.500 ± 0.263 mg in plants that did not (Fig. 5). Difference between the 2 groups was significant at the 0.1% level, although standard deviation was large. The sum of dry weight of new rhizome and fruit per milligram of leaf was 0.520 ± 0.177 mg in plants that set fruit. The dry weight of new rhizome per square centimeter of leaf was 1.289 ± 0.596 mg in plants that set fruit and 1.689 ± 1.003 mg in plants that did not. The difference between the 2 groups was significant at the 0.1% level. The dry weight sum of new rhizome and fruit per square centimeter of leaf was 1.683 ± 0.579 mg in plants that set fruit. This value was similar to that of plants that did not set fruit. Thus, the productivity of leaves was the same in both plant groups, and sexual reproduction suppressed rhizome growth in this *P. humile* population. If

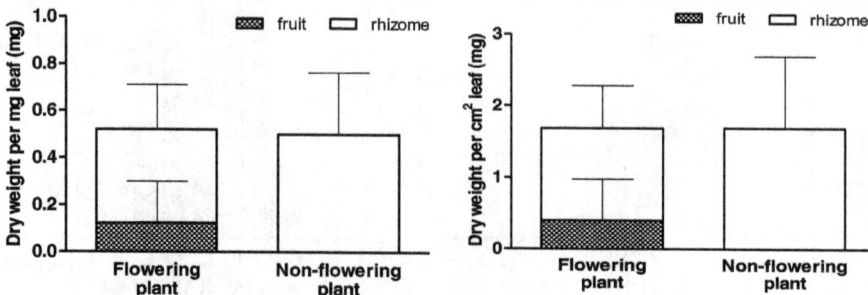

Fig. 5 Dry weight (mg) of fruit and/or new rhizome per milligram leaf dry weight (*left*) or per square centimeter leaf area (*right*) in fruit plants and non-fruit plants

rhizome and seed reproduction are not clearly temporally separated processes, a trade-off between both reproductive modes might be expected, especially when plants grow under severe resource limitation (Stears 1989). As mentioned above, this *P. humile* population growing in a forest gap did not have sufficient sunlight and probably invested the resources in asexual rather than sexual reproduction. As a result, the fruit set was very low below 10%. Generally, the proximate factor that limits fruit and seed production is the energy supply from leaf photosynthesis (Stephenson 1981). Thus, it was thought that the low fruiting rate caused more energy to be invested in the rhizomes, leading to a longer rhizome. A longer rhizome was thought to be more advantageous than a short one to avoid the shading.

Conclusions

Relationship between the sexual organ and the plant size
The minimum sizes the plant needed for flowering (at least 1 flower a stem) were 9 in leaf number, 100 cm^2 in total leaf area, and 200 mg in total leaf dry weight, and the flowering rate increased with the plant sizes. Of the 3 size factors, total leaf dry weight was thought to be the most proper indicator of the relationship between the plant size and flowering.

Leaf ranks (or nodes) with flower
There was no flower at 2 nodes (to L_2) from the bottom and at 2~4 nodes from the top in all plants. And flowering rate was highest at the L_5 in 9~12-leaf plants and decreased toward the bottom (in the direction of the L_1) and to the top (in the direction of the last leaf). Thus, the sex organ and new rhizome might be assured of resources from the several leaves at the end and at the anterior of the stem, respectively.

Resource budget of new rhizome and fruit
Fruiting rate was 6.1% of the total flowers. And the dry weight of new rhizome per milligram of leaf was 0.397 ± 0.190 mg in fruiting plants and 0.500 ± 0.263 mg in non-fruiting plants. Thus, the low fruiting rate was thought to be a strategy for shade avoidance by allocating resources in the rhizome in a forest gap.

Abbreviation
a.s.l.: Above sea level; CC: Correlation coefficient; L_1: The 1st leaf on a stem (the closest leaf to soil surface); L_2–L_{13}: The 2nd leaf to the 13th leaf; L_R: The leaf which was the closest to the last one and had a reproductive organ; R_0: New (this year) rhizome; R_1: The last year's rhizome; R_2: Two-years-ago rhizome; R_3: Three-years-ago rhizome; SLA: Specific leaf area

Acknowledgements
Not applicable

Funding
Not applicable

Competing interests
The author declares no competing interests.

References
Augusto, L., Dupouey, J.-L., & Ranger, J. (2003). Effects of tree species on understory vegetation and environmental conditions in temperate forests. *Annals of Forest Science, 60,* 823–832.

Bierzydek, P. (1982). Life histories and demography of shade-tolerant temperate forest herbs: a review. *The New Phytologist, 90,* 757–776.

Choung, Y.-S. (1991). Growth characteristics and demography of *Polygonatum involucratum* and *Polygonatum humile* ramet population. *J Ecol Environ, 14,* 305–316.

Cook, R. E. (1983). Clonal plant population. *American Scientist, 71,* 244–253.

De Kroon, H., & Schieving, F. (1991). Resource allocation patterns as a function of clonal morphology: a general model applies to a foraging clonal plant. *Journal of Ecology, 79,* 519–530.

Ehrlén, J., & van Groenendael, J. (2001). Storage and the delayed costs of reproduction in the understorey perennial *Lathyrus vernus. Journal of Ecology, 89,* 237–246.

Emborg, J. (1998). Understorey light conditions and regeneration with respect to the structural dynamics of a near-natural temperature deciduous forest in Denmark. *Forest Ecology and Management, 106,* 83–95.

Hartnett, D. C. (1990). Size-dependent allocation to sexual and vegetative reproduction in four clonal composites. *Oecologia, 84,* 254–259.

Hasegawa, T., & Kudo, G. (2005). Comparisons of growth schedule, reproductive property and allocation pattern among three rhizomatous *Polygonatum* species with reference to their habitat types. *Plant Species Biology, 20,* 23–32.

Hirose, T., & Kachi, N. (1982). Critical plant size for flowering in biennials with special reference to their distribution in a sand dune system. *Oecologia, 55,* 281–284.

Houle, G. (2002). The advantage of early flowering in the spring ephemeral annual plant *Floerkea proserpinacoides. The New Phytologist, 154*(3), 689–694.

Jang, C.-C. (2002). A taxonomic review of Korean *Polygonatum* (Ruscaceae). *Korean Journal of Plant Taxonomy, 32,* 417–447.

Jang, K. H., Park, J. M., Kang, J. H., & Lee, S. T. (1998). Growth and flowering characteristics of *Polygonatum* spp. Journal of Medicinal Crop Science, 6, 142-148.

Jerling, L. (1988). Clone dynamics, population dynamics and vegetation pattern of *Glaux maritima* on a Baltic sea shore meadow. *Vegetatio, 74,* 171–185.

Lee, T. B. (2003). *Coloured flora of Korea* (p. 676). Seoul: Hyangmoon Publishing Co..

Lee, W. T. (1996). *Coloured standard illustrations of Korean plants* (p. 399). Seoul: Academybook Co..

Legner, N., Fleck, S., & Leuschner, C. (2013). Low light acclimation in five temperate broad-leaved tree species of different successional status: the significance of a shade canopy. *Annals of Forest Science, 70,* 557–570.

Klinkhamer, P. G. L., de Jong, T. M., & Meelis, E. (1990). How to test for proportionality in the reproductive effort of plants. *Amer Natl, 135,* 291–300.

Klinkhamer, P. G. L., Meelis, E., de Jong, T. J., & Weiner, J. (1992). On the analysis of size-dependent reproductive output in plant. *Functional Ecology, 6,* 308–316.

Reekie, E. G., & Bazzaz, F. A. (1987). Reproductive effort in plants. 2. Does carbon reflect allocation of other resources? *Am Natl, 129,* 897–906.

Samson, D. A., & Werk, K. S. (1986). Size-dependent effects in the analysis of reproductive effort in plants. *The American Naturalist, 127,* 667–680.

Schemske, D. W., Willson, M. F., Melampty, M. N., Miller, L. J., Verner, L., Schemske, K. M., & Best, L. B. (1978). Flowering ecology of some spring woodland herbs. *Ecol, 59,* 351–366.

Schmid, B., & Weiner, J. (1993). Plastic relationships between reproductive and vegetative mass in *Solidago altissima. Evolution, 47,* 61–74.

Seo, Y.-S., Park, W.-H., & Cha, Y.-Y. (2011). Effects of *Polygonatum odoratum* on lowering lipid and antioxidation. *J Oriental Rehab Med, 21,* 49–62.

Silvertown, J. W. (1982). *Introduction to population ecology* (p. 209). New York: Longman Group Ltd.

Stephenson, A. G. (1981). Flowers and fruit abortion: proximate causes and ultimate functions. *Ann Rev Ecol Syst, 12,* 253–279.

Stears, S. C. (1989). Trade-offs in life-history evolution. *Functional Ecology, 3,* 259–268.

Verburg, R. W., Kwant, R., & Werger, M. J. A. (1996). The effect of plant size on vegetative reproduction in a pseudo-annual. *Vegetatio, 125,* 185–192.

Worley, A. C., & Harder, L. D. (1996). Size-dependent resource allocation and costs of reproduction in *Pinguicula vulgaris* (Lentibulariaceae). *Journal of Ecology, 84,* 195–206.

22

Distribution characteristic of invasive alien plants in Jeju Island

Tae-Bok Ryu, Mi-Jeoung Kim, Chang-Woo Lee, Deok-Ki Kim, Dong-Hui Choi, Hyohyemi Lee, Hye-Ran Jeong, Do-Hun Lee and Nam-Young Kim[*]

Abstract

Background: This study was undertaken to analyze the distribution and ecological characteristics of invasive alien plant species on Jeju Island, and to provide basic data for their management and control.

Results: A field research was conducted at 436 locations on Jeju Island. The field research identified nine species of invasive alien species growing on Jeju Island. Based on the distribution pattern, *Hypochaeris radicata* L., *Rumex acetosella*, and *Ambrosia artemisiifolia* L. were found to be distributed horizontally throughout Jeju Island, with vertical growth in two or more vegetation zones, from warm temperate to the subalpine zone. Widely distributed species penetrate various habitats, such as grasslands, ranches, roadsides, farmlands, and empty lots, and have an immensely negative impact on the ecosystem, including declining biodiversity on Jeju Island. *Paspalum distichum* var. *indutum Shinners*, *Paspalum distichum* L., *Solanum carolinense* L., and *Aster pilosus* Willd. were distributed in some areas as a biased distribution species, whereas *Lactuca scariola* L. and *Solidago altissima* L. were found only in certain areas as centralized distribution species.

Conclusions: The centralized distribution species and biased distribution species of the invasive plants in the ecosystem of Jeju Island should be physically eliminated, keeping in mind the short- and mid-term perspectives and monitoring, and by considering expansion of additional distribution areas. Due to limitations of physical/chemical elimination, time, and cost, widely distributed species require to be eliminated and managed, mainly to restore the integrity of the ecosystem, by planting native species to reestablish the habitat.

Keywords: Invasive alien plants, Jeju Island, Naturalized plants, Alien species

Background

Increasing global exchanges and climate change have led to an increase in national and regional naturalized species pools (Lee et al. 2011). The rapid increase in naturalized plants positively or negatively influences human life and the local ecosystems. Invasive alien species are a threat to local ecosystems and biodiversity, and alien species expertise groups, such as international union for conservation of nature (IUCN) and invasive species specialist group (ISSG), require focused management of invasive alien species (Lowe et al. 2000). In Korea, the group of species that have a negative impact on the local ecosystem, such as *Sicyos angulatus* L. which dominates the water layer and simplifies the lower layer vegetation

* Correspondence: nykim@nie.re.kr
National Institute of Ecology, Seo-Cheon Gun, Chungcheongnam Province
325-813, South Korea

(National Institute of Environmental Research, 2005), is designated as an ecosystem-invasive species (Chapter 5, Article 21 of the Act on the Protection and Utilization of Biodiversity).

Jeju Island has recently had a rapid influx of naturalized plants, accompanied by an increase in tourism demand and intensive land development (Yang and Kim, 2005; Yang 2007a). Since the study of naturalized plants on Jeju Island reported by Nakai (Nakai, 1914), the naturalized plants seem to have gradually increased from 170 taxa (Yang and Kim, 1998) to 183 taxa (Yang et al. 2001a). Until recently, Jeju island was a primary naturalization center for newly naturalized plant species, including *Sisyrinchium micranthum* Cav. (Shin et al. 2016), *Gamochaeta pensylvanica* Willd. *Cabrera* (Ji et al. 2014) and 254 taxa (Kim et al. 2006), that account for about 76% of the 334 domestically introduced species (National Institute of Ecology, 2015).

Preceding literature analysis on the ecological information of naturalized plant species on Jeju Island is inadequate, compared to the results of abundant studies on newly introduced species.

In addition, research of the naturalized plant species pool in Jeju Island has been conducted locally since 2007 (Yang 2007a) (about 10 years ago), but existing distribution information on each plant species is limited. Ecosystem-invasive plant species have been described in the Jeju Island Naturalized Plant Species Pool, but research on regional ecological characteristics, such as the distribution status in Jeju Island and the habitat types, is lacking. This study therefore aims to find out information about ecological and distribution characteristics of ecosystem-invasive alien plants growing in Jeju Island. The information on ecosystem-invasive alien plant species in Jeju Island confirmed through this study will provide basic information for management, prevention, control, biodiversity conservation area, and future global climate change scenario.

Methods
Materials and methods
Located at the southernmost part of the Korean peninsula, Jeju Island is a volcanic island formed between the end of the third and the fourth Cenozoic era (Haraguchi 1931). It has an average annual temperature of 15.8 °C (1981 to 2010; Korea Meteorological Administration, 2016), and average annual precipitation of 1497.6 mm. It has a horizontal warm, evergreen, broad-leaved forest zone according to geographical and climatic factors, and various vegetation zones such as cool temperature zone and subalpine zone are distributed vertically (Kim, 2006). Jeju Island is divided into five ecoregions: Daejeong ecoregion, Seogwipo ecoregion, Sungsanpo ecoregion, Jeju ecoregion, and Halla ecoregion (Kim and Choi, 2012), which are classified as per the micro climate conditions such as climate and precipitation, and each ecoregion represents a unique environmental characteristic. In 2015, 13,600,000 tourists have visited Jeju Island, which depicts increased tourism by 11.3%, as compared to the previous year (Jeju Special Self-Governing Province, 2016). This has therefore resulted in increased pressure for land development. As a result, a variety of naturalized plant species are growing around the damaged habitat, in response to the warm and humid climatic environment and high human intervention.

This study has collected previously published literature information to identify the invasive alien plants in the Jeju Island region. The survey was conducted from March 2015 to October 2016, and a total of 436 sites were examined.

The collected geographical coordinates were mapped using the Arc GIS, and the national names and scientific

names of plant species followed the Korean Plant Names Index (2016) and Korean naturalized plants (Park 2009).

The study a field research on the ecosystem-invasive organisms in Jeju Island conducted in the area for ecosystem-invasive organisms used by the National Ecology Institute of Ministry of Environment (National Institute of Ecology, 2015). It records the distribution information (GPS), habitat types and ecological characteristics of each plant species. A total of 436 sites were examined. The collected geographical coordinates were mapped using the Arc GIS, and the national names and scientific names of plant species followed the Korean plant names index (2016) and Korean naturalized plants (Park 2009).

Result and discussion
Distribution characteristics of invasive alien species of the ecosystem
The results of the field research confirmed the growth of nine species of ecosystem-invasive alien plants designated by the Ministry of Environment: *Lactuca scariola* L., *Ambrosia artemisiifolia* L., *Aster pilosus* Willd, *Hypochaeris radicata* L., *Solidago altissima* L., *Rumex acetosella*, *Solanum carolinense* L., *Paspalum distichum* L., and *Paspalum distichum* var. *indutum* Shinners. Based on the distribution of each plant species, they were divided into three types: wide type, biased type, and centralized type. The wide distribution type is distributed horizontally throughout Jeju Island and grows vertically in one or more climate zones. *Hypochaeris radicata* L., *Rumex acetosella*, and *Ambrosia artemisiifolia* L. belong to this type. The biased distribution type shows a biased distribution in some parts of Jeju Island horizontally and is a species that grows vertically only in one climate zone. *Paspalum distichum* var. *indutum* Shinners, *Paspalum distichum* L., *Solanum carolinense* L., and *Aster pilosus* Willd belong to this type. The centralized distribution type has a limited distribution in some areas in Jeju Island, and *Lactuca scariola* L. and *Solidago altissima* L. belong to this type. Each type of plant species has unique distribution characteristics based on their natural environment and habitat characteristics of Jeju Island.

Hypochaeris radicata L. is distributed horizontally throughout Jeju Island (Fig. 1). It is uniformly distributed in five ecological stations, and large-scale habitats over 1000 m^2 are observed throughout Jeju Island (Fig. 1). It grows vertically mainly in warm temperate regions and also in the cooler temperate and the subalpine zones (Fig. 2). In Jeju Island, *Hypochaeris radicata* L. mainly grows in areas with frequent aperiodic physical disturbance (Lee et al., 2001). The frequent aperiodic disturbance were observed not only at the grassland (52 points; 23.6%), which is the distribution center of *Hypochaeris radicata* L., but also at the edge of the road (110 points; 50%) maintained by

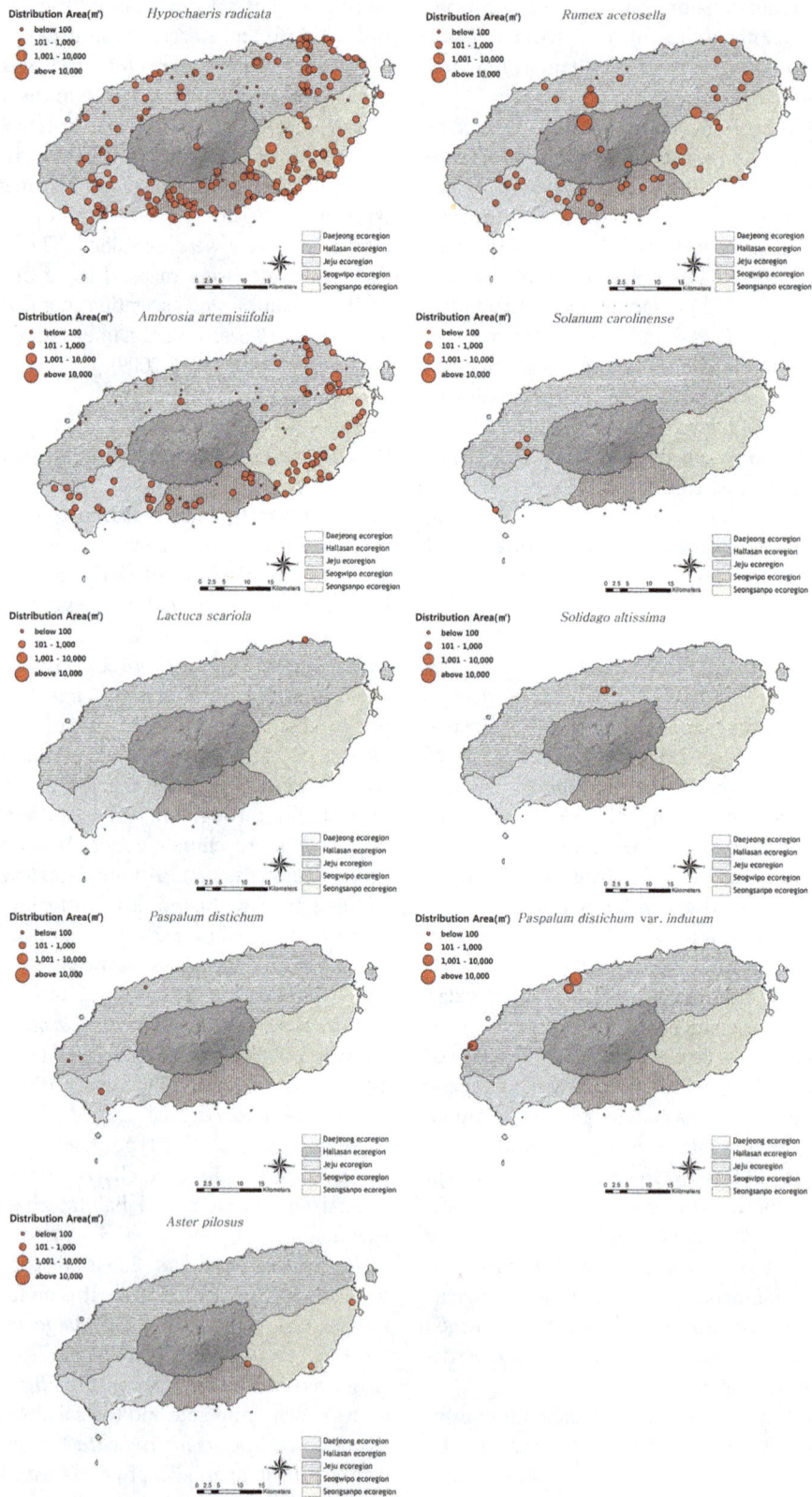

Fig. 1 Distribution map of invasive alien species by ecoregions

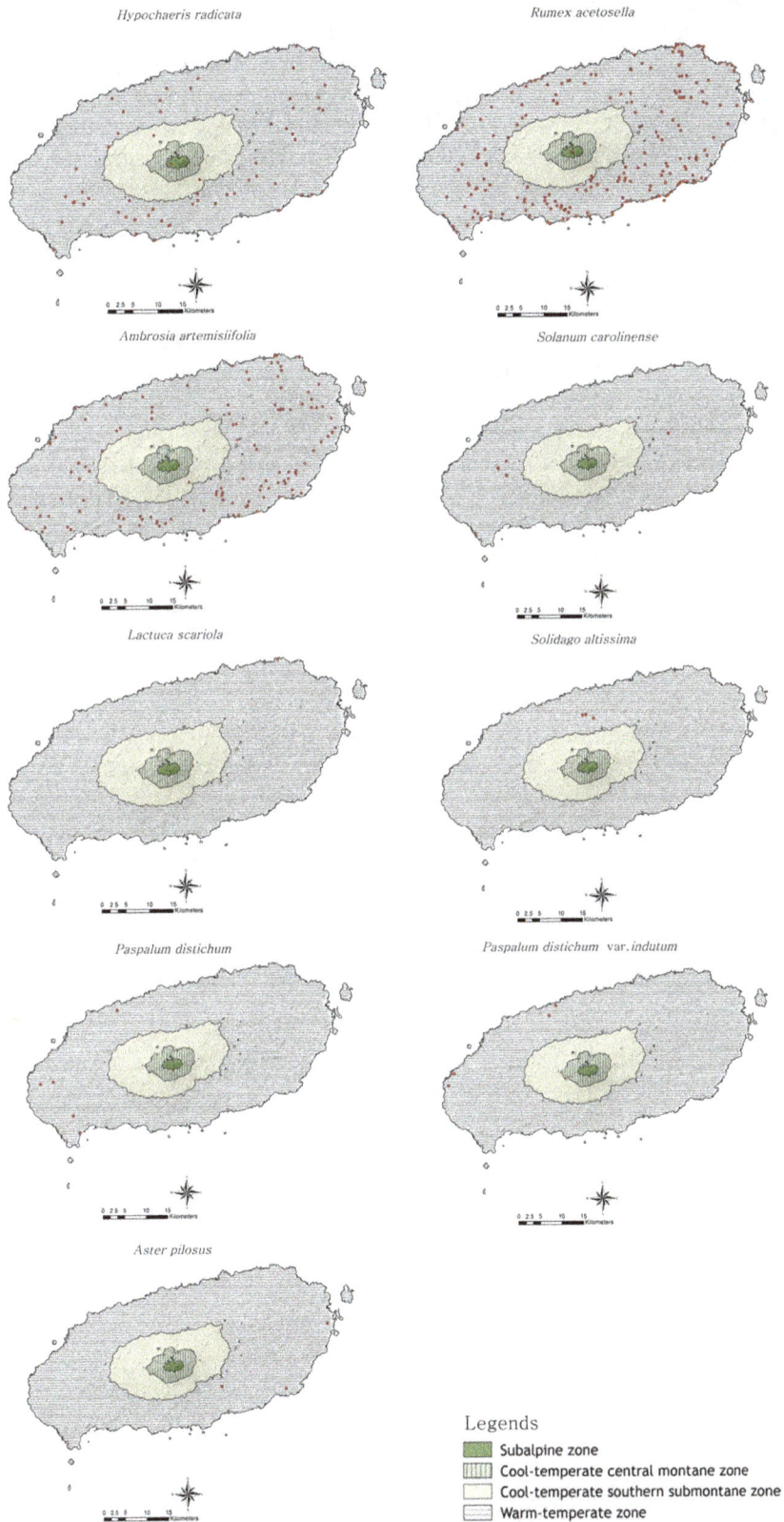

Fig. 2 Distribution map of invasive alien species by temperate zone

aperiodic thinning (Table 1). *Hypochaeris radicata* L. is widely distributed in the temperate climatic regions throughout the world (Turkington and Aarssen, 1983) and is commonly found in temperate grasslands in neighboring Japan (Doi et al. 2006). It is widely distributed in the southern coast of the Korean peninsula or along the coastline to the central region, and has a distribution center vertically in the temperate and southern lowland of cool temperate zones (National institute of Environmental Research, 2006a). In the subalpine region, the invasion of *Hypochaeris radicata* L. causes serious problems in the regional ecosystem species diversity. In the Kosciuszko National Park area in Australia, *Hypochaeris radicata* L. invades and disturbs the ecosystem in the natural vegetation in the alpine and subalpine regions, as well as the secondary vegetation (McDougall et al. 2005; Bear et al. 2006). Especially in the Australian Alps grassland vegetation at over 1200 m above sea level, it results in a qualitative and quantitative imbalance of species diversity (Walsh and McDougall 2004). The subalpine region is a unique ecosystem characterized by low temperature and poor nutrition, which is currently being evaluated as a vulnerable ecosystem due to climate changes (Lee and Kim, 2007). Along with climatic changes, the human intervention in the subalpine regions has resulted in a high possibility of quality decline of the natural ecosystem caused by habitat encroachment, and decrease of biodiversity by *Hypochaeris radicata* L. in Korea. Although this is notable only in the disturbed areas of the subalpine regions due to the continuous elimination project of the National Park Management Corporation, due attention should be given to the penetration into potential natural life. In addition, it is necessary to continue

monitoring the inflow of uncontrolled seeds due to climbers and transportation of supply for resting places, and to prepare adequate countermeasures.

Rumex acetosella was found to be similar in its distribution and ecological properties to *Hypochaeris radicata* L.. It is horizontally distributed throughout Jeju Island and is classified as a wide distribution type grown in five ecological regions (Fig. 1). Vertical distribution is located in the warm temperate zone, southern part of cool temperature zone, and the subalpine area (Fig. 2). It is more frequent in the open grasslands, such as the pasture grassland, as compared to *Hypochaeris radicata* L.; this was similar with a previous study which showed its concentration in grasslands of the Korean peninsula. However, *Rumex acetosella* was most frequently observed at the edge of the road where human intervention, such as aperiodic weeding and trampling, is constant (30 sites; 48.4%). *Rumex acetosella* also affects alpine and subalpine areas such as the Australian alpine region (Johnston and Pickering, 2001) and the European Alps (Den Nijs, 1976). Economic loss due to decline in grass productivity of grasslands of Daekwanlyeong Gangwon-do (Kim et al. 1999) has been actively reported due to indigenization, resulting in a negative influence on human life. It is therefore necessary to establish a control and regulation plan for *Rumex acetosella*, which invades various vegetation environments such as warm temperate, southern cool temperate zone, and subalpine zones, and negatively affects not only the natural ecosystem but also human life.

Ambrosia artemisiifolia L. grow in all five ecoregions and are distributed vertically from the warm temperate zone to the southern part of cool temperate zone (Figs. 1

Table 1 Number of survey areas of invasive alien plants according to the habitat types

	Grassland	Road side	Ruderal	Cultivated land	Wetland	River
Hypochaeris radicata	52 (23.64%)	110 (50%)	45 (20.45%)	13 (5.91%)	.	.
Rumex acetosella	18 (29.03%)	30 (48.39%)	12 (19.35%)	2.00 (3.23%)	.	.
Ambrosia artemisiifolia	26 (19.85%)	65 (49.62%)	29 (22.14%)	10 (7.63%)	.	1 (0.76%)
Solanum carolinense	3 (42.86%)	3 (42.86)	.	1 (14.29)	.	.
Solidago altissima	.	1 (33.33%)	2 (66.67%)	.	.	.
Lactuca scariola	.	.	1 (100%)	.	.	.
Aster pilosus	.	3 (100%)	.	.	.	.
Paspalum distichum	.	.	.	.	5 (100%)	.
Paspalum distichum var. indutum	.	.	.	.	4 (8%)	1 (20%)

and 2). It is categorized into the wide distribution type, which is distributed evenly across Jeju Island. Unlike *Hypochaeris radicata* L. and *Rumex acetosella* which are distributed on the grasslands and the edge of the road, *Ambrosia artemisiifolia* L. is a plant species that recognizes dumping grounds and ruderal vegetation. Although frequently observed alongside the road and in the grasslands at Jeju Island, it is frequently found in the eutrophic sites such as ruderal vegetation and cultivated land (Table 1). Studies have confirmed its distribution in the southern regions of cool temperate zones, and this has been ascertained as a result of its spreading distribution along the edge of the road. Rapid spreading of the distribution area is expected, proportionate to the development of the middle mountain region of Jeju Island. Countermeasures are thus needed for the control and elimination of the roads in the mountainous regions of Jeju Island.

Paspalum distichum L. and *Paspalum distichum* var. *indutum* Shinners are distributed in the Jeju ecoregion and Daejung ecoregion (Fig. 1). Vertically, they grow only in the warm temperate zone (Fig. 2) and are classified as a biased distribution type showing a horizontal biased distribution in the west. Among the 175 inland wetlands (Kang, 2013) of Jeju Island, they only grow in the wetland on the western side. These areas have the lowest rainfall in Jeju Island, being less than 1500 mm (Korea Meteorological Administration, 2012). Considering the distribution characteristics of *Paspalum distichum* L. and *Paspalum distichum* var. *indutum* Shinners that are highly frequent in the southern provinces having more than 1500 mm of annual precipitation, research for distribution in various environments, such as water gates, human interference, microgeomorphology, and microclimate of *Paspalum distichum* L. and *Paspalum distichum* var. *indutum* Shinners in the Jeju Island, is required. Both *Paspalum distichum* L. and *Paspalum distichum* var. *indutum* Shinners infest the wetlands and rivers in the southern part of the Korean peninsula, but they grow mainly in the wetlands of Jeju Island. Since the river in Jeju Island shows physical characteristics of a temporary river (Yang et al. 2014), the habitat of *Paspalum distichum* L. and *Paspalum distichum* var. *indutum* Shinners is inevitably limited, and the direct risk of river ecosystem will not be great. However, there is a need for long-term monitoring of the possibility of the distribution spread to permanent rivers such as the Chungmun stream and the Hyodon stream (Yang 2007b), and the wetlands located on the eastern side of Jeju Island.

Solanum carolinense L. and *Aster pilosus* Willd. are distributed opposingly at Jeju Island. Both species were classified into a biased distribution type with limited distribution in some areas. The distribution of *Solanum*

carolinense L. shows a biased distribution on the eastern side of Jeju Island (Figs. 1 and 2). Depending on the habitat type, it penetrated into the edge of the road and the edge of grasslands and was confirmed to affect the local ecosystem. This is similar to the distribution characteristics (roads, barelands, and pastures) of *Solanum carolinense* L. distributed on the Korean Peninsula (National institute of Environmental Research, 2006b). As mentioned above, this region has the lowest annual precipitation and shows relatively higher temperature in Jeju Island (KMA, 2012). Conversely, distribution of *Aster pilosus* Willd. is limited to the Sungsanpo ecoregion of the southeastern part of Jeju Island. The Sungsanpo ecoregion has relatively high rainfall and warm microclimate characteristics (Korea Meteorological Administration, 2012). Due to Mt. Halla, it is relatively less affected by the northwest monsoon in winter (Korea Meteorological Administration, 2012). However, the association with climatic factors due to small populations and distribution sites should be analyzed through further studies. It develops around the edge of severely disturbed roads.

Lactuca scariola L. and *Solidago altissima* L. were classified as centralized distribution type, growing in a limited region on Jeju Island. *Lactuca scariola* L. was found in less than 30 individuals in the area of Kimneong-li, Gujwa-eup, Jeju City. It was found growing in an empty lot near a wind power plant and barelands, and it is presumed that it has entered Jeju Island along with construction equipment and materials for the wind power plant. *Solidago altissima* L. was observed in Ora-dong, Jeju City. It grows in empty lots around the village and the edge of the road. Although the information on distribution characteristics of *Lactuca scariola* L. and *Solidago altissima* L. could not be confirmed due to its narrow distribution area and small number of species in Jeju Island, a careful approach for the distribution spread is required in the future.

State of invasive alien plants in Jeju Island
Although 10 species of invasive alien plants are listed in the preceding literature of Jeju Island, a total of nine species have been identified through field research (Table 2). As a result of the analysis of preceding literature, some problems describing the invasive alien species in Jeju Island were revealed. In 1998, nine species of invasive alien species were recorded for the first time in Jeju Island (Yang and Kim, 1998). While *Lactuca scariola* L. was first described in 1998 (Yang and Kim, 1998), it was not described in 2001 (Yang et al. 2001a; Yang et al. 2001b), 2003 (Yang 2003), and 2005 (Yang and Kim, 2005). *Lactuca scariola* L. was described in a species pool by Kim et al., but the distribution information is unclear since it failed to describe whether *Lactuca*

Table 2 List of invasive alien plants designated by the Ministry of Environment in Jeju Island

Scientific name	References							This study
	1	2	3	4	5	6	7	
Lactuca scariola	O	.	.	.	.	O	.	O
Solanum carolinense	O	O	O	O	O	O	O	O
Ambrosia artemisiifolia	O	O	O	O	O	O	O	O
Paspalum distichum	O	O	O	O	O	O	O	O
Aster pilosus	O	.	O	O	O	O	O	O
Hypochaeris radicata	O	O	O	O	O	O	O	O
Rumex acetosella	O	O	O	O	O	O	O	O
Solidago altissima	O	O	O	O	O	O	O	O
Paspalum distichum var. indutum	O	O	O	O	O	O	O	O
Sicyos angulatus	.	.	.	.	.	.	O	.

No.1 Yang and Kim 1998, No.2 Yang, Park and Kim. 2001a, No.3 Yang, Park and Kim. 2001b, No.4 Yang 2003, No.5 Yang and Kim 2005, No.6 Kim et al. 2006, No.7 Yang 2007a, 2007b

scariola L. is related to the preceding literature or actual field research results (Kim et al. 2006); there is no mention later in the species pool (Yang 2007a). In conclusion, there is a high possibility that Lactuca scariola L. in Jeju Island was misrepresented and misidentified initially, but it is also likely that it disappeared after its introduction into Jeju Island. These facts raise doubt on the reliability of the initial distribution of Lactuca scariola L. on Jeju Island, but it is necessary to monitor the distribution of the existing Lactuca scariola L.. Aster pilosus Willd. has not been described in one of the two papers by the same author (Yang et al. 2001a), published in 2001 after the first description in 1998 (Yang and Kim, 1998). Sicyos angulatus L. was recorded for the first time in Jeju Island in 2006 and sampling information collected from Ora-dong, Jeju City, is recorded in the National Biological Species Information System. However, this study did not confirm the existence of Sicyos angulatus L. in Jeju Island. Jeju Island has a temporary river in the form of an ephemeral stream for most time of the year, and has a poor development of rivers characterized by short flow channel. Most of the year-round precipitation is concentrated in summer (Korea Meteorological Administration, 2016), and the physical pressures on river channels and spill of water during intensive rainfall are huge. The river bed is composed of sand and gravel having high water permeability, but despite being rivers, they are very dry and have poor nutritional characteristics. On the other hand, Sicyos angulatus L. is preferred at eutrophic sites with high nitrogen content, growing on well-watered river banks and the terrace lands of the river (Kim 2013). As a result, the growth of Sicyos angulatus L. is inevitably limited to the physical system of the river in Jeju, which shows dryness, poor nutrition, and extreme physical disturbance. In

conclusion, it is considered that Sicyos angulatus L. in Jeju Island has disappeared, but there is also the possibility of redistribution by the seed bank in the soil. It is therefore considered to be ephemerophyten, which is not settled in the environment in Jeju Island, and its impact on the river system of Jeju Island is not significant.

Conclusions
Risk and management plans

It is difficult to know precisely about the epicenter and main introduction routes of invasive alien plants in Jeju Island. However, it is estimated that it has been imported through overseas grain imports, tourists, and ports.

The management of alien species at the local level is inevitably different from the national level, and the distribution and ecological characteristics of the alien species should be taken into account. Typically, Rumex acetosella and Ambrosia artemisiifolia L. were considered to be national wide distribution species (WS) in the intrusion grade considering the range of distribution, and were considered to have a negative impact on the overall regional and national ecosystem (Jung 2014). Aster pilosus Willd. is an acutely spreading species and regarded to negatively affect the entire ecosystem of the country. It is considered that the spread of Hypochaeris radicata L. should also be a concern, although the impact on the entire ecosystem is not great (Jung 2014). However, while Hypochaeris radicata L. is one of the invasive alien species that has the widest growth range horizontally and vertically in Jeju Island, and has the greatest influence on the local ecosystem, the ecological status of Aster pilosus Willd. is not excessive, considering the distribution regions in Jeju Island. Hence, it is necessary to establish a management plan for alien species at the national as well as the local level. The elimination and management of invasive alien plants in Jeju Island should be established at the short-, medium-, and long-term levels, corresponding to their centralized distribution species, biased distribution species, and wide distribution species, depending on the distribution regions.

The short-term elimination management plan needs to be implemented on centralized distribution species such as Lactuca scariola L. and Solidago altissima L.. It is estimated that it requires low cost and less time, as these are distributed only in some areas, and should be considered as a short-term priority elimination of species, considering the potential of distribution spread. Considering the size of the species population on Jeju Island, the goal requires the complete elimination by physical removal in less than 5 years, with special attention to prevent the re-introduction and spread after the elimination project. The mid-term elimination plan should be carried out on the biased distribution species such as

Paspalum distichum L., *Paspalum distichum* var. *indutum* Shinners, *Solanum carolinense* L., and *Aster pilosus* Willd.. In addition to the physical elimination, monitoring and spread prevention programs should be carried out considering the distribution regions. Particular attention should be paid to the spreading and distribution outside the current area. After the physical elimination, incineration and treatment of the plant should be undertaken inside the distribution area to prevent the potential spread of seeds. Elimination projects should be carried out from the mid-term point of view, such as 10 years or less, including monitoring the redistribution. The long-term elimination management plan should be carried out on wide distribution species such as *Rumex acetosella*, *Aster pilosus* Willd., and *Ambrosia artemisiifolia* L.. Considering the existing distribution and populations, it is necessary to establish a long-term management system of more than 10 years, which involves a high cost and longer duration for the elimination. In addition to the physical/chemical elimination project, monitoring and spread prevention program, it is necessary to focus on restoration of the primary habitat and ecosystem. This is because in the case of wide distribution species, complete eradication using only physical/chemical elimination is considered impossible due to buried seeds in the soil and distribution region. Hence, it is critical that the restoration of disturbed habitat and improvement of habitat integrity and restoration through native plant planting should be induced. It is supposed that the spread distribution and growth of invasive alien species should be controlled by managing the potential habitat of invasive alien species.

Acknowledgements

This study was supported by "Ecological Studies of Alien Species and The study on the Inhabitation status of Nutria (*Myocastor coypus*) through Ministry of Environment" (NIE-2016-8). The authors are grateful to the reviewers for their constructive comments for the improvement of the earlier version of the manuscript.

Funding

This study was supported by Ecological Studies of Alien Species and The study on the Inhabitation status of Nutria (*Myocastor coypus*) through Ministry of Environment (NIE-2016-8).

Authors' contributions

TBR carried out the field studies and drafted the manuscript. MJK participated in the design of the study. CWL, DGK, DHC, HRJ, HHL, and DHL researched the field. NYK conceived of the study, participated in its design, and helped to draft the manuscript. All authors read and approved the final manuscript.

Competing interests

The authors declare that they have no competing interests.

References

Bear, R., Hill, W., & Pickering, C. M. (2006). Distribution and diversity of exotic plant species in montane to alpine areas of Kosciuszko National Park. *Cunninghamia, 9*(4), 559–570.

Den Nijs, J. C. M. (1976). Biosystematic studies of the Rumex acetosella complex II. The alpine region. *Acta Botanica Neerlandica, 25*(6), 417–447.

Doi, M., Ito, M., & Auld, B. A. (2006). Growth and reproduction of Hypochoeris radicata L. *Weed Biol Manag, 6*(1), 18–24.

Haraguchi, K. (1931). Geology of Saishu (Jeju) Island. *Bull Geol Surv Korea, 10,* 1–34. in Japanese.

Jeju Special Self-Governing Province(2016) Research for tourist information. Accessed Oct. 1. 2016. http://www.jeju.go.kr/(Jeju Special Self-Governing Province)

Ji, S. J., Jung, S. Y., Hong, J. K., Hang, H. S., Park, S. H., Yang, J. C., Chang, K. S., Oh, S. H., & Lee, Y. M. (2014). Original article: two newly naturalized plants in Korea: *Euthamia graminifolia* (L.) Nutt. and *Gamochaeta pensylvanica* (Willd.) Cabrera. *Kor J Plant Taxon, 44,* 13–17 (in Korean with English abstract).

Johnston, F. M., & Pickering, C. M. (2001). Alien plants in the Australian Alps. *Mt Res Dev, 21,* 284–291.

Jung, S. Y. (2014). *A study on the distribution characteristics of Invasive Alien Plant (IAP) in South Korea* (p. 234). Andong: Ph. D. Dissertation, Univ. of Andong. in Korean with English abstract.

Kang, D. H. (2013). *Flora of Aquatic and Wetland habitats in Jeju Island, MS thesis.* Jeju: Univ. of Jeju. 200 pp. (in Korean with English abstract).

Kim, C. S., Koh, J. G., Song, G. P., Moon, M. O., Kim, J. E., Lee, E. J., Hwang, S. I., & Jeong, J. H. (2006). Distribution of naturalized plants in Jeju Island, Korea. *Korean J Plant Res, 19*(5), 640–648 (in Korean with English abstract).

Kim, J. W. (2006). *Vegetation ecology* (2nd ed.). Seoul: Worldscience Press. 340 pp. (in Korean).

Kim, J. W. (2013). *The plant book of Korea Vol. 1* (Plants living close to the village, p. 1200). Seoul: Econature. in Korean.

Kim, J. W., & Choi, B. K. (2012). *Discovering the essence of the Korean vegetation for filed excursion: sprit of place, Korea.* Seoul: Worldscience Press. 164 pp.

Kim, S. M., Kim, Y. H., Hwang, K. H., Ahn, M. S., & Hur, J. H. (1999). Response of red sorrel (*Rumex acetosella* L.) to several soil-and foliar-applied herbicides. *Korean J Pestic Sci, 3*(3), 45–53 (In Korean with English abstract).

Korea Meteorological Administration. (2012). *The climate atlas of Korea* (p. 174). Seoul: KMA. in Korean.

Korea Meteorological Administration. (2016) Research for Climate information. Accessed Oct. 1. 2016. http://www.kma.go.kr.

Korean plant names index(2016) Research for Plant Information. Accessed Oct. 1. 2016. http://www.nature.go.kr/kpni/index.do

Lee, D. G., & Kim, J. U. (2007). Vulnerability assessment of sub-alpine vegetations by climate change in Korea. *Korea Soc Environ Restor Revegetation Technol, 10,* 110–119 (In Korean with English abstract).

Lee, I. Y., Park, J. H., Onet, S. M., Kim, C. S., Moon, B. C., Kim, S. T., & Jeong, J. W. (2001). Geographical distribution and characteristics of seed germination and rhizomes growth of *Rumex acetosella*. *Korean J Weed Sci, 21,* 259–267. In Korean with English abstract.

Lee, Y. M., Park, S. H., Jung, S. Y., Oh, S. H., & Yang, J. C. (2011). Original article: study on the current status of naturalized plants in South Korea. *Korean J Plant Taxon, 41,* 87–101 (In Korean with English abstract).

Lowe, S., Browne, M., Boudjelas, S., & De Poorter, M. (2000) 100 of the world's worst invasive alien species: a selection from the global invasive species database. Invasive species specialist group (ISSG), A specialist group of the Species Survival Commission (SSC) International union for conservation union (IUCN), University of Auckland, New Zealand (http://www.issg.org/booklet.pdf).

McDougall, K. L., Morgan, J. W., Walsh, N. G., & Williams, R. J. (2005). Plant invasions in treeless vegetation of the Australian Alps. *Perspect Plant Ecol, Evol Syst, 7*(3), 159–171.

Nakai, T. (1914). *Flora of Quelpaert and Wando Island.* Chosen, Seoul: Govern. 164 pp. (in Japanese).

National Institute of Ecology. (2015). *Guidebook on nationwide survey of non-native species in Korea* (p. 87). Seocheon: NIE. in Korean.

National institute of Environmental Research. (2005). *Effects of ecosystem disturbance wildplants on ecosystem and their management(1).* Incheon: National institute of Environmental Research. 76 pp. (in Korean with English abstract).

National institute of Environmental Research. (2006a). *A study of detailed survey on invasive alien species in Korea and designation of invasive alien species in foreign countries.* Incheon: National institute of Environmental Research. 408 pp. (in Korean with English abstract).

National institute of Environmental Research. (2006b). *Spread and management scheme of solanum carolinense as an ecosystem disturbance wildplant.* Incheon: National institute of Environmental Research. 178 pp. (in Korean with English abstract).

Park, S. H. (2009). *New illustrations and photographs of naturalized plants of Korea* (p. 575). Seoul: Ilchokak. in Korean.

Shin, H. W., Kim, M. J., & Lee(, N. S. (2016). First report of a newly naturalized Sisyrinchium micranthum and a taxonomic revision of Sisyrinchium rosulatum in Korea. *Korean J Plant Tacon, 46*(3), 295–300. in Korean with English abstract.

Turkington, R., & Aarssen, L. W. (1983). Biological flora of the British Isles. No. 156. *Hypochoeris radicata* L. (*Achyrophorus radicatus* (L.) Scop.). *J Ecol, 71*(3), 999–1022.

Walsh, N. G., & McDougall, K. L. (2004). Progress in the recovery of treeless subalpine vegetation in Kosciuszko National Park after the 2003 fires. *Cunninghamia, 8,* 439–452.

Yang, S. (2007a). River management and improvement plan in Jeju island. *Riv Cult, 3*(4), 105–115 (in Korean).

Yang, S. K., Kim, D. S., & Jung, W. Y. (2014). Rainfall-Runoff characteristics in a Jeju stream considering antecedent precipitation. *J Environ Sci, 23,* 553–560 (in Korean with English abstract).

Yang, Y. H. (2003). *Studies on the distribution and vegetation of naturalized plants on Jeju Island* (p. 108). Jeju: Ph. D. Dissertation, Univ. of Jeju. in Korean with English abstract.

Yang, Y. H (2007a) Studies on the Vegetation of Naturalized Plants in Jeju Island. Korean Journal of Weed Science 27: 112-121. (in Korean with English abstract)

Yang, Y. H., & Kim, M. H. (1998). A study on the naturalized plants in Cheju Island. *Cheju J Life Sci, 1,* 49–58 (in Korean with English abstract).

Yang, Y. H., & Kim, M. H. (2005). The restudying of naturalized plants in Jeju Island. *Korean J Plant Res, 18,* 325–336 (in Korean with English abstract).

Yang, Y. H., Park, S. H., & Kim, M. H. (2001a). The restudying of naturalized plants in Jejudo. *The J Basic Sci Cheju Natl Univ, 14,* 53–62 (in Korean with English abstract).

Yang, Y. H., Park, S. H., & Kim, M. H. (2001b). The flora of naturalized plants in Jeju Island. *Korean J Plant Res, 14*(3), 277–285 (in Korean with English abstract).

23

Comparison of IgE induction in mice by pollens from three pine tree species

Seo-Yoong Kim[1], In-Bo Oh[2] and Kee-Ryong Choi[1]*

Abstract

Background: Over the years, pine pollens have been excluded as an allergen due to its relatively large size, low protein content, and waxy hydrophobic layer, despite their abundance. However, recent studies suggest the possibilities of pine pollens being allergens, and it has been reported that allergy symptoms were highly prevalent in areas with considerably large pine forests and high possibility of exposure to the pollen. Therefore, we conducted a comparative analysis of the allergenicities of the pollens from the dominant species of Korean pines, red pine (*Pinus densiflora*), black pine (*Pinus thunbergii*), and pitch pine (*Pinus rigida*), in mice.

Methods: The protein composition of the pollens from the three pine species was compared via sodium dodecyl sulfate polyacrylamide gel electrophoresis (SDS-PAGE). The pine pollens and proteins extracted from the pollens were introduced to BALB/c mice by nasal inhalation and application to exposed skin and the IgE produced by the mice were extracted from blood and analyzed via ELISA.

Results: SDS-PAGE showed differing protein compositions of the pollens of the three pine species. Analysis of blood IgE compositions showed a similar amount of IgE produced when pollens were applied to skin. In contrast, when mice inhaled the pollens, *P. densiflora* was shown to induce significantly more IgE production than those of the other two species.

Conclusions: The experimental results demonstrate that the pollens of all three South Korean pine species induce IgE production, and this production was more pronounced when the pollens were inhaled than when they were applied to the skin. Of the three species, the pollen of *P. densiflora* was found to induce the highest level of IgE production.

Keywords: Pollinosis, Allergen, IgE, Red pine, Black pine, Pitch pine

Background

Several characteristics of pine pollen, such as its relatively large size (45–65 μm), low protein content, and waxy hydrophobic layer, have led researchers to exclude it as a possible cause of allergic reactions, despite its abundance (Howlett et al. 1981; Pettyjohn and Levetin 1997). It has even been used as a negative control in inhalation challenge tests (Frølund et al. 1986). However, some studies have suggested the possibility of pine pollens being allergens (Walker 1921; Rowe 1939; Newmark and Itkin 1967; Harris and German 1985; Cornford et al. 1988; Kalliel and Settipane 1988; Cornford et al. 1990). In addition, significantly high rates of allergy symptoms have been reported in areas with considerably large pine forests and, hence, a high possibility of exposure to pine pollen (Farnham and Vaida 1982; Farnham 1988; Fountain and Cornford 1991; Freeman 1993; Gastaminza et al. 2009). In South Korea, it has been reported that 7.31% of asthmatic patients and 16.9% of allergic coryza and conjunctivitis patients test positive to pine pollen as an allergen (Hong 2015).

There are over 90 species belonging to the *Pinus* genus worldwide (Lawrence 1951). Korean forests are composed of 40.5% coniferous forests, 27% broadleaf forests, and 29.3% mixed stand forests. These coniferous forests are primarily dominated by pines, including *Pinus densiflora* (56.1%), *Pinus rigida* (15%), *Pinus koraiensis* (8.3%), *Cryptomeria japonica*, and *Chamaecyparis obtusa* (3.4%) (Korean Forest Service 2011). It has also been

* Correspondence: pollen@ulsan.ac.kr
[1]Department of Biological Science, University of Ulsan, Ulsan 44610, South Korea
Full list of author information is available at the end of the article

determined that among the airborne pollens in South Korea, pine pollen disperses for the longest period at the highest abundance (Oh et al. 2000; Choi et al. 2011; Jung and Choi 2013; Choi et al. 2014). Studies have reported differences in the antigenicities of the pollens of various allergenic plants. Therefore, detailed information regarding the antigenicity of each species is needed (Calenoff et al. 1990; Park et al. 1999; Shahali et al. 2007; Cox et al. 2009, 2011). The purpose of this study was to investigate the antigenicity of pine pollen, especially *P. densiflora*, *P. thunbergii*, and *P. rigida* pollen, derived from Korean pine plants. To analyze the protein allergens of these pollens, three types of proteins were compared by SDS-PAGE. We also observed the changes in IgE expression induced by pine pollens in mice after nasal inhalation and skin application (Tamura et al. 1986; Tordesillas et al. 2015).

Methods
Collection of pine pollens from *P. densiflora*, *P. thunbergii*, and *P. rigida* and preparation of pollen antigens
The pine pollens used in this study were collected from *P. rigida*, *P. densiflora*, and *P. thunbergii* samples from the University of Ulsan (St. A), Mt. Munsu (St. B), Guyu-dong (St. C), and Yeonam-dong (St. D), South Korea, respectively from April to May, 2014 (Fig. 1).

Male flowers were collected and dried in the shade for 7 days. From these, pollen was extracted through a sieve with a pore size of 150 µm. Next, 5 g of each pollen sample was added to 5 volumes of ethyl ether (w/v 1:5), stirred twice every 12 h at 4 °C, and air-dried for 12 h to completely remove ethyl ether. The completely defatted sample was then added to phosphate-buffered saline (PBS; pH 7.8) in a w/v ratio of 1:5 and stirred for 24 h at 4 °C, after which it was centrifuged for 30 min at 21,000g at 4 °C. The protein-containing supernatant was extracted and dialyzed by adding the sample to a dialysis tube and placing the tube in distilled water at 4 °C for 48 h, during which the water was changed 2–3 times. Then, the samples were filtered with a 0.45-µm Millipore filter and stored at − 20 °C. Using the Thermo Scientific Pierce bicinchoninic acid (BCA) Protein Assay, the presence of proteins in the sample was confirmed by a change in color, and the amount of protein was quantified and standardized based on the absorbance of each sample.

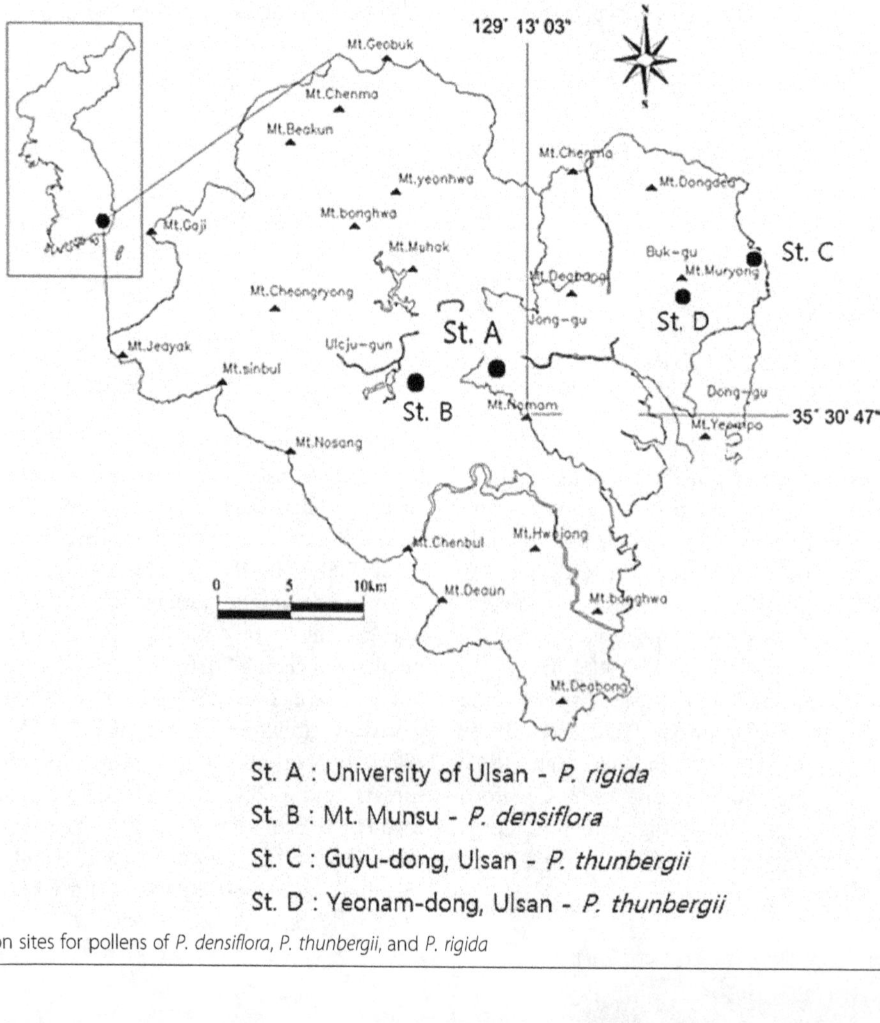

St. A : University of Ulsan - *P. rigida*

St. B : Mt. Munsu - *P. densiflora*

St. C : Guyu-dong, Ulsan - *P. thunbergii*

St. D : Yeonam-dong, Ulsan - *P. thunbergii*

Fig. 1 The collection sites for pollens of *P. densiflora*, *P. thunbergii*, and *P. rigida*

SDS-PAGE of pine pollen antigens

Polyacrylamide gels were made at 10%, 15% (separating gels), and 5% (stacking gels), and the protein samples were mixed with SDS gel-loading buffer (50 mM Tris·Cl, 2% SDS, 0.1% bromophenol blue, 10% glycerol, and 1% β-mercaptoethanol) and boiled for 3 min to prepare for electrophoresis. Standard size markers and the samples were added to the wells and run at 80 V for 3 h in a Tris-glycine electrophoresis buffer system (25 mM Tris, 250 mM glycine, and 0.1% SDS). Then, the gel was stained in a staining solution (0.25% Coomassie Brilliant Blue R250 and 10% glacial acetic acid in methanol: H_2O 1:1 v/v) and destained with a destaining solution (methanol: glacial acetic acid:H_2O 3:1:6) that was changed 3 times. SDS-PAGE was used to conduct a comparative analysis of the proteins of the pine pollens of *P. densiflora*, *P. thunbergii*, and *P. rigida*. Gels of different concentrations (5, 10, and 15%) were used to broaden the size range of the proteins that could be analyzed.

Mice

Female Balb/c mice aged 7 weeks were purchased from the Dae-Han Bio Link Co. Ltd. They were maintained under specific pathogen-free (SPF) conditions in the animal facility of the University of Ulsan and were used at 8–10 weeks.

Detection of changes in IgE expression induced by pine pollens in mice

In order to assess the induction of IgE expression in a mouse system by pine pollens, the procedure described in Fig. 2 was carried out. That is, for nasal inhalation tests, 0.05 g of pine pollen from *P. densiflora*, *P. thunbergii*, or *P. rigida* was added to 1 mL of sterilized distilled water and mixed by vortexing. This produced a homogeneous mixture of a 0.05 g/mL solution; 25 μL (1.25 g) was injected directly into the nasal cavity of each mouse (*n* = 5) daily for 15 days. Another method

involved a skin test during which each anesthetized mouse had its back fur removed with clippers. Then, 100 μL of pine pollen proteins from *P. densiflora*, *P. thunbergii*, or *P. rigida* was applied to the skin daily for 15 days. In the control group, 25 μL of sterilized distilled water was injected into the nasal cavity of each mouse once daily for 15 days. On days 5, 10, and 15 after introduction of pine pollens, capillary tubes were used to extract blood from the capillary vessels of the eyes of mice in the control group, nasal inhalation test group, and skin test group. The blood samples were centrifuged at 800*g* for 15 min, and the supernatant was extracted to

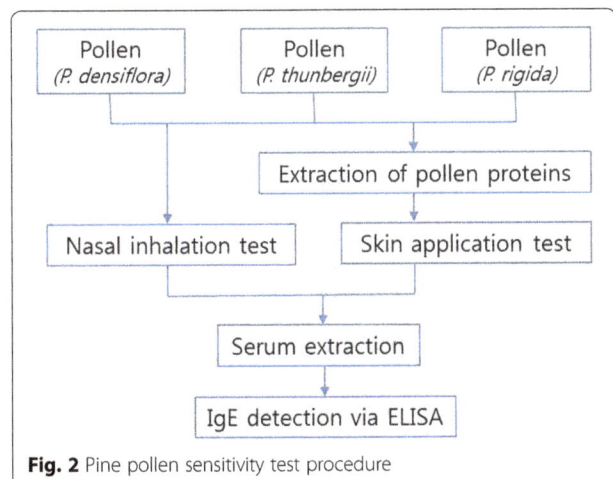

Fig. 3 Analysis of South Korean pine pollen extracts using 5% polyacrylamide gel. Lane 1, SDS-PAGE marker; lane 2, *P. densiflora* extract; lane 3, *P. thunbergii* extract; lane 4, *P. rigida* extract

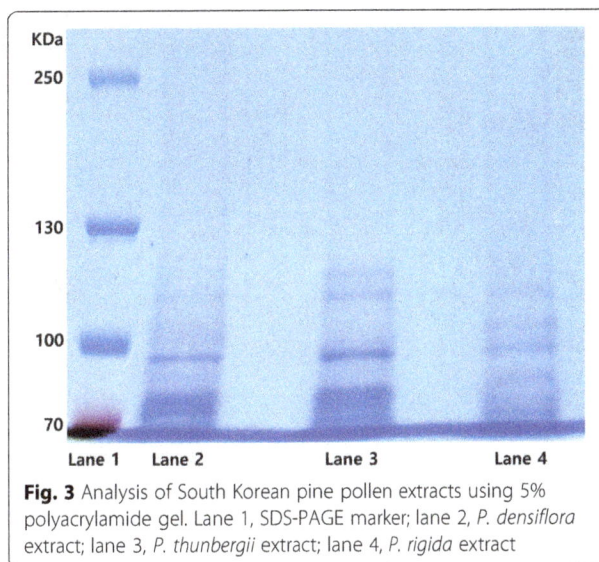

Fig. 2 Pine pollen sensitivity test procedure

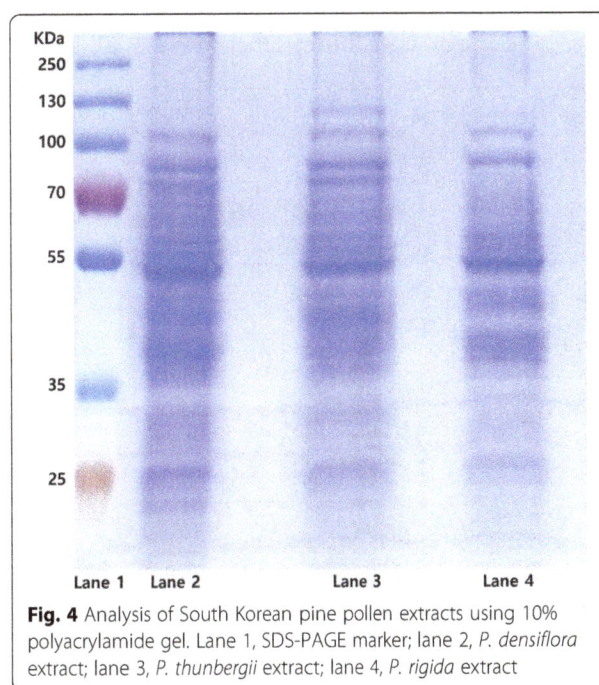

Fig. 4 Analysis of South Korean pine pollen extracts using 10% polyacrylamide gel. Lane 1, SDS-PAGE marker; lane 2, *P. densiflora* extract; lane 3, *P. thunbergii* extract; lane 4, *P. rigida* extract

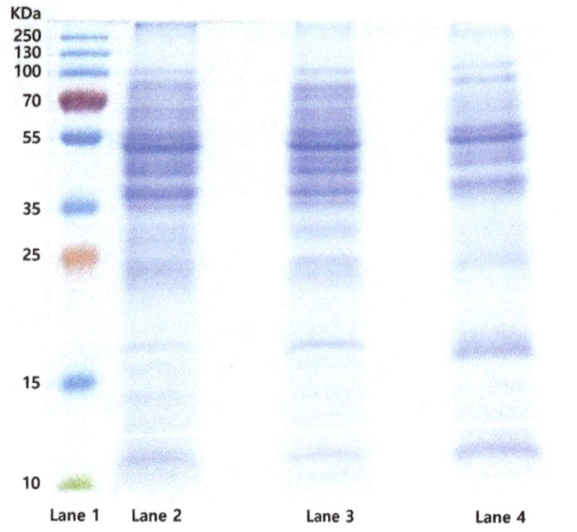

Fig. 5 Analysis of South Korean pine pollen extracts using 15% polyacrylamide gel. Lane 1, SDS-PAGE marker; lane 2, *P. densiflora* extract; lane 3, *P. thunbergii* extract; lane 4, *P. rigida* extract

isolate serum, which was stored at − 20 °C. A Mouse IgE ELISA kit (ebioscience mouse IgE ELISA Ready-SET-Go) was used to detect changes in IgE composition in blood serum samples of each group. A 96-well microplate was coated with IgE-capture antibody (Ab) and incubated at 4 °C for 18 h. To each well, 200 µL of PBS-T (Tween 20, 0.05%) was added and removed three times for washing. Then, 100 µL of 2% bovine serum albumin (BSA) was added to each well and incubated at 20 °C for 2 h to prevent unspecific binding. After washing each well three times with 200 µL PBS-T, 50 µL uniformly diluted mouse serum was added to each well and incubated for 2 h. After washing each well three times with 200 µL PBS-T,

IgE-detection Ab was added and incubated at 20 °C for 1 h. After washing each well three times with 200 µL PBS-T, horseradish peroxidase-bound secondary Ab was added and incubated at 20 °C for 30 min. After washing each well three times with 200 µL PBS-T, 100 µL of a substrate for horseradish peroxidase, TMB (3,3′,5,5′-tetramethylbenzidine), was added per well to induce a color change. Then, 100 µL of 0.16 M sulfuric acid stop solution was added to each well to stop the color change, and the absorbance was read at 450 nm using a microplate reader.

Results and discussion
Analysis of pine pollen protein
After 5% SDS-PAGE, analysis of protein bands above 70 kDa showed different band patterns at approximately 80 and 120 kDa for *P. densiflora*, *P. thunbergii*, and *P. rigida* (Fig. 3). Next, 10% SDS-PAGE was run to analyze proteins of 35–70 kDa. Even though an exact comparison of the protein bands of the pine species was difficult owing to the large number of proteins in this size range, it was clear that *P. densiflora*, *P. thunbergii*, and *P. rigida* all showed similar band patterns (Fig. 4). Finally, 15% SDS-PAGE was run to analyze proteins under 35 kDa. The pollens from the three species showed different band patterns at approximately 30 and 17 kDa (Fig. 5). According to previous data, the major antigens in pine pollen are of 42 and 6–8 kDa (Weber 2005). In addition, in the case of *Pinus radiata*, it has been reported that the main allergens are of 140, 85, 70, 55, 42, 32, 22, 19, and 6–8 kDa (Cornford et al. 1988; Gastaminza et al. 2009). The three species that were used in this experiment showed different band patterns at approximately 17, 30, 80, and 120 kDa. This indicates that the three

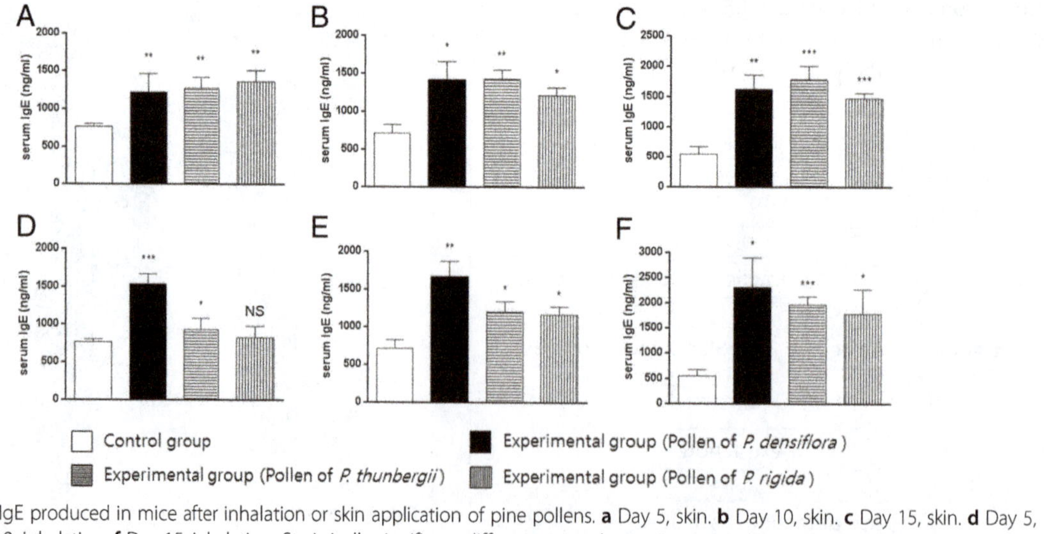

Fig. 6 Analysis of IgE produced in mice after inhalation or skin application of pine pollens. **a** Day 5, skin. **b** Day 10, skin. **c** Day 15, skin. **d** Day 5, inhalation. **e** Day 10, inhalation. **f** Day 15, inhalation. Statistically significant differences are shown. *$p < 0.05$, **$p < 0.01$, ***$p < 0.001$ vs control. *NS*, not significant

species exhibit differences in antigenicity. It is necessary to identify the region of each of these proteins that acts as an allergen by binding to IgE. Future studies should also identify which protein band patterns cause allergic reactions using sera extracted from patients showing allergic reactions after exposure to pine pollens.

Comparison of IgE induction by pine pollens

Considering that the dispersion of pine pollens is most intense for about 2 weeks, IgE induction experiments were conducted on mice for 15 days to induce allergic reactions via inhalation and skin application (Tamura 1986; Conford et al. 1990; Maejima et al. 2000; Mahler et al. 2000; Seitzer et al. 2003a, b; Mondoulet et al. 2010, 2012; Tordesillas et al. 2015). Control mice showed no change in the IgE compositions of blood sera collected on days 5, 10, and 15 (Fig. 6). The experimental groups that underwent nasal inhalation and skin tests showed higher levels of IgE than the control group, and these levels increased with the duration of exposure (Fig. 6). In the skin test group, pollen from all three species induced IgE production in mice on the 5th day, and levels increased by similar amounts until the 15th day (Fig. 6a–c). In the nasal inhalation test, mice that inhaled the pollen of *P. densiflora* showed significant increases in IgE production on the 5th and 10th days, but those that inhaled the pollens of the other two species showed only very small increases in IgE production (Fig. 6d, e). However, mice that inhaled each of the three species showed significant increases in IgE production by the 15th day, such that all three groups showed a similar amount of IgE produced by the end of the experiment (Fig. 6f) In addition, on the 15th day, all groups that had inhaled pine pollen exhibited higher IgE levels than those in which pollen was applied to the skin (Fig. 6c, f). Thus, our results indicate that *P. densiflora*, *P. thunbergii*, and *P. rigida* pollen from South Korea can cause allergic reactions in animal models.

Conclusions

The experimental results demonstrate that the pollens of all three South Korean pine species induce IgE production, and this production was more pronounced when the pollens were inhaled than when they were applied to the skin. Of the three species, the pollen of *P. densiflora* was found to induce the highest level of IgE production.

Abbreviations

Ab: Antibody; BCA: Bicinchoninic acid; BSA: Bovine serum albumin; ELISA: Enzyme-linked immunosorbent assay; IgE: Immunoglobulin E; *P. densiflora*: Pinus densiflora; *P. rigida*: Pinus rigida; *P. thunbergii*: Pinus thunbergii; PBS: Phosphate-buffered saline; PBS-T: Phosphate-buffered saline Tween 20; SDS: Sodium dodecyl sulfate; SDS-PAGE: Sodium dodecyl sulfate polyacrylamide gel electrophoresis; SPF: Specific pathogen-free; TMB: 3,3',5,5'-tetramethylbenzidine; *w/v*: Weight-to-volume

Acknowledgements
This research was supported by the Environmental Health Center of Ulsan University Hospital of the Ministry of Environment, Korea.

Funding
This work was supported by the 2016 Research Fund of the University of Ulsan.

Authors' contributions
SY conducted the experiments, analyzed the data, and wrote the manuscript. IB participated and designed the research. KR mainly designed the research and revised the paper totally. All authors read and approved the final manuscript.

Competing interests
The authors declare that they have no competing interests.

Author details
[1]Department of Biological Science, University of Ulsan, Ulsan 44610, South Korea. [2]Environmental Health Center, University of Ulsan College of Medicine, Ulsan 44033, South Korea.

References
Calenoff, E., Beigler, M. A., Friesen, G. L., & Nichols, J. L. (1990). *U.S. patent no. 4,963,356* (pp. 1–30). Washington, DC: U.S. Patent and Trademark Office.

Choi, S. H., Jung, I. Y., Kim, D. Y., Kim, Y. H., Lee, J. H., Oh, I. B., & Choi, K. R. (2011). Seasonal distribution of airborne pollen in Ulsan, Korea in 2009-2010. *Journal of Ecology and Field Biology, 34*(4), 371–379.

Choi, S. W., Lee, J. H., Kim, Y. H., Oh, I. B., & Choi, K. R. (2014). Association between the sensitization rate for inhalant allergens in patients with respiratory allergies and the pollen concentration in Ulsan, Korea. *The Korean Journal of Medicine, 86*(4), 453–461.

Cornford, C. A., Fountain, D. W., & Burr, R. G. (1990). IgE-binding proteins from pine (*Pinus radiata* D. Don) pollen: evidence for cross- reactivity with ryegrass (*Lolium perenne*). *International Archives of Allergy and Immunology, 93*, 41–46.

Cornford, C. A., Fountain, D. W., Burr, R. G., & O'leary, L. (1988). Hayfever in university students (letter). *The New Zealand Medical Journal, 101*, 520.

Cox, L., Esch, R. E., Corbett, M., Hankin, C., Nelson, M., & Plunkett, G. (2011). Allergen immunotherapy practice in the United States: guidelines, measures, and outcomes. *Annals of Allergy, Asthma & Immunology, 107*, 289–302.

Cox, L., & Jacobsen, L. (2009). Comparison of allergen immunotherapy practice patterns in the United States and Europe. *Annals of Allergy, Asthma & Immunology, 103*, 451–459.

Farnham, J. E. (1988). A new look at conifer allergy (editorial). *Allergy and Asthma Proceedings, 9*(3), 237–238.

Farnham, J. E., & Vaida, G. A. (1982). A new look at New England tree pollen. *Allergy and Asthma Proceedings, 3*(2), 320–326.

Fountain, D. W., & Cornford, C. A. (1991). Aerobiology and allergenicity of *Pinus radiata* pollen in New Zealand. *Grana, 30*, 71–75.

Freeman, G. L. (1993). Pine pollen allergy in northern Arizona. *Annals of Allergy, 70*(6), 491–494.

Frølund, L., Madsen, F., Gerner Svendsen, U., & Weeke, B. (1986). Reproducibility of standardized bronchial allergen provocation test. *Allergy, 41*(1), 30–36.

Gastaminza, G., Lombardero, M., Bernaola, G., Antepara, I., Munoz, D., Gamboa, P. M., Audicana, M. T., Marcos, C., & Ansotegui, I. J. (2009). Allergenicity and cross-reactivity of pine pollen. *Clinical & Experimental Allergy, 39*, 1438–1446.

Harris, R. M., & German, D. F. (1985). The incidence of pine pollen reactivity in an allergic atopic population. *Annals of Allergy, 55*, 678–679.

Hong, C. S. (2015). Pollen allergy plants in Korea. *Allergy, Asthma & Respiratory Disease, 3*(4), 239–254.

Howlett, B. J., Vithanage, H. I. M. V., & Knox, R. B. (1981). Pollen antigens, allergens, and enzymes. *Commentaries in plant science, 2*(19), 191–207.

Jung, I. Y., & Choi, K. R. (2013). Relationship between airborne pollen concentrations and meteorological parameters in Ulsan, Korea. *Journal of Ecology and Environment, 36*(1), 65–71.

Kalliel, J. N., & Settipane, G. A. (1988). Eastern pine sensitivity in New England. *Allergy and Asthma Proceedings, 9*(3), 233–235.

Korea forest service. (2011). *Statistical yearbook of forest 41* (pp. 32–33).

Lawrence, G. H. (1951). *Taxonomy of vascular plants.* New York: Macmillan.

Maejima, A., Tamura, K., Taniguchi, Y., & Saito, S. (2000). Study on the inhalation exposure conditions of pollen for inducing the production of Japanese cedar (Cryptomeria Japonica) pollen-specific IgE antibody in mice. *Jpn J. Palynol, 46*(2), 155–161.

Mahler, V., Diepgen, T. L., KuB, O., Leakakos, T., Truscott, W., Schuler, G., Kraft, D., & Valenta, R. (2000). Mutual boosting effects of sensitization with timothy grass pollen and latex glove extract on IgE antibody responses in a louse model. *The Journal of Investigative Dermatology, 114*(5), 1039–1043.

Mondoulet, L., Dioszeghy, V., Ligouis, M., Dhelft, V., Dupont, C., & Benhamou, P. H. (2010). Epicutaneous immunotherapy on intact skin using a new delivery system in a murine model of allergy. *Clinical & Experimental Allergy, 40*(4), 659–667.

Mondoulet, L., Dioszeghy, V., Ligouis, M., Dhelft, V., Puteaux, E., Dupont, C., & Benhamou, P. H. (2012). Epicutaneous immunotherapy compared with sublingual immunotherapy in mice sensitized to pollen (*Phleum pratense*). *International Scholarly Research Network Allergy, 2012,* 1–8.

Newmark, F. M., & Itkin, I. H. (1967). Asthma due to pine pollen. *Annals of Allergy, 25,* 251–252.

Oh, J. W., Lee, H. R., Kim, J. S., Lee, K. I., Kang, Y. J., Kim, S. W., Kook, M. H., Kang, H. Y., Kim, J. S., Lee, M. H., Lee, H. B., Kim, K. E., Pyun, B. Y., Lee, S. I., & Han, M. J. (2000). Aerobiological study of pollen and mold in the 10 states of Korea. *Pediatric Allergy And Respiratory Disease, 10,* 22–33.

Park, J. W., Ko, S. H., Kim, C. W., Jeoung, B. J., & Hong, C. S. (1999). Identification and characterization of the major allergen of the *Humulus japonicus* pollen. *Clinical & Experimental Allergy, 29*(8), 1080–1086.

Pettyjohn, M. E., & Levetin, E. (1997). A comparative biochemical study of conifer pollen allergens. *Aerobiologia, 13,* 259–267.

Rowe, A. H. (1939). Pine pollen allergy. *Journal of Allergy, 10,* 377–378.

Seitzer, U., Bussler, H., Kullmann, B., Petersen, A., Becker, W. M., & Ahmed, J. (2003a). Characterization of immunoglobulin E responses in Balb/c mice against the major allergens of timothy grass (*Phleum pratense*) pollen. *Clinical & Experimental Allergy, 33*(5), 669–675.

Seitzer, U., Bussler, H., Kullmann, B., Petersen, A., Becker, W. M., & Ahmed, J. (2003b). Quantitative assessment of immediate cutaneous hypersensitivity in a mouse model exhibiting an IgE response to Timothy grass allergens. *Medical Science Monitor, 9*(12), BR407–BR412.

Shahali, Y., Majd, A., Pourpak,Z., Tajadod G., Haftlang, M., & Moin, M. (2007). Comparative study of the pollen protein contents in two major varieties of *Cupressus arizonica* planted in Tehran. Iranian journal of allergy, asthma and immunology, 6*(3),* 123-127.

Tamura, S., Kobayashi, T., Kikuta, K., Nakagawa, M., Sakaguchi, M., & Inouye, S. (1986). IgE antibody responses against Japanese cedar pollen in the mouse. *Microbiology and Immunology, 30*(9), 883–891.

Tordesillas, L., Goswami, R., Benedé, S., Grishina, G., Dunkin, D., Järvinen, K. M., Maleki, S. J., Sampson, H. A., & Berin, M. C. (2015). Skin exposure promotes a Th2-dependent sensitization to peanut allergens. *The Journal of Clinical Investigation, 124,* 4965–4975.

Walker, I. C. (1921). Frequent causes and the treatment of seasonal hay-fever. *Archives of Internal Medicine, 28,* 71–118.

Weber, R. W. (2005). Cross-reactivity of pollen allergens: recommendations for immunotherapy vaccines. *Current opinion in allergy & clinical immunology, 5*(6), 563–569.

24

A study of low-temperature and mountain epilithic diatom community in mountain stream at the Han River system, Korea

Yong Jin Kim[1,2] and Ok Min Lee[1*]

Abstract

Background: This study was conducted to assess the physicochemical water quality and the altitudinal distribution of low-temperature and mountain epilithic diatom (LTMD) community in Buk and Hangae streams that are located in Seorak Mountain with the height of 1708 m in Korea. And the community characteristics of LTMD found in the Buk and Hangae streams were compared to that of LTMD from the Han River system.

Results: The physicochemical water qualities of Buk and Hangae streams were determined to be very clean. As a result of analyzing the community composition, 135 taxa of epilithic diatoms were determined, and 22 taxa appeared including *Hannaea arcus* var. *subarcus* which are known to have low-temperature and mountain ecological characteristics in the literatures. The relative frequencies of LTMD were 37.0~0.9% range from the upper to lower regions. Although *Diatoma tenuis*, *Eunotia minor*, and *Gomphonema affine* are known to be ubiquitous in streams and lakes, in this research, the three taxa were added into low-temperature and mountain epilithic diatom, since *D. tenuis* and *E. minor* appeared only in altitudes above 600 m, and *G. affine* had the highest relative frequency during spring and fall in altitudes above 700 m, when water temperature was around 10 °C.

Conclusions: Among the 24 taxa of low-temperature and mountain epilithic diatom (LTMD) (including the 3 taxa added in this study), 14 taxa (*Diatoma hyemalis*, *D. mesodon*, *D. tenuis*, *Hannaea arcus*, *H. arcus* var. *subarcus*, *Ulnaria inaequalis*, *Eunotia bilunaris*, *E. implicata*, *E. minor*, *E. muscicola*, *E. silvahercynia*, *E. septena*, *Delicata delicatula*, and *Gomphonema affine*) represented the characteristics of LTMD very well; they grow best in water temperatures below 15 °C in Buk and Hangae streams and Han River system.

Keywords: Altitudinal distribution, Buk and Hangae streams, Epilithic diatom, Han River system, Low-temperature and mountain

Background

Mountain streams are fragile ecosystems (Kim et al. 2012; O'Driscoll et al. 2012), sensitive to human influences such as road construction and environmental changes such as those due to climate change (Lang and Murphy 2012). Species richness in these streams is usually limited by extreme environmental conditions (Chapin and Körner 1995) and communities are therefore particularly sensitive to disturbances. Because of their sensitivity, freshwater ecosystems in these areas are useful for biodiversity and biological indicator research. Diatoms are an important component in these communities. However, although diatoms play key roles in inorganic nutrient cycling involving phosphorus and silica, studies have not been conducted on these diatoms due to accessibility and economic constraints (Loeb et al. 1983; Sánchez-Castillo et al. 2008).

Diatoms respond to changes in water quality, making them useful as biological indicators (Descy 1979; Watanabe et al. 1990). Diatoms also recover quickly after habitat disturbances, compared to other organisms (Descy 1979). Additionally, changes in their biomass can indicate shifts in other environmental metrics, such as feeding by herbivores, water temperature, substrate, nutrients, flow velocity, and flow rate (Allan 1995).

---

* Correspondence: omlee@kyonggi.ac.kr
[1]Department of Life Science, Kyonggi University, Iui-dong, Suwon-si 443-760, Korea
Full list of author information is available at the end of the article

Studies on diatoms in mountain streams have been conducted in the headwaters of montane to alpine regions in Canada (Antoniades and Douglas 2002; Lauriol et al. 2006), Russia (Potapova 1996; Medvedeva 2001), Poland (Kownacki et al. 2006), and Bolivia (Servant-Vildary 1982). Recently, studies related to climate change and biological water assessment have been conducted, including research on the integrity of freshwater ecosystems (Falasco et al. 2012), physicochemical factors of upland blanket peat catchments (O'Driscoll et al. 2012), aquatic ecosystem monitoring, and on monitoring the integrity of biological resources and habitats. Until now, the reported diatomic indices were developed through indexing nutrients and organic matter such as the trophic diatom index (TDI) or the diatom assemblage index to organic pollution (DAIpo); however, because high altitude streams generally have relatively clean water, Falasco et al. (2012) have suggested modifying the diatomic indicators.

In Korea, only research on fishes and benthic macroinvertebrates of mountain streams has been conducted (Chung et al. 2011; Son et al. 2011), and very little research has been done on epilithic diatoms (Kim et al. 2012). Currently in Korea, nationwide biological assessments of water quality are being performed. The sites of the biological assessments of water quality are mostly in mid-lower streams; thus, the research and monitoring of mountain stream are only being done in limited areas. In addition, the appropriate indicators and indices to monitor mountain streams have not been developed; research on community composition or habitat characteristics of epilithic diatoms is insufficient.

In this study, we investigated the epilithic diatom community and physicochemical water quality parameters in the Buk and Hangae streams, the highest altitude streams in the Han River system. We focused on the distribution of low-temperature and mountain epilithic diatoms according to temperature and altitude from the literature. The aims of our research were to (i) select the indicator species that can monitor mountain stream and (ii) evaluate the applicability of the selected indicator species.

Methods
Study areas and sampling sites
The 20 sites of investigation for this research were set every 100 m between the Buk and Hangae streams, which flow into Soyang Lake and the Soyang River on Seorak Mountain (1708 m). We sampled four times in the sampling site (total of 80 samples) from May to November 2011.

The Buk stream rises from Seorak Mountain, which is the highest mountain in Korea's Han River system. The Buk stream flows into Soyang Lake, joining the Soyang River and the Naerin stream past Baekdam (573 m), Gugokdam (710 m), and Gayangdong Valley (778 m). The distance from Gayangdong Valley to the Soyang Dam (100 m) is 97 km, and 13 sites (BC01~BC13) were investigated between the two. The Hangae stream rises from Hangaeryung (881 m), passes through Hangaeryung Valley (764 m), Jangsudae (652 m), and Oknyutang (365 m), and flows into the Buk stream. The distance from Hangaeryung to the point where the Buk stream joins is 16 km, and seven sites (HG01~HG07) were investigated between the two points (Fig. 1).

The community characteristics of LTMD found in the Buk and Hangae streams were compared to that of LTMD from the Han River system. Data from the Han River system the research results of 19 sites (220–871 m) from Kim et al. (2012).

Sampling and analysis methods
The altitude at the sampling sites was measured by GPS (Triton, Magellan) and was also based on referenced data from the National Geographic Information Institute (www.ngii.go.kr). A portable meter (Horiba D-55, Orion 5-Star) was used to measure the physiochemical factors: water temperature, dissolved oxygen (DO), pH, conductivity, and turbidity in the field. Water velocity was measured by the Craig method (Craig 1987). Chlorophyll a, biological oxygen demand (BOD), total nitrogen (TN), and total phosphorus (TP) were analyzed by standard methods (Greenberg et al. 2000). Rock or gravel was selected 20–30 cm under the water surface for attached algae sampling. Rock (25–100 cm^2) was brushed and scraped, and the scraped material was diluted in 300 ml distilled water. Samples were stored after fixation in Lugol's solution (Greenberg et al. 2000). Permanent slide samples were made for analysis of diatoms. Epilithic diatoms in the samples were identified using × 200 to × 1000 magnification under light and phase-contrast microscopy (Olympus BX41, Japan). Species were identified in accordance with Krammer and Lange-Bertalot (1986, 1988, 1991), Krammer (2002), Lange-Bertalot et al. (2010), Round et al. (1990) and AlgaeBase (www.algaebase.org).

Illustrations, including those by Joh et al. (2010), Chung (1993), Krammer (2002), Lange-bertalot et al. (2010), Round et al. (1990); and literature, including Antoniades and Douglas (2002), Kim et al. (2009), Lee et al. (1992), Medvedeva (2001), O'Driscoll et al. (2012), Torne's et al. (2012), were referenced to make the list of LTMD that live in low-temperature or mountainous areas.

Detrended correspondence analysis (DCA) and correlation analysis were performed to assess the relationships among the number of cells, relative frequency, physicochemical factors, and altitude (n = 80) using PC-ORD (MjM Software, Gleneden Beach, OR, USA) and SPSS ver. 12.0 software (SPSS, Inc., Chicago, IL, USA).

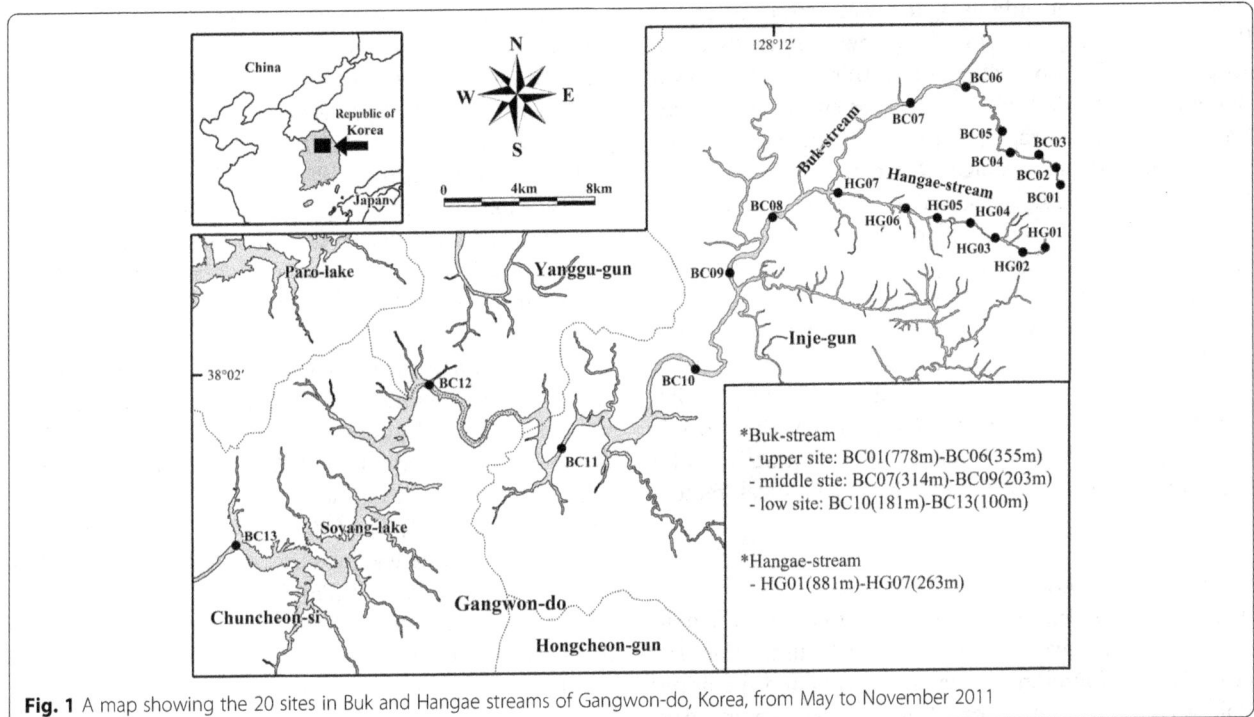

Fig. 1 A map showing the 20 sites in Buk and Hangae streams of Gangwon-do, Korea, from May to November 2011

Results

Physicochemical factors

All physicochemical factors, excluding water temperature, were subdivided into upper, middle, and lower stream reaches based on altitude and habitat environment. All Hangae stream sites were classified as upper stream reaches.

The water temperature in the Buk and Hangae streams was 8.3–26.9 °C and increased as altitude decreased (Fig. 2). The greatest difference between the highest and lowest water temperature in the Buk stream was measured in May, and the least difference was measured in August. Water temperatures at sites BC01 to BC04, with an average altitude of 500 m or greater, averaged < 15 °C, decreasing by 1.4 °C (r^2 = 0.893) per 100 m of altitude. In contrast, the greatest and least differences between the highest and lowest water temperatures were measured in August and November, respectively, in the Hangae stream. The average water temperature was < 15 °C from HG01 to HG04, sites with elevations of > 500 m. For every 100 m increase in elevation, water temperature dropped by 1.1 °C (r^2 = 0.964). The pH of the Buk stream was slightly acidic at 6.5, whereas the Hangae stream had a pH of 7.4 – 8.0. The conductivity in the upper part of the Buk stream was very low with an average of

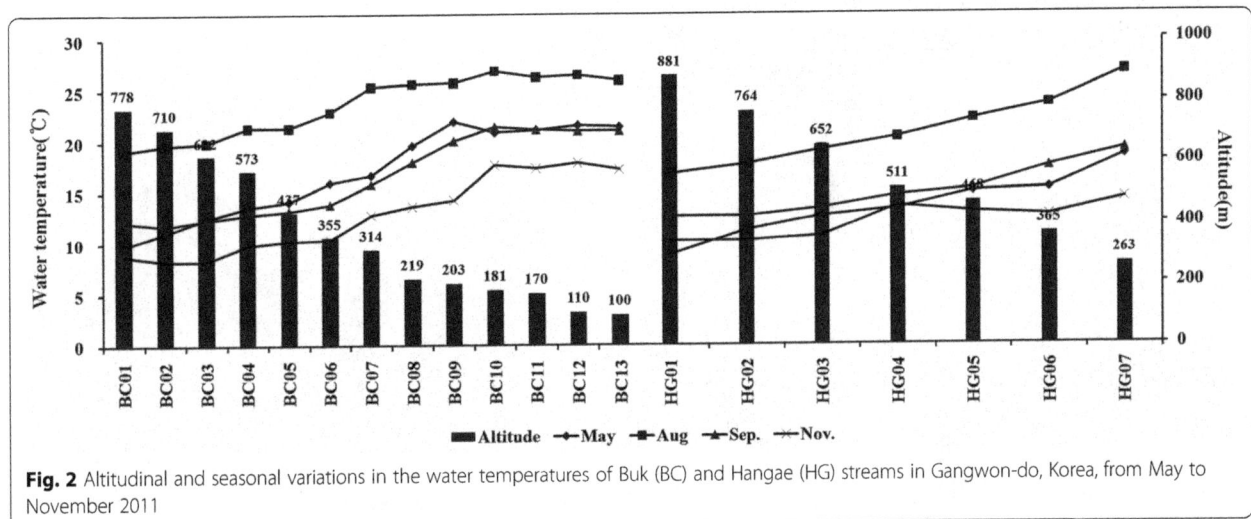

Fig. 2 Altitudinal and seasonal variations in the water temperatures of Buk (BC) and Hangae (HG) streams in Gangwon-do, Korea, from May to November 2011

25.9 μS cm^{-1}. The middle and lower stretches of the Buk and Hangae streams had low conductivity of 60 μS cm^{-1}. Chlorophyll a concentration in the lower Buk stream was 13.3 μg L^{-1}. In other sites, the concentration was low, averaging ≤ 1 μg L^{-1}. The water quality was clean with an average DO consumed in 5 days (BOD$_5$) of ≤ 2 mg L^{-1} at all sites. The average TN was 1.283–1.891 mg L^{-1} and did not show much seasonal variation. TP was 17.1–22.1 μg L^{-1}, increasing downstream (Table 1).

Altitude was highly correlated with velocity ($r = 0.73$), pH ($r = -0.66$), water temperature ($r = -0.64$), and TP ($r = -0.55$). Chlorophyll a showed an extremely high correlation with turbidity ($r = 0.82$) and a high correlation with TP. TN was correlated with conductivity and turbidity but not with other physicochemical factors (Table 2).

Community composition and LTMD

A total of 135 epilithic diatom taxa were identified in the Buk and Hangae streams. The least number of taxa (10–12) were found in the Hangae stream and the upper Buk stream. An average of 22 and 21 taxa appeared in the middle and lower stretches of the Buk stream, respectively. Site BC08, in the middle of the stream, showed the most diversity with 33 taxa collected in May. The number of taxa decreased at site BC10, at the inlet of Soyang Lake (Fig. 3).

From the literature in Korea, low-temperature and mountain-dwelling algae of 39 taxa were identified. Of these 39 taxa, 32 taxa were diatoms, 5 taxa were Crysophyceae, and 2 taxa were Chlorophyceae. Of the 32 taxa of diatoms in Korea, 21 taxa appeared in this study. These include *Diatoma hyemalis, D. mesodon, D. moniliformis, Hannaea arcus, H. arcus* var. *subarcus, Ulnaria inaequalis,* and *Tabellaria flocculosa* in Fragilariaceae;

Eunotia bilunaris, E. implicata, E. muscicola, E. silvahercynia, and *E. septena* in Eunotiaceae; *Eucocconeis laevi* and *Planothidium lanceolatum* of Achnanthaceae; and *Cymbella aspera, C. leptoceros, C. tumida, Delicata delicatula, Reimeria. sinuata, Gomphonema acuminatum,* and *G. clavatum* in Naviculaceae. Additionally, *Diatoma tenuis, Eunotia minor,* and *Gomphonema affine* were classified as LTMD due to their limited appearance at altitudes > 500 m. The relative frequencies of LTMD ranged from 0.0 to 94.6%, appearing more frequently in altitudes > 500 m (Fig. 4).

The five *Eunotia* species usually appeared at altitudes of > 500 m; in particular, *E. septena* and *E. implicata* were limited to altitudes > 700 m. *Gomphonema affine* showed a similar trend to that of *Eunotia* with a high relative frequency at altitudes > 700 m and a decreasing relative frequency below 600 m. It also showed decreasing relative frequency in summer and increasing frequency in spring and fall (Fig. 5).

Hannaea arcus and *H. arcus* var. *subarcus* showed differences in habitat range. The appearance of *H. arcus* was limited to altitudes of 200–300 m, whereas *H. arcus* var. *subarcus* showed its highest relative frequency at 400 m, and decreased in either the upper or lower parts of the streams. *Ulnaria inaequalis* had high relative frequency at about 300 m and showed a similar range to that of *H. arcus*. *Ulnaria inaequalis* and *H. arcus* had a high relative frequency during spring, when they appeared at < 200 m (Fig. 5).

Similar to *Eunotia, Diatoma hyemalis* only appeared in altitudes > 700 m, whereas *D. tenuis* was limited to about 600 m. *D. mesodon* only appeared at altitudes > 700 m during summer when water temperature was high and appeared at around 300 m in the fall and spring (Fig. 5). Excluding *Delicata delicatula*, most taxa in Cymbellaceae appeared at low altitudes. Regardless of

Table 1 Mean values of physicochemical factor values in Buk (BC) and Hangae (HG) streams of Gangwon-do, Korea, from May to November 2011

Parameter	Buk stream (mean ± SD)			Hangae stream (mean ± SD)
	Upper (BC01–06)	Middle (BC07–09)	Lower (BC10–13)	
Altitude (m)	579.2 ± 160.6	351.2 ± 60.0	185.9 ± 41.2	503.8 ± 219.9
Velocity (cm s^{-1})	65.8 ± 19.5	35.0 ± 16.1	13.7 ± 12.8	54.6 ± 4.8
Dissolved oxygen (mg L^{-1})	9.3 ± 2.1	9.1 ± 0.1	10.1 ± 0.3	8.8 ± 0.4
pH	6.5 ± 0.4	7.8 ± 0.7	8.0 ± 0.1	7.4 ± 0.2
Conductivity (μS cm^{-1})	25.9 ± 4.2	61 ± 12.8	60.9 ± 6.3	64.1 ± 19.4
Turbidity (NTU)	0.8 ± 1.0	1.2 ± 1.2	12.0 ± 4.1	0.6 ± 0.6
Chlorophyll a (μg L^{-1})	0.2 ± 0.1	1.2 ± 0.4	13.3 ± 5.6	0.6 ± 0.2
BOD$_5$ (mg L^{-1})	0.6 ± 0.1	1.6 ± 1.0	2.0 ± 0.8	0.5 ± 0.0
Total nitrogen (mg L^{-1})	1.386 ± 0.140	1.283 ± 0.138	1.891 ± 0.152	1.323 ± 0.302
Total phosphorus (μg L^{-1})	17.7 ± 8.3	22.1 ± 3.4	49.5 ± 6.7	17.1 ± 1.1

Table 2 Correlation comparison of physicochemical factors in Buk (BC) and Hangae (HG) streams of Gangwon-do, Korea, from May to November 2011

	WT	DO	pH	EC	Tur.	Chl.a	BOD$_5$	TN	TP	Altitude	Velocity
WT	1.00	−0.29**	0.57**	0.15	0.51**	0.44**	0.44**	0.03	0.49**	−0.64**	−0.51**
DO		1.00	0.20	0.19	0.14	0.07	0.04	0.07	0.00	−0.28*	−.024*
pH			1.00	0.46**	0.44**	0.43**	0.35**	0.15	0.41**	−0.66**	−0.62**
Con.				1.00	0.07	0.00	0.09	0.40**	0.03	−0.40**	−0.36**
Tur.					1.00	0.82**	0.57**	0.35**	0.55**	−0.39**	−0.41**
Chl.a						1.00	0.45**	0.21	0.56**	−0.38**	−0.44**
BOD$_5$							1.00	0.09	0.40**	−0.38**	−0.30**
TN								1.00	−0.02	−0.02	−0.16
TP									1.00	−0.55**	−0.60**
Altitude										1.00	0.73**
Velocity											1.00

WT water temperature, *DO* dissolved oxygen, *EC* electric conductivity, *Tur.* turbidity, *TN* total nitrogen, *TP* total phosphorus

*$P < 0.05$, **$P < 0.01$, $n = 80$

season, the relative frequency of *Reimeria sinuata* appeared in altitudes 200 m. *Cymbella tumida* did not show a relationship with altitude or water quality, appearing at altitudes < 300 m. Along with *Reimeria sinuata*, *Planothidium lanceolatum* showed the highest relative frequency at altitudes of 200 m (Fig. 5).

DCA was conducted by assigning all LTMD and all other species with a relative frequency higher than 1% to a main matrix, and assigning physicochemical factors to the second matrix. As a result, the eigenvalue of axis 1 was 0.599 and was highly correlated with factors such as water temperature, pH, TP, and BOD$_5$, but was negatively correlated with altitude ($r = -0.809$) and velocity ($r = -0.747$). The eigenvalue of axis 2 was 0.346, and it was positively correlated with water temperature, pH, and conductivity but negatively correlated with altitude (Fig. 6).

In the graph, taxa with positive values on axis 1 show increased relative frequency when the altitude and water velocity decreased and the water temperature increased. These taxa had the greatest correlation to changes in physicochemical factors such as pH, BOD$_5$, and TP. The taxa on the positive side of axis 2 are those in which relative frequencies increased as pH increased. Taxa on the negative side of axis 2 decreased in relative frequency as pH increased. Therefore, taxa placed in the third quadrant best represent the characteristics of LTMD (Fig. 6).

LTMD at other mountain streams in the Han River system (Kim et al. 2012)

The average water temperature of the streams in the Han River system was 14.5 °C, which shows the weak relationship between altitude and water temperature ($r^2 = 0.106$). The relationship between altitude and water temperature seems to be greater in an area where the riverbed is rarely changed, and in riparian vegetation

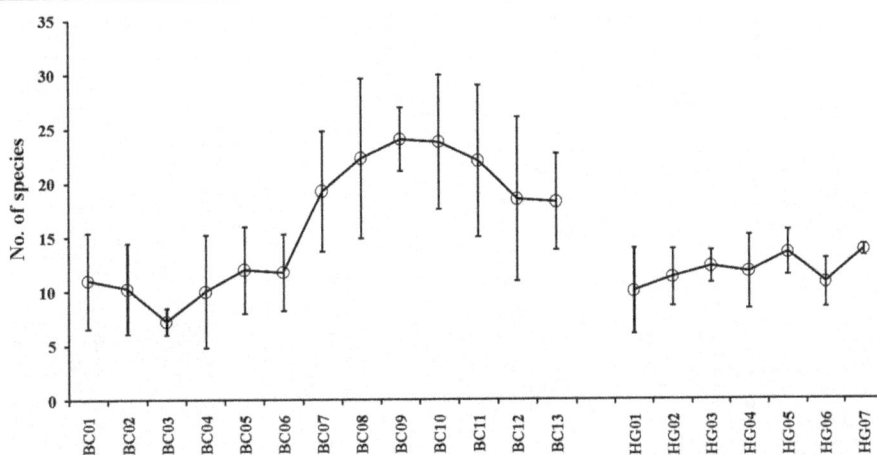

Fig. 3 The species numbers of epilithic diatom in Buk (BC) and Hangae (HG) streams of Gangwon-do, Korea, from May to November 2011

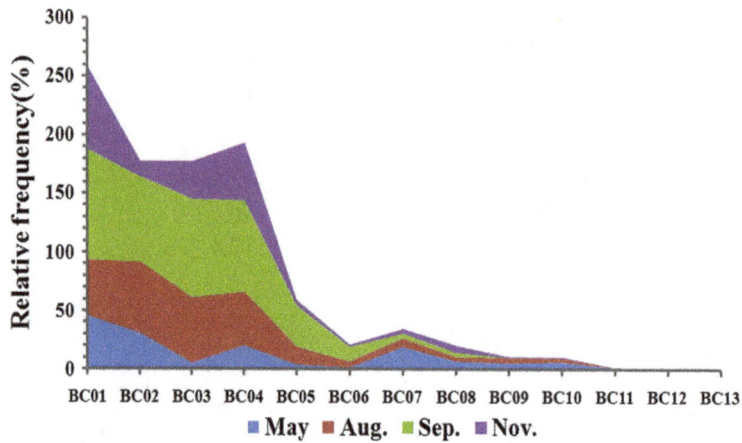

Fig. 4 Relative frequency of LTMD in Buk stream of Gangwon-do from May to November 2011

environments, such as the Buk and Hangae streams (Kim et al. 2012). The pH ranged from 6.8 to 9.5, with a large variation, depending on the area. A total of 14 taxa of LTMD were found, including *Cymbella affine, C. tumida, Diatoma mesodon, D. vulgaris, Eunotia minor, E. pectinalis, Hannaea arcus, H. arcus* var. *subarcus, Planothidium lanceolatum, Psammothidium oblongellum, Reimeria sinuata,* and *Ulnaria inaequalis.* The relative frequencies of LTMD ranged from 2.1 to 80.9%, with a high average of 33.9% at sites where water temperature

was 15 °C or less; the site S6 had the highest relative frequency, at 80.9%. The relative frequency of *Hannaea arcus* var. *subarcus* at site S6 was 56.6%; *H. arcus* var. *subarcus* showed the highest relative frequency among the LTMD ranging from 27.1 to 40.5% Fig. 7.

Discussion
Physicochemical factors
Of the physiochemical environmental factors examined, water temperature, pH, TP, and velocity were found to

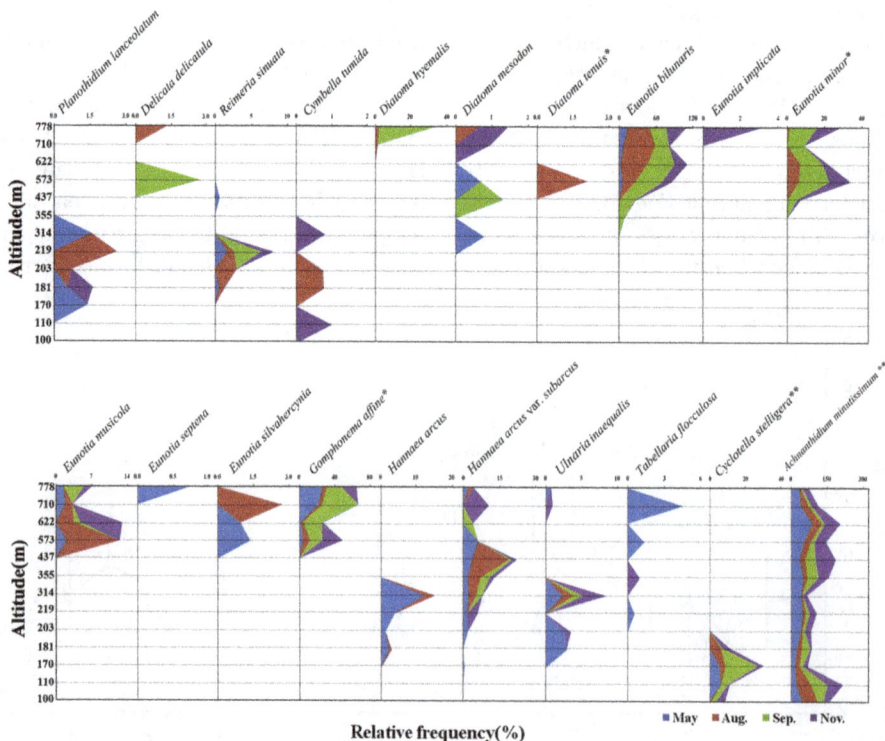

Fig. 5 The distribution of LTMD in Buk stream of Gangwon-do, Korea, from May to November 2011 (*added in this study; **taxa of widespread and lower altitude)

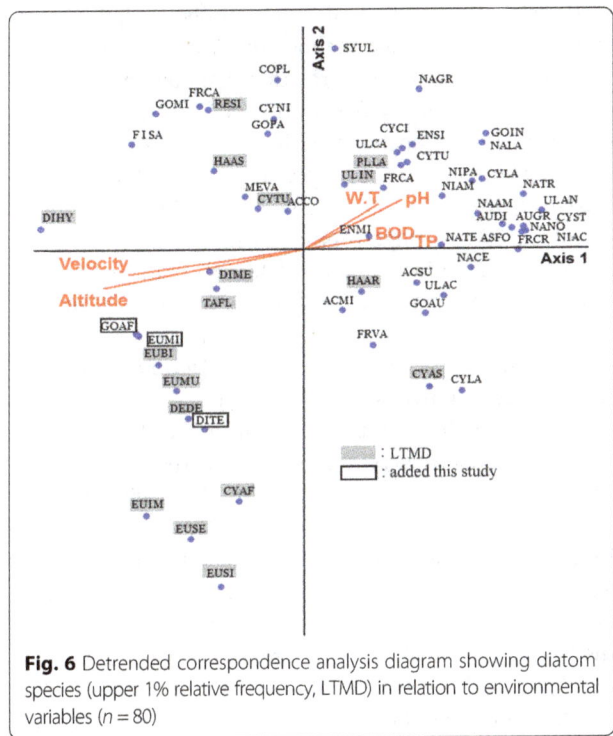

Fig. 6 Detrended correspondence analysis diagram showing diatom species (upper 1% relative frequency, LTMD) in relation to environmental variables (n = 80)

Community composition and LTMD

The 135 taxa of epilithic diatoms from the Buk and Hangae streams were similar to the 138 taxa found in high altitude streams in the Maritime Alps National Park (Falasco et al. 2012). The taxa that appeared in the upper region were not diverse. An increase in taxa due to nutrients (TP and TN), pH, and water temperature was observed beginning in the middle sections of the streams. Among the Pennales, diatoms in Naviculaceae and Fragilariaceae families were most diverse. In the upper region, saproxenous taxa (Watanabe et al. 2005) including *Encyonema minutum*, *Hannaea arcus* var. *subarcus*, and *Diatoma mesodon* were detected. In the middle and lower parts of the streams, the abundance of saprophilous taxa (Kelly and Whitton 1995; Watanabe et al. 2005) such as *Nitzshcia amphibian*, *N. palea*, and *Navicula gregaria* of Naviculaceae and Bacillariaceae increased. Generally, the relative frequency of taxa in *Navicula* and *Nitzschia* were higher in the lower regions than those in the upper regions (Atazadeh et al. 2007; Kim et al. 2009), which was thought to be due to pH and nutrient concentrations (Ginn et al. 2007; Passy, 2007).

Among the 32 taxa of mountain diatoms in Korea, 21 taxa were identified in this study, and 3 additional taxa were added. The genus *Eunotia* best represented the characteristics of LTMD. Most taxa in the genus *Eunotia* appear in low pH, alpine headwaters (Joh et al. 2010, Lange-Bertalot et al. 2010). This is thought to be due to the greater effect of pH than water temperature on this genus. *Eunotia implicata* was limited to altitudes > 700 m (Fig. 5). The seasonal average pH was slightly acidic at ≤ 6.5 (Table 1). *Eunotia bilunaris* had a decreasing relative frequency starting at an altitude of 500 m, and did not appear at 300 m, where the pH was 7.0. All *Eunotia* taxa were placed in the third quadrant of the DCA chart, which means they appeared in the high altitude, low-pH environment (Fig. 6). The seven taxa of *Eunotia* including *E. minor* showed these characteristics clearly. The newly added LTMD from this study,

have strong relationships with altitude. For the Buk and Hangae streams, the change in factors such as water temperature, pH, and velocity can be sorted based on altitude of 500 m. Based on the altitude of 500 m, water temperature fluctuated around 15 °C, pH ranged from neutral to slightly acidic, and velocity showed about 30 cm s^{-1} difference. The TP was greater in lower sites and showed little difference in concentration at upper and middle sites (Table 1).

The negative correlation between altitude and water temperature can only be applied to streams with ecologically similar environments (Kim et al. 2012), because they are affected by groundwater and riparian vegetation. Complementary data are needed for each ecological environment to compare between streams.

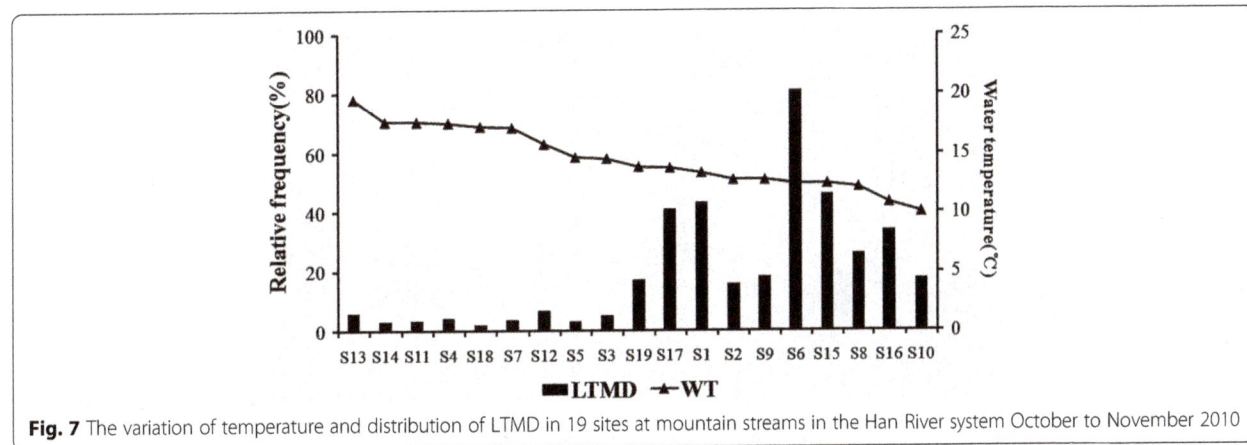

Fig. 7 The variation of temperature and distribution of LTMD in 19 sites at mountain streams in the Han River system October to November 2010

Gomphonema affine and *D. teunis*, showed trends similar to those of *Eunotia*. In particular, during each of the four samples, *G. affine* showed less than 10% frequency only in August, when the water temperature was high. In altitudes above 700 m, the frequency of appearance of *G. affine* did not significantly differ, based on factors such as pH, water velocity, TN, and TP; its growth seems to be limited by water temperature.

Hannaea arcus appears as a major dominant species in the Sophia River and Sophia Lake in Canada (Antoniades and Douglas 2002). Sophia Lake and its catchment area are a low temperature water system (4–11 °C), located at the most northern part of Cornwallis Island (75° 06′ N, 93° 36′ W). *H. arcus* appears at 15 sites in the Sophia River and Sophia Lake catchment area, and has a high negative correlation with water temperature ($r^2 = 0.710$). The dominance ratio changes greatly with water temperature. Although *H. arcus* was detected in the present study at relatively low altitudes of 200–300 m, it was mostly in May, when the water temperature was about 15 °C. Growth of this genus was inhibited in the Buk stream, where pH was ≤ 7.0 and altitudes were > 300 m, which contrasted with the Antoniades and Douglas (2002) study, where it was found in water with a pH of around 8.0.

Hannaea arcus var. *subarcus*, a variety of *H. arcus*, was detected at the highest altitudes and had its greatest relative frequency at 400 m. The presence of this species was positively correlated with altitude ($r^2 = 0.539$) and negatively correlated with water temperature ($r^2 = 0.338$) in a study by Kim et al. (2012). *H. arcus* var. *subarcus*, along with *H. arcus*, clearly exhibited the characteristics of LTMD. *H. arcus* var. *subarcus* seems to have greater growth range compared to that of *H. arcus*.

Ulnaria inaequalis, of the family Fragilariaceae, showed similar trends to those of *H. arcus*, although it seemed to have a greater growth range compared to that of *H. arcus*. *Diatoma hyemalis* and *D. tenuis* were detected in similar locations as those of *Eunotia*, with limited appearance in the upper regions of the streams. *Diatoma hyemalis* is usually found in alpine low-temperature and oligotrophic water systems (Joh et al. 2010) such as in Switzerland (Robinson et al. 2010). *Diatoma mesodon* appeared at altitudes of 300–600 m during the spring and fall, and was seen at the highest altitude site in the summer and fall. *Diatoma mesodon*, as a representative mountain diatom along with *H. arcus*, is the dominant species in the Colorado River (Vavilova and Lewis 1999), the Willamette River in Oregon (Carpenter and Waite 2000), the Swiss Alps in Switzerland (Robinson and Kawecka 2005), and the Tatra River in the National Park of Poland (Kawecka and Robinson 2008). *Delicata delicatula* is usually found in oligotrophic mountains (Chung 1993) and was found in the upper region of the Buk stream during the summer and fall (Fig. 5). It was identified at the highest site during the summer and found at 600 m during the fall. *Delicata delicatula* was located near *E. bilunaris* and *E. musicola* on the DCA diagram; its range of growth seemed to be limited by pH.

Every epilithic diatom community found in streams has different characteristics due to differences in physicochemical factors, including velocity, riverbed, turbidity, flow rate, habitat environment, pH, conductivity, and nutrient concentrations. In the case of the Buk and Hangae streams, the physicochemical factors that affected the epilithic diatom community the most were altitude, water temperature, pH, and TP (Fig. 7). The

Table 3 The list of low temperature and mountain diatoms (LTMD) and ecological character which appeared in this study

Taxa	Code	Altitude	Season	Trophic state	Habitat character
Delicata delicatula	DEDE	U/M/L	S/Su/A/W	Oligo	Mt
Diatoma hyemalis	DIHY	U	Su/A	Oligo	Low pH/Mt
Diatoma mesodon	DIME	U/M/L	S/Su/A/W	Oligo/Meso/Eu	Mt/Lt
Diatoma tenuis	DITE	U/M	W	Oligo/Meso	Mt
Eunotia bilunaris	EUBI	U/M	S/Su/A/W	Oligo/Meso	Low pH/ Mt
Eunotia minor	EUMI	U/M	Su/A/W	Oligo/Meso	Lt
Eunotia musicola	EUMU	U	S/Su/A/W	Oligo/Meso	Mt
Eunotia implicate	EUIM	U/M/L	W	Oligo	Low pH/Mt
Eunotia silvahercynia	EUSI	U	S/A	Oligo	Low pH/Mt
Eunotia septena	EUSE	U/M	S	Oligo	Mt
Gomphonema affine	GOAF	U	S/Su/A/W	Oligo/Meso	Low pH/Mt
Hannaea arcus	HAAR	M/L	S/Su	Meso/Eu	pH 7.0~8.5/Lt
Hannaea arcus var. *subarcus*	HAAS	U/M/L	S/Su/A/W	Oligo/Meso/Eu	Lt
Ulnaria inaequalis	ULIN	M/L	S/Su/A/W	Meso/Eu	pH 7.0~8.5/Lt

U upper, *M* middle, *L* low, *S* spring, *Su* Summer, *A* autumn, *W* winter, *Mt* mountain, *Lt* low temperature

relative frequencies of LTMD were 37, 5.6, and 0.9% from the upper, middle, and lower regions, respectively (Fig. 4). The average relative frequency was ≥ 50% at altitudes > 700 m. Additionally, LTMD appeared in the upper region, which is to be expected, as *Eunotia. bilunaris*, *E. implicata*, *H. arcus*, *Diatoma hyemalis*, *D. mesodon*, *D. tenuis*, and *Tabellaria flocculosa* also appear as dominant species in high altitude streams from places such as Italy (Falasco et al. 2012), Canada (Antoniades and Douglas 2002), and Ireland (O'Driscoll et al. 2012).

Among the LTMD, *H. arcus*, *H. arcus* var. *subarcus*, *D. mesodon*, and *S. inaequalis* had similar frequency of occurrence in 19 sites at mountain streams of the Han River system, in the Buk and Hangae streams. Although a direct comparison of genus *Eunotia* was difficult due to low abundance, most appeared in upper streams of 19 sites in mountain streams. Genus *Cymbella* did not demonstrate the general characteristics of LTMD in either the Buk and Hangae streams, or mountain streams of the Han River system.

Conclusions

Among the 24 taxa of LTMD (excluding the 10 taxa with low occurrence tendencies and relative frequencies, such as genus *Achnanthes* and *Cymbella*), 14 taxa represented the characteristics of LTMD very well; they grow best in water temperatures below 15 °C. These are *Diatoma hyemalis*, *D. mesodon*, *D. tenuis*, *Hannaea arcus*, *H. arcus* var. *subarcus*, *Ulnaria inaequalis*, *Eunotia bilunaris*, *E. implicata*, *E. minor*, *E. muscicola*, *E. silvahercynia*, *E. septena*, *Delicata delicatula*, and *Gomphonema affine* (Table 3).

In all, *Hannaea arcus* var. *subarcus* was found in most mountain streams; it is thought to be a useful taxon as the most representative of Korean LTMD. The occurrence of five species in the genus *Eunotia* were found to have a more significant relationship to pH than to water temperature. Therefore, they were not useful as indicator species. However, if physicochemical factors such as oligotrophy, low conductivity, and weakly acidic pH are considered the primary characteristics of mountain streams or headwaters, the *Eunotia* species could be used as indicators. Although *Hannaea arcus* and *D. mesodon* are representative of the mountain-based epilithic diatom genera found in Korea and international studies, they can be useful as indicators for increases in water temperature; however, more studies are needed to identify additional ecological characteristics. More diverse indicators are needed to obtain accurate results monitoring mountain streams using LTMD.

Abbreviations
DCA: Detrended correspondence analysis; LTMD: Low-temperature and mountain epilithic diatom

Acknowledgements
Not applicable.

Funding
This research was supported by the foundation research project of National Research Foundation of Korea (2011–0005974).

Authors' contributions
KYJ carried out the design of the study, performed the fieldwork, and drafted the manuscript. LOK participated in the design and coordination of manuscript and helped draft the manuscript. All authors read and approved the final manuscript.

Competing interests
The authors declare that they have no competing interests.

Author details
[1]Department of Life Science, Kyonggi University, Iui-dong, Suwon-si 443-760, Korea. [2]Han River Environment Research Center, National Institute of Environmental Research, 68 42, Dumulmeorigil Yangseo-myeon, Yangpyong-gun, Gyunggi Province 473-823, Korea.

References
Allan, J. D. (1995). *Stream ecology-structure and function of running waters.* London: Chapman and Hall.

Antoniades, D. M., & Douglas, M. S. V. (2002). Characterization high arctic stream diatom assemblages from cornwallis island, Nunavut, Canada. *Canadian Journal of Botany, 80,* 50–58.

Atazadeh, I., Sharifi, M., & Kelly, M. G. (2007). Evaluation of the trophic diatom index for assessing water quality in river Gharasou, western Iran. *Hydrobiologia, 589,* 165–173.

Carpenter, K. D., & Waite, I. R. (2000). Relations of habitat-specific algal assemblages to land use and water chemistry in the Willamette basin, Oregon. *Environmental Monitoring and Assessment, 64,* 247–257.

Chapin, F. S. III & Körner, C. (1995). Patterns, causes, changes and consequences of biodiversity in arctic and alpine ecosystems. In: Arctic and alpine biodiversity: Patterns, causes and consequences (Ed. by F.S. Chapin III and C. Körner), pp. 313–320. Springer–Verlag, Berlin.

Chung, J. (1993). *Illustration of the freshwater algae of Korea.* Seoul: Academy Publishing Company.

Chung, N. I., Park, B. K., & Kim, K. H. (2011). Potential effect of increased water temperature on fish habitats in Han-river watershed. *Journal of Korean Society on Water Environment, 27,* 314–321.

Craig, D. A. (1987). Some of what you should know about water. *Journal of North American Benthological Society, 4,* 178–182.

Descy, J. P. (1979). New approach to water quality estimation using diatoms. *Nova Hedwigia, 64,* 305–323.

Falasco, E., Ector, L., Ciaccio, E., Hoffmann, L., & Bona, F. (2012). Alpine freshwater ecosystems in a protected area: A source of diatom diversity. *Hydrobiologia, 695,* 233–251.

Ginn, B. K., Cumming, B. F., & Smol, J. P. (2007). Diatoms-based environmental inferences and model comparisons from 494 northeastern north American lakes. *Journal of Phycology, 43,* 647–661.

Greenberg, A. E., Clesceri, L. S., & Eaton, A. N. (2000). *Standard method for the examination of water and wastewater* (21st ed.). Washington, D.C., USA: American Public Health Association.

Joh, G. J., Lee, J. H., Lee, K., & Yoon, S. K. (2010). *Algal flora of Korea, Volume 3 Number 2, Freshwater Ditoms II (Chrysophyta, Bacillariophyceae, Pennales, Araphidineae, Diaomaceae)* (p. 153). Korea: National Institute of Biological Resources.

Kawecka, B., & Robinson, C. T. (2008). Diatom communities of lake/stream networks in the Tatra Mountains, Poland, and the Swiss alps. *Oceanological and Hydrobiological Studies, 37,* 21–35.

Kelly, M. G., & Whitton, B. A. (1995). The trophic diatom index: A new index for monitoring eutrophication in rivers. *Journal of Applied Phycology, 7,* 433–444.

Kim, Y. J., Kong, D. S., & Lee, O. M. (2012). The community of cryophilic and mountain periphyton at high altitude streams in the Han-river system. *Journal of Environmental Impact Assessment, 21,* 143–160.

Kim, Y. J., Shin, K. A., & Lee, O. M. (2009). Water quality assessed by DAIpo and TDI of Bokha stream and Dal stream in south Han-river, Korea. *Journal of Environmental Biology, 27,* 414–424.

Kownacki, A., Dumnicka, E., Kwandras, J., Galas, J., & Ollik, M. (2006). Benthic communities in relation to environmental factors in small high mountain ponds threatened by air pollutants. *Boreal Environment Research, 11,* 481–492.

Krammer, K. (2002). Cymbella. In Diatoms of Europe 3(Ed. By H. Lange-berralot), A. R. G. Gantner Verlag K. G., Ruggell.

Krammer, K. & Lange-Bertalot, H. (1986). *Bacillariophyceae.* 1. Teil: Naviculaceae. In Süßwasserflora von Mitteleuropa(Ed. By H. Ettl, J. Gerloff, H. Heynig and D. Mollenhauer), Vol. 2/1. G. Fischer, Stuttgart and New York.

Krammer, K. & Lange-Bertalot, H. (1988). *Bacillariophyceae.* 2. Teil: Bacillariaceae, Epithemiaceae, Surirellaceae. In Süßwasserflora von Mitteleuropa(Ed. By H. Ettl, J. Gerloff, H. Heynig and D. Mollenhauer), Vol. 2/2. G. Fischer, Stuttgart and New York.

Krammer, K. & Lange-Bertalot, H. (1991). *Bacillariophyceae. 4. Teil: Achnanthaceae, Kritische Ergänzungen zu Navicula*(Lineolatae) und *Gomphonema*,Gesamtliteraturverzeichnis. In Süßwasserflora von (Ed. By H. Ettl, J. Gerloff, H. Heynig and D. Mollenhauer), Vol. 2/4. G. Fischer, Stuttgart and New York.

Lang, P., & Murphy, J. (2012). Environmental drivers, life strategies and bioindicator capacity of bryophyte communities in high-latitude headwater streams. *Hydrobiologia, 679,* 1–17.

Lange-Bertalot, H., Bak, M. & Witkowski, A. (2010). Eunotia and some related genera. In Diatoms of Europe 6(Ed. by H. Lange-Bertalot), A. R. G. Gantner Verlag K. G., Ruggell.

Lauriol, B., Préevost, C., & Lacelle, C. (2006). The distribution of diatom flora in ice caves of the northern Yukon territory, Canada: Relationship to air circulation and freezing. *International Journal of Speleology, 35,* 83–92.

Lee, J. H., Gotoh, T., & Chung, J. (1992). Diatoms of Yungchun dam reservoir and its tributaries. Kyung pook prefecture, Korea. *Diatom, 7,* 45–70.

Loeb, S. L., Reuter, J. E. & Goldman, C. R. (1983). Littoral zone production of oligotrophic lakes. In Periphyton of freshwater ecosystems (Ed. by R.G. Wetzel). Junk W. Publishers, The Hague.

Medvedeva, L. A. (2001). Biodiversity of aquatic algal communities in the Sikhote-Alin biosphere reserve (Russia). *Cryptogamie Algologie, 22,* 65–100.

O'Driscoll, C., Eyto, E., Rodgers, M., O'Connor, M., Asam, Z. Z., & Xiao, L. (2012). Diatom assemblages and their associated environmental factors in upland peat forest rivers. *Ecological Indicators, 18,* 443–451.

Passy, S. I. (2007). Diatom ecological guilds display distinct and predictable behavior along nutrient and disturbance gradients in running waters. *Aquatic Botany, 86,* 171–178.

Potapova, M. (1996). Epilithic algal communities in rivers of the Kolyma mountains, NE Siberia, Russia. *Nova Hedwigia, 63,* 3–4.

Robinson, C. T., & Kawecka, B. (2005). Benthic diatoms of an alpine stream/lake network in Switzerland. *Aquatic Science, 67,* 492–506.

Robinson, C. T., Kawecka, B., Füreder, L., & Peter, A. (2010). Biodiversity of flora and fauna in alpine waters. *Alpine Waters, 6,* 193–223.

Round, F. E., Crawford, R. M., & Mann, D. G. (1990). *The diatoms.* New York, USA: University of Cambridge.

Sánchez-Castillo, P. M., Linares-Cuesta, J. E., & Fernández-Moreno, D. (2008). Changes in epilithic diatom assemblages in a Mediterranean high mountain lake(Laguna de la caldera, sierra Nevada, Spain) after a period of drought. *Journal of Limnology, 76,* 49–55.

Servant-vildary, S. (1982). Altitudinal zonation of mountainous diatom flora in Bolivia: Application to the study of the quaternary. *Acta Geologica Academiae Scientiarum Hungaricae, 25,* 179–210.

Son, S. H., Kim, J. Y., Jo, J. J., & Kong, D. S. (2011). Altitudinal distribution aspect of benthic macroinvertebrates in a mountain stream of Seoraksan. *Journal of Korean Society on Water Environment, 27,* 680–688.

Tornés, E., Leira, M., & Sabater, S. (2012). Is the biological classification of benthic diatom communities concordant with ecotypes? *Hydrobiologia, 695,* 44–51.

Valvilova, V. V., & Lewis, W. M. J. (1999). Temporal and altitudinal variations in the attached algae of mountain streams in Colorado. *Hydrobiologia, 193,* 81–93.

Watanabe, T., Asai, K., & Houki, A. (1990). Numerical simulation of organic pollution in flowing waters. *Hazardous Waste Containment and Treatment, 4,* 251–281.

Watanabe, T., Ohtsuka, T., Tuji, A., & Houki, A. (2005). *Picture book and ecology of the freshwater diatoms.* Tokyo: Uchida-rokakuho.

Permissions

The contributors of this book come from diverse backgrounds, making this book a truly international effort. This book will bring forth new frontiers with its revolutionizing research information and detailed analysis of the nascent developments around the world.

We would like to thank all the contributing authors for lending their expertise to make the book truly unique. They have played a crucial role in the development of this book. Without their invaluable contributions this book wouldn't have been possible. They have made vital efforts to compile up to date information on the varied aspects of this subject to make this book a valuable addition to the collection of many professionals and students.

This book was conceptualized with the vision of imparting up-to-date information and advanced data in this field. To ensure the same, a matchless editorial board was set up. Every individual on the board went through rigorous rounds of assessment to prove their worth. After which they invested a large part of their time researching and compiling the most relevant data for our readers.

The editorial board has been involved in producing this book since its inception. They have spent rigorous hours researching and exploring the diverse topics which have resulted in the successful publishing of this book. They have passed on their knowledge of decades through this book. To expedite this challenging task, the publisher supported the team at every step. A small team of assistant editors was also appointed to further simplify the editing procedure and attain best results for the readers.

Apart from the editorial board, the designing team has also invested a significant amount of their time in understanding the subject and creating the most relevant covers. They scrutinized every image to scout for the most suitable representation of the subject and create an appropriate cover for the book.

The publishing team has been an ardent support to the editorial, designing and production team. Their endless efforts to recruit the best for this project, has resulted in the accomplishment of this book. They are a veteran in the field of academics and their pool of knowledge is as vast as their experience in printing. Their expertise and guidance has proved useful at every step. Their uncompromising quality standards have made this book an exceptional effort. Their encouragement from time to time has been an inspiration for everyone.

The publisher and the editorial board hope that this book will prove to be a valuable piece of knowledge for researchers, students, practitioners and scholars across the globe.

List of Contributors

Neha P. Ingole and Kwang-Guk An
Department of Biological Sciences, College of Biological Sciences and Biotechnology, Chungnam National University, Daejeon 34134, South Korea

Hui Seong Ryu, Ra Young Shin and Jung Ho Lee
Department of Biology Education, Daegu University, Gyeongbuk 38453, South Korea

Jeom-Sook Lee
Department of Biology, Kunsan National University, Gunsan 54150, South Korea

Seung Ho Lee
Marine & Environmental Research Laboratory, Ansan 15486, South Korea

Hyeon-Ho Myeong
Division of Ecosystem Research, National Park Research Institute, Wonju 26441, South Korea

Jung-Yun Lee and Jong-Wook Kim
Department of Biological Science, Mokpo National University, Muan-gun 58554, South Korea

Jang Sam Cho
Division of Ecological Assessment, National Institute of Ecology, Seocheon 33657, South Korea

Md Niamul Haque
Department of Ocean System Engineering, College of Marine Science, Gyeongsang National University, Cheondaegukchi-Gil 38, Tongyeong, Gyeongnam 650-160, South Korea

Sung-Hyun Kwon
Department of Marine Environmental Engineering, College of Marine Science, Engineering Research Institute (ERI), Gyeongsang National University, Cheondaegukchi-Gil 38, Tongyeong, Gyeongnam 650-160, South Korea

Bok Yeon Jo and Han Soon Kim
Department of Biology, Kyungpook National University, Daegu 41566, South Korea

Sungmin Jung, Yunkyoung Lee, Jaeyong Lee, Yukyong Cheong, Arif Reza, Jaiku Kim and Bomchul Kim
Department of Environmental Science, Kangwon National University, Chuncheon 24341, South Korea

Kiyong Kim
Department of Hydrology, University of Bayreuth, 95447 Bayreuth, Germany

Jeffrey S. Owen
Department of Environmental Science, Hankuk University of Foreign Studies, Yongin 17053, South Korea

Seung-Yeon Lee, Kyu-Tae Cho, Rae-Ha Jang and Young-Han You
Department of Biology, Kongju National University, Gongju, South Korea

Uhram Song and Han Eol Kim
Department of Biology, Jeju National University, Jeju 63243, South Korea

Kenji Sugimoto and Yoichi Nakano
National Institute of Technology, Ube College, Ube, Japan.

Tetsuji Okuda
Faculty of Science and Technology, Ryukoku University, Kyoto, Japan

Satoshi Nakai
Department of Chemical Engineering, Hiroshima University, Hiroshima, Japan

Wataru Nishijima
Environmental Research and Management Center, Hiroshima University, Hiroshima, Japan

Mitsumasa Okada
The Open University of Japan, Chiba, Japan

Bo Eun Nam and Jong Min Nam
Department of Biology Education, Seoul National University, Seoul 08826, South Korea

Jae Geun Kim
Department of Biology Education, Seoul National University, Seoul 08826, South Korea
Center for Education Research, Seoul National University, Seoul 08826, South Korea

Sun-Ja Cho
Department of Microbiology, Pusan National University, Busan 46269, South Korea

Mi-Hee Kim
Busan Metropolitan City Institute of Health and Environment, Busan 46616, South Korea

Young-Ok Lee
Department of Biological Sciences, Daegu University, Daegu 38453, South Korea

Uhram Song
Department of Biology and Research Institute for Basic Sciences, Jeju National University, Jeju 63243, South Korea

Hun Park
Institute for Legal Studies, Yonsei University, Seoul 03722, South Korea

Ji-Woong Choi and Kwang-Guk An
Department of Biological Science, College of Biosciences and Biotechnology, Chungnam National University, Daejeon 305-764, South Korea

Jong-Yun Choi and Se-Hwan Son
National Institute of Ecology, Seo-Cheon Gun, Chungcheongnam province 325-813, South Korea

Gea-Jae Joo
Department of Biological Sciences, Pusan National University, Busan 609-735, South Korea

Kwang-Seuk Jeong
Department of Biological Sciences, Pusan National University, Busan 609-735, South Korea
Institute of Environmental Technology and Industry, Pusan National University, Busan 609-735, South Korea

Seong-Ki Kim
Nakdong River Environment Research Center, Goryeong-Gun, Gyeongsangbuk-do, South Korea

Yoori Cho and Dowon Lee
Department of Environmental Planning, Seoul National University, 1 Gwanak-ro, Seoul 08826, South Korea

SoYeon Bae
Division of Ecological Survey Research, NIE, 1210 Geumgang-ro, Seocheon-gun 33657 Seoul, South Korea

Seongjun Kim, Seung Hyun Han, Guanlin Li and Yowhan Son
Department of Environmental Science and Ecological Engineering, Graduate School, Korea University, Seoul 02841, South Korea

Tae Kyung Yoon
Environmental Planning Institute, Seoul National University, Seoul 08826, South Korea

Sang-Tae Lee
Forest Practice Research Center, National Institute of Forest Science, Pocheon 11186, South Korea

Choonsig Kim
Department of Forest Resources, Gyeongnam National University of Science and Technology, Jinju 52725, South Korea

Heon-Mo Jeong
Division of Ecosystem Services and Research Planning, National Institute of Ecology, Seocheon, South Korea

Rae-Ha Jang and Young-Han You
Department of Biology, Kongju National University, Gongju, South Korea

Hae-Ran Kim
Division of Education Planning and Management, Nakdonggang National Institute of Biological Resources, Sangju, South Korea

Sookyung Shin and Hyesoon Kang
Department of Biology, Sungshin University, Seoul 01133, South Korea

Sang Gil Lee
Hankang Tree Hospital, Seongnam 13631, South Korea

Seon-Mi Lee, Jae-Gyu Cha and Ho-Gyung Moon
Division of Ecological Conservation, National Institute of Ecology, 1210 Geumgang-ro, Maseo-myeon, Seocheon-gun, Chungcheongnam-do 33657, Republic of Korea

Seo Hyeon Kim and Jong Min Nam
Department of Biology Education, Seoul National University, Seoul 08826, South Korea

Byeong-Mee Min
Department of Science Education, Dankook
University, Yongin 16890, South Korea

**Tae-Bok Ryu, Mi-Jeoung Kim, Chang-Woo Lee,
Deok-Ki Kim, Dong-Hui Choi, Hyohyemi Lee, Hye-
Ran Jeong, Do-Hun Lee and Nam-Young Kim**
National Institute of Ecology, Seo-Cheon Gun,
Chungcheongnam Province 325-813, South Korea

Seo-Yoong Kim and Kee-Ryong Choi
Department of Biological Science, University of
Ulsan, Ulsan 44610, South Korea

In-Bo Oh
Environmental Health Center, University of Ulsan
College of Medicine, Ulsan 44033, South Korea

Ok Min Lee
Department of Life Science, Kyonggi University,
Iui-dong, Suwon-si 443-760, Korea

Yong Jin Kim
Department of Life Science, Kyonggi University,
Iui-dong, Suwon-si 443-760, Korea
Han River Environment Research Center, National
Institute of Environmental Research, 68 42,
Dumulmeorigil Yangseo-myeon, Yangpyong-gun,
Gyunggi Province 473-823, Korea

Index

Phytoremediation, 85-90
Pollination Service, 112
Pollinosis, 183
Polygonatum Humile, 166-167, 173
Post-remediation, 85-89
Precipitation, 43, 83, 92, 100, 113, 127-129, 131-133, 136, 147, 161, 167, 175, 179-180, 182
Pyrosequencing, 76-78, 80-83

Q
Quercus Glauca, 127-128

R
Rapd-pcr, 71, 73
Regeneration Niche, 155, 164
Rhizome, 162, 164, 166-169, 171-173

S
S. Conopea, 32-34, 36-40
S. Japonica, 20-22
S. Truttae, 32-33, 35-40
Salt Marsh Plant, 20, 23
Seasonality, 1
Seedling Establishment, 155-160, 162
Six Deciduous Oaks, 51
Soil Factor, 20
Soil Organic Matter, 120, 122-125, 157
Soil Temperature, 127-133, 165
Spatial Pattern, 134, 137, 139, 144-145

Spatial Segregation of Sexes, 134-135, 141, 144
Sprouting Ability, 51-53, 55-56
Stable Isotope Analysis, 102, 105, 107, 109
Suaeda Glauca, 21
Successional Stage, 146
Synura Americana, 32-33, 36

T
Thinning Intensity, 120-121, 123
Torreya Nucifera, 134-135, 137, 144-145
Trophic State Deviation, 1

U
Ultrastructure, 32-33, 36, 40
Upper Dam Construction, 1, 5-9

V
Vegetation Structure, 90, 112-114, 118, 133, 145, 148, 153
Vegetation Succession, 146

W
Waterway Tunnel, 91-100

Z
Zonation, 20, 23, 164-165, 198
Zooplankton, 42-50, 103-104, 106, 108, 110-111
Zostera Japonica, 62, 69-70
Zostera Marina, 62, 67, 69-70

www.ingramcontent.com/pod-product-compliance
Lightning Source LLC
Chambersburg PA
CBHW082025190326
41458CB00010B/3276